EXISTENCE THEORY FOR GENERALIZED NEWTONIAN FLUIDS

Mathematics in Science and Engineering

EXISTENCE THEORY FOR GENERALIZED NEWTONIAN FLUIDS

Dominic Breit
Mathematical & Computer Sciences; Mathematics
Heriot-Watt University
Edinburgh, Midlothian, United Kingdom

Series Editor
Goong Chen

ELSEVIER

AMSTERDAM • BOSTON • HEIDELBERG • LONDON
NEW YORK • OXFORD • PARIS • SAN DIEGO
SAN FRANCISCO • SINGAPORE • SYDNEY • TOKYO

Academic Press is an imprint of Elsevier

ACADEMIC PRESS

Academic Press is an imprint of Elsevier
125 London Wall, London EC2Y 5AS, United Kingdom
525 B Street, Suite 1800, San Diego, CA 92101-4495, United States
50 Hampshire Street, 5th Floor, Cambridge, MA 02139, United States
The Boulevard, Langford Lane, Kidlington, Oxford OX5 1GB, United Kingdom

Library of Congress Cataloging-in-Publication Data
A catalog record for this book is available from the Library of Congress

British Library Cataloguing-in-Publication Data
A catalogue record for this book is available from the British Library

ISBN: 978-0-12-811044-7

For information on all Academic Press publications
visit our website at https://www.elsevier.com

Publisher: Nikki Levy
Acquisition Editor: Graham Nisbet
Editorial Project Manager: Susan Ikeda
Production Project Manager: Paul Prasad Chandramohan
Designer: Maria Ines Cruz

Typeset by VTeX

Dedicated

to

Janine

CONTENTS

PREFACE

In continuum mechanics, the motion of an incompressible, homogeneous fluid is described by the velocity field and the hydrodynamical pressure. The time evolution of the fluid is governed by the Navier–Stokes system of partial differential equations which describes the balance of mass and momentum. In the classical formulation – which goes back to C.-L. Navier and G.G. Stokes – the relation between the viscous stress tensor and the symmetric gradient of the velocity field is linear (i.e. we have a Newtonian fluid). This system is already quite challenging from a mathematical point of view and has fascinated many mathematicians. However, it can only model fluids with a very simple molecular structure such as water, air and several oils. In order to study more complex fluids one has to deal with a generalized Navier–Stokes system. In this model for generalized Newtonian fluids, the viscosity ν is assumed to be a function of the shear-rate $\dot{\gamma}$. Very popular among rheologists is the power-law model in which the generalized viscosity function is of power-type $\nu \sim \dot{\gamma}^{p-2}$ with $p > 1$. For a specific fluid physicists can identify this power by experiments. Instead of the Laplacian, as in the classical Navier–Stokes system, the main part of the mathematical model is a power-type nonlinear second order differential operator. So, in addition to the convective term, a second highly nonlinear term appears. The only framework which is available today is the concept of weak solutions. These solutions belong to appropriate Sobolev spaces: Derivatives have to be understood in the sense of distributions and singularities may occur.

The mathematical observation of power-law fluids began in the late sixties in the pioneering work of J.-L. Lions and O.A. Ladyzhenskaya. The cases they could handle were restrictive for real world situations. Nevertheless, it was a breakthrough in the theory of partial differential equations. Since then, there has been huge progress in the mathematical theory of generalized Newtonian fluids. The first systematic study was initiated by the group around J. Nečas in the 1990s. Situations which are realistic for physical and industrial applications could finally be handled. In the 2000s the Lipschitz truncation (the approximation of a Sobolev function by a Lipschitz continuous one done in such a way that both are equal outside of a small set whose size can be controlled) was shown to be a powerful tool in the analysis of generalized Newtonian fluids. A very wide range of non-Newtonian fluids could finally be included in the mathematical theory. The best known bound $p > \frac{6}{5}$ for three dimensional flows of power-law fluids was achieved.

A drawback of the Lipschitz truncation is its nonlinear and nonlocal character. In fact, the property of a function to be solenoidal (that is, divergence-free) is lost by truncating it. So, one has to introduce the pressure function in the weak formulation which results in additional technical difficulties. An advanced pressure decomposition via singular integral operators is necessary in the non-stationary case. An improved version, the "solenoidal Lipschitz truncation", was developed only recently. It allows the existence of weak solutions to the generalized Navier–Stokes system to be shown without the appearance of the pressure function and therefore highly simplifies the proofs. Moreover, it allows the Prandtl–Eyring fluid model to be studied which was out of reach before. In this model the power-growth is replaced by some logarithmic function: the law $\nu \sim \ln(1 + \dot{\gamma})/\dot{\gamma}$ was introduced in 1936 based on a molecular theory. This leads to a limit case in the functional analytical setting of generalized Newtonian fluids. It is not possible to introduce the pressure in the expected function space. Neither can the divergence be corrected.

The aim of this book is to present a complete and rigorous mathematical existence theory for generalized Newtonian fluids – for stationary, non-stationary and stochastic models. The balance laws are formulated in all situations via a generalized Navier–Stokes system. The proofs are presented as self-contained as possible and require from the reader only basic knowledge of nonlinear partial differential equations.

The heart of this book is the construction of the "solenoidal Lipschitz truncation". It has numerous applications and is of interest for future research beyond the scope of this monograph. The stationary truncation is presented in Chapter 3 and the non-stationary version in Chapter 6. Based on the "solenoidal Lipschitz truncation" the existence of weak solutions to generalized Navier–Stokes equations is shown. The existence proof itself is only slightly more complicated than the classical monotone operator theory and easy to follow.

In Chapter 4 we study the stationary Prandtl–Eyring model in two dimensions. Here, several important tools like the Bogovskiĭ operator and Korn's inequality loose some of their continuity properties. Optimal results, which might allow for a loss of integrability, can be achieved in the framework of Orlicz spaces. They are flexible enough to study fine properties of measurable functions which are required in the Prandtl–Eyring fluid model. We present optimal versions of the Bogovskiĭ operator, Nečas' negative norm theorem and Korn's inequality in this framework in Chapter 2. The background about Orlicz- and Orlicz–Sobolev spaces is revised in Chapter 1.

In Chapter 7 we deal with non-stationary flows of power law fluids. A first step is to approximate the equations by a system whose solutions are known to exist. In order to pass to the limit in the regularization parameter, one has to apply compactness of the velocity, the "solenoidal Lipschitz truncation" and arguments from monotone operator theory.

In the last part of the book we study stochastic partial differential equations in fluid mechanics. Probabilistic models have become more and more important for applications and earned a strong interest amongst mathematicians. They can, for instance, take into account physical uncertainties and model turbulence in the fluid motion. We present first existence theorems for generalized Navier–Stokes equations under random perturbations. The results and methods build a basis for future research on stochastic partial differential equations in the analysis of generalized Newtonian fluids. All probabilistic tools are presented as well. The Chapter can be viewed as an introduction to stochastic partial differential equations from an analytical point of view. Thus, the proofs of the main result given in Chapter 10 are accessible to analysts without prior knowledge in stochastic analysis.

ACKNOWLEDGMENT

This monograph is based on my habilitation thesis [26]. It contains results from the papers [27,28,30,31,33,34]. Note that the results from [31] are an improvement on [32] and are not yet contained in [26]. Thanks goes to J. Frehse, J. Málek and E. Süli for being referees of [26].

Special thanks go to all the people without whom the writing of this book would have been impossible: to M. Fuchs for introducing me to mathematical fluid mechanics, to L. Diening for being a constant source of inspiration, for much valuable advise and his unique view on the Lipschitz truncation method, to A. Cianchi for sharing his deep knowledge of Orlicz spaces and to M. Hofmanová for opening the door for me to stochastic partial differential equations. I would like to extend my sincere appreciation to all the people with whom I have discussed research related to this monograph, in particular: M. Bildhauer, J. Frehse, S. Schwarzacher. Moreover, thanks goes to all people who carefully read parts of it: L. Diening, M. Hofmanová, M.V. Lawson, M. Ottobre, S. Paulus, O. Penrose, B.P. Rynne, Z.B. Wyatt.

Dominic Breit
Edinburgh
August 2016

NOTATION

Formula	Meaning
C	Continuous functions
C^α	α-Hölder continuous functions
C^k	k-times continuously differentiable functions
C^∞	∞-times continuously differentiable functions
$\operatorname{spt}(\varphi)$	Support of φ
C_0^∞	C^∞-functions with compact support
C_0^k	C^k-functions with compact support
$\mathcal{D}'$	Dual of C_0^∞
$C_{0,\mathrm{div}}^\infty$	C_0^∞-functions with vanishing divergence
$\mathcal{D}'_{\mathrm{div}}$	Dual of $C_{0,\mathrm{div}}^\infty$
$C_{0,\perp}^\infty$	C_0^∞-functions with vanishing mean value
X'	Dual space of X
$\Vert \cdot \Vert_X$	Norm on X
$\langle \cdot, \cdot \rangle_X$	Inner product on X
$\mathcal{L}^n$	n-dimensional Lebesgue measure
$\mathcal{H}^s$	s-dimensional Hausdorff measure
$\mathfrak{B}$	Borelian σ-algebra
L^p	Lebesgue-space of p-integrable functions
L^p_{loc}	Lebesgue-space of locally p-integrable functions
L^p_{div}	L^p-functions with vanishing divergence
$L^p_\perp$	L^p-functions with vanishing mean
p'	Dual exponent of p: $p' = p/(p-1)$
$L^A/A(L)$	Orlicz-space generated by A
$L^A_\perp/A(L)_\perp$	L^A-functions with vanishing mean
$\widetilde{A}$	Young conjugate of A
u^*	Decreasing rearrangement of u
$u * v$	Convolution of u and v
Mf	Hardy-Littlewood maximal operator applied to f
$\mathcal{B}_r$	Ball with radius r
$\kappa\mathcal{B}_r$	Ball with same center as $\mathcal{B}_r$ and radius κr
Π^k	Averaged Taylor polynomial of order k
$W^{k,p}$	Sobolev functions with differentiability k and integrability p

Formula	Meaning
$W^{k,p}_{loc}$	Sobolev functions with differentiability k and local integrability p
$W^{k,p}_{\mathrm{div}}$	$W^{k,p}$-functions with vanishing divergence
$W^{k,p}_0$	$W^{k,p}$-functions with vanishing trace
$W^{-k,p}$	Dual space of $W^{k,p}_0$
$W^{k,p}_{\mathrm{div}}$	$W^{k,p}$-functions with vanishing divergence
$W^{k,p}_{0,\mathrm{div}}$	$W^{k,p}_0$-functions with vanishing divergence
$W^{-k,p}_{\mathrm{div}}$	Dual space of $W^{k,p}_{0,\mathrm{div}}$
$W^{1,A}$	Orlicz-Sobolev space generated by A
$W^{1,A}_0$	$W^{1,A}$-functions with vanishing trace
$\mathbb{R}^{d\times d}_{sym}$	Symmetric $d \times d$ matrices
$\odot$	Symmetric tensor product
E^p	L^p-functions with symmetric gradient in L^p
E^p_0	E^p-functions with vanishing trace
$E^p_{0,\mathrm{div}}$	E^p_0-functions with vanishing divergence
E^A	L^A-functions with symmetric gradient in L^A
E^A_0	E^A-functions with vanishing trace
$E^A_{0,\mathrm{div}}$	E^A_0-functions with vanishing divergence
BD	Functions of bounded deformation
$\mathcal{M}$	Bounded signed measures
$H^A/H^{A(L)}$	L^A-functions with divergence in L^A
$H^A_0/H^{A(L)}_0$	H^A-functions with vanishing normal trace
Bog	Bogovskiĭ-operator
Δ^{-1}	Solution operator to the Laplace equation
Δ^{-2}	Solution operator to the bi-Laplace equation
curl^{-1}	Solution operator to the curl equation
Q_r	Parabolic cube with radius r
κQ_r	Cube with same center as Q_r and radius κr
$\mathcal{M}^\alpha$	α-parabolic maximal function
$\mathcal{M}^\alpha_\sigma$	α-parabolic maximal function with power σ
$L^p(0,T;\mathcal{V})$	Bochner measurable functions with values in $\mathcal{V}$ and integrability p
$C([0,T];\mathcal{V})$	Continuous functions with values in $\mathcal{V}$
$C^\alpha([0,T];\mathcal{V})$	α-Hölder continuous functions with values in $\mathcal{V}$
$C_w([0,T];\mathcal{V})$	Weakly continuous functions with values in $\mathcal{V}$

Formula	Meaning
$W^{k,p}(0,T;\mathscr{V})$	k-times weakly differentiable functions with values in $\mathscr{V}$ and integrability p
$(\Omega,\mathcal{F},\mathbb{P})$	Probability space with sample space Ω, σ-algebra $\mathcal{F}$ and probability measure $\mathbb{P}$
$(\mathcal{F}_t)_{t\geq 0}$	filtration
$L^p(\Omega,\mathcal{F},\mathbb{P};\mathscr{V})$	Random variable over $(\Omega,\mathcal{F},\mathbb{P})$ with values in $\mathscr{V}$ and moments of order p
$\mathscr{M}_2$	Quadratically integrable martingales
$\mathscr{M}_2^c$	Quadratically integrable continuous martingales
$\mathscr{M}_2^{c,loc}$	Locally in time quadratically integrable continuous martingales
$L_2(\mathscr{H}_1,\mathscr{H}_2)$	Hilbert-Schmidt operators from $\mathscr{H}_1\to\mathscr{H}_2$
$\big(\langle X\rangle_t\big)_{t\geq 0}$	Quadratic variation process of $(X_t)_{t\geq 0}$
$\big(\langle X,Y\rangle_t\big)_{t\geq 0}$	Covariation process of $(X_t)_{t\geq 0}$ and $(Y_t)_{t\geq 0}$
$\mathbb{A}_q$	Stokes operator on L^q
$\mathscr{A}_q$	$\mathcal{A}$-Stokes operator on L^q
D	Domain of an operator

PART 1

Stationary problems

CHAPTER 1

Preliminaries

Contents

Abstract

In this chapter we present some preliminary material which will be needed in order to study stationary models for generalized Newtonian fluids. We begin with the functional analytic framework. In particular, we define Lebesgue-, Sobolev- and Orlicz-spaces and describe their basic properties. After this we present the Lipschitz truncation method in its classical framework and present two applications. Finally, we discuss some modelling aspects concerning power law fluids and provide a historical overview on the mathematical theory of weak solutions for stationary flows.

1.1 LEBESGUE & SOBOLEV SPACES

In this section we define various function spaces. For proofs, further details and references we refer to [5].

Definition 1.1.1 (Classical function spaces)**.** Let $G \subset \mathbb{R}^d$ be open and $k \in \mathbb{N}$. We define

$$C(G) := \{u : G \to \mathbb{R} : u \text{ is continuous}\},$$
$$C^k(G) := \left\{u : G \to \mathbb{R} : \nabla^i u \text{ is continuous for } i = 0, ..., k\right\},$$
$$C^\infty(G) := \left\{u : G \to \mathbb{R} : \nabla^i u \text{ is continuous for all } i \in \mathbb{N}_0\right\}.$$

Remark 1.1.1. In Definition 1.1.1 if we replace G by the closed set $\overline{G}$ we are considering functions whose derivatives are continuous up to the boundary of G.

Definition 1.1.2. Let $G \subset \mathbb{R}^d$ be open and $\alpha \in (0, 1]$. We define

$$C^\alpha(\overline{G}) := \left\{u : G \to \mathbb{R} : \sup_{x \neq y} \frac{|u(x) - u(y)|}{|x - y|^\alpha} < \infty\right\},$$
$$C^\alpha(G) := \left\{u : G \to \mathbb{R} : u \in C^\alpha(\overline{K}) \text{ for all } K \Subset G\right\}$$

as the set of (locally) α-Hölder continuous functions.

Existence Theory for Generalized Newtonian Fluids.
DOI: http://dx.doi.org/10.1016/B978-0-12-811044-7.00002-1

Remark 1.1.2. In the case $\alpha = 1$ we obtain the set of Lipschitz continuous functions.

Definition 1.1.3. Let $G \subset \mathbb{R}^d$ be open and $u : G \to \mathbb{R}$. We define the support of u by

$$\operatorname{spt} u := \overline{\{x \in G : u(x) \neq 0\}}.$$

By $C_0^0(G)$ and $C_0^\infty(G)$ we denote subclasses of $C^0(G)$ and $C^\infty(G)$ whose elements satisfy $\operatorname{spt} u \Subset G$.

Definition 1.1.4 (Lebesgue spaces). Let $(\mathfrak{X}, \Sigma, \mu)$ be a measure space. We define

$$L^p(\mathfrak{X}, \Sigma, \mu) := \left\{ u : \mathfrak{X} \to \mathbb{R} : u \text{ is } \mu\text{-measurable}, \int_{\mathfrak{X}} |u|^p \, d\mu < \infty \right\}, \ 1 \leq p < \infty,$$

$$L^\infty(\mathfrak{X}, \Sigma, \mu) := \left\{ u : \mathfrak{X} \to \mathbb{R} : u \text{ is } \mu\text{-measurable}, \inf_{\mu(N)=0} \sup_{x \in \mathfrak{X} \setminus N} |u(x)| < \infty \right\}.$$

The elements of $L^p(\mathfrak{X}, \Sigma, \mu)$ are equivalence classes. Two functions u and v belong to the same equivalence class if $u = v$ μ-almost everywhere in $\mathfrak{X}$, i.e. if

$$\mu\{x \in \mathfrak{X} : u(x) \neq v(x)\} = 0.$$

Remark 1.1.3.

- $L^p(\mathfrak{X}, \Sigma, \mu)$ is a Banach space together with the natural norm

$$\|u\|_p := \|u\|_{L^p(\mathfrak{X})} := \left(\int_{\mathfrak{X}} |u|^p \, d\mu \right)^{\frac{1}{p}} \quad \text{for} \quad 1 \leq p < \infty,$$

$$\|u\|_\infty := \|u\|_{L^\infty(\mathfrak{X})} := \inf_{\mu(N)=0} \sup_{x \in \mathfrak{X} \setminus N} |u(x)|.$$

- We set $p' = \frac{p}{p-1}$ for $1 < p < \infty$, $p' = \infty$ for $p = 1$ and $p' = 1$ for $p = \infty$. Let $1 \leq p < \infty$ and $v \in L^{p'}(\mathfrak{X}, \Sigma, \mu)$. Then the mapping

$$L^p(\mathfrak{X}, \Sigma, \mu) \ni u \mapsto \int_G uv \, d\mu$$

belongs to the dual space $L^p(\mathfrak{X}, \Sigma, \mu)'$. On the other hand each element of $L^p(\mathfrak{X}, \Sigma, \mu)'$ can be represented via a function $v \in L^{p'}(\mathfrak{X}, \Sigma, \mu)$. In particular, the mapping

$$L^{p'}(\mathfrak{X}, \Sigma, \mu) \ni v \mapsto \left(L^p(\mathfrak{X}, \Sigma, \mu) \ni u \mapsto \int_{\mathfrak{X}} uv \, d\mu \right) \in L^p(\mathfrak{X}, \Sigma, \mu)'$$

is an isomorphism.

- The space $L^p(\mathfrak{X}, \Sigma, \mu)$ is reflexive if and only if $1 < p < \infty$.
- If $\mu(\mathfrak{X}) < \infty$ we have that $L^q(\mathfrak{X}, \Sigma, \mu) \subset L^p(\mathfrak{X}, \Sigma, \mu)$ for $1 \leq p \leq q \leq \infty$. In particular, the following holds

$$\|u\|_{L^p(\mathfrak{X})} \leq c(d, p, q, \mu(\mathfrak{X}))\, \|u\|_{L^q(\mathfrak{X})}$$

  for every $u \in L^q(\mathfrak{X}, \Sigma, \mu)$.
- We can define vector- or matrix-valued L^p-spaces component-wise. For simplicity we do not mention this in the notation. However, when considering functions with values in an infinite dimensional spaces X we denote this by writing $L^p(\mathfrak{X}, \Sigma, \mu; X)$.

The most important special case is $(\mathfrak{X}, \Sigma, \mu) = (G, \mathcal{B}(\mathbb{R}^d), \mathcal{L}^d)$ where $G \subset \mathbb{R}^d$, $\mathcal{B}(\mathbb{R}^d)$ is the Borel σ-algebra on $\mathbb{R}^d$ and $\mathcal{L}^d$ the d-dimensional Lebesgue measure. In this case we abbreviate the notation by setting $L^p(G) := L^p(G, \mathcal{B}(\mathbb{R}^d), \mathcal{L}^d)$. If $1 \leq p < \infty$ then $C_0^\infty(G)$ is dense in $L^p(G)$. In particular, for every $u \in L^p(G)$ there is $(u_m) \subset C_0^\infty(G)$ such that $u_m \to u$ in $L^p(G)$.

Definition 1.1.5 (Local Lebesgue spaces)**.** Let $G \subset \mathbb{R}^d$ be open. We define

$$L^p_{loc}(G) := \{u : G \to \mathbb{R} :\ u \in L^p(K) \text{ for all } K \Subset G\} \quad \text{for} \quad 1 \leq p \leq \infty.$$

Definition 1.1.6 (Weak derivative)**.** Let $G \subset \mathbb{R}^d$ be open and $u \in L^1_{loc}(G)$.

- u is called weakly differentiable with respect to the i-th variable if there is a function $v_i \in L^1_{loc}(G)$ such that

$$\int_G u\, \partial_i \varphi \,\mathrm{d}z = -\int_G v_i \varphi \,\mathrm{d}z \quad \text{for all} \quad \varphi \in C_0^\infty(G).$$

  We call v_i the weak derivative of u with respect to the i-th variable. If weak derivatives with respect to all variables exist, we call u weakly differentiable and denote by $\nabla u = (v_1, \ldots, v_n)$ the weak gradient of u.
- Let $\alpha \in N_0^d$ and $|\alpha| := \alpha_1 + \ldots + \alpha_n$. For a smooth function u we denote by $D^\alpha u := \frac{\partial^{|\alpha|} u}{\partial^{\alpha_1} x_1 \ldots \partial^{\alpha_n} x_n}$ its α-th classical derivative. A function $u \in L^1_{loc}(G)$ is called k-times weakly differentiable if for all $\alpha \in N_0^d$ with $|\alpha| \leq k$ there is a function $v_\alpha \in L^1_{loc}(G)$ such that

$$\int_G u\, D^\alpha \varphi \,\mathrm{d}z = (-1)^{|\alpha|} \int_G v_\alpha \varphi \,\mathrm{d}z \quad \text{for all} \quad \varphi \in C_0^\infty(G).$$

We call $D^\alpha u = v_\alpha$ the α-th weak derivative.

Definition 1.1.7 (Sobolev spaces). For $k \in \mathbb{N}$ and $1 \leq p \leq \infty$ we define

$$W^{k,p}(G) := \left\{ u : G \to \mathbb{R},\ D^\alpha u \in L^p(G),\ \text{for all } \alpha \in \mathbb{N}_0^d,\ |\alpha| \leq k \right\},$$
$$W_0^{k,p}(G) := \overline{C_0^\infty(G)}^{W^{k,p}(G)}.$$

Remark 1.1.4.
- $W^{k,p}(G)$ $(1 \leq p \leq \infty)$ is a Banach space together with the norm

$$\|u\|_{k,p} := \|u\|_{W^{k,p}(G)} := \left(\sum_{|\alpha| \leq k} \int_G |D^\alpha u|^p \, dz \right)^{\frac{1}{p}} \quad \text{for} \quad 1 \leq p < \infty,$$
$$\|u\|_{k,\infty} := \|u\|_{W^{k,\infty}(G)} := \sum_{|\alpha| \leq k} \inf_{\mathcal{L}^d(N)=0} \sup_{x \in G \setminus N} |D^\alpha u(x)|.$$

- The space $W^{k,p}(G)$ is reflexive iff $1 < p < \infty$.
- A sequence $(u_k) \subset W^{k,p}(G)$ $(1 < p < \infty)$ converges weakly to u in $W^{k,p}(G)$ if

$$\int_G D^\alpha u_k \, v \, dz \longrightarrow \int_G D^\alpha u \, v \, dz, \quad k \to \infty,$$

for all $v \in L^{p'}(G, \mathbb{R}^N)$ $(p' := \frac{p}{p-1})$ and for all $\alpha \in \mathbb{N}_0^d$ with $|\alpha| \leq k$.
- The dual space of $W_0^{k,p}(G)$ will be denoted by $W^{-k,p'}(G)$.

Theorem 1.1.1 (Smooth approximation). *Let $G \subset \mathbb{R}^d$ be open and bounded with Lipschitz boundary (i.e. $\partial\Omega$ can be locally parametrized by Lipschitz continuous functions of $d-1$ variables) and $1 \leq p < \infty$. Then $C^\infty(\overline{G})$ is dense in $W^{k,p}(G)$. In particular, for every $u \in W^{k,p}(G)$ there is $(u_m) \subset C_0^\infty(G)$ such that $u_m \to u$ in $W^{k,p}(G)$.*

Theorem 1.1.2 (Sobolev). *Let $G \subset \mathbb{R}^d$ be open.*

a) *The embeddings*

$$W_0^{1,p}(G) \hookrightarrow L^{\frac{dp}{d-p}}(G) \quad \text{if} \quad p < d,$$
$$W_0^{1,p}(G) \hookrightarrow C^{1-\frac{d}{p}}(\overline{G}) \quad \text{if} \quad p > d,\ \mathcal{L}^d(G) < \infty,$$

are continuous.

b) *Let $G \subset \mathbb{R}^d$ be open and bounded with Lipschitz-boundary. The embeddings*

$$W^{1,p}(G) \hookrightarrow L^{\frac{dp}{d-p}}(G) \quad \text{if} \quad p < d,$$
$$W^{1,p}(G) \hookrightarrow C^{1-\frac{d}{p}}(\overline{G}) \quad \text{if} \quad p > d,$$

are continuous.

Theorem 1.1.3 (Kondrachov). *Let $G \subset \mathbb{R}^d$ be open and bounded. The embedding*

$$W_0^{1,p}(G) \hookrightarrow L^q(G), \quad q < \frac{pd}{d-p},$$

is compact for all $p < d$.

In Definition 1.1.7 we have interpreted the boundary values of a Sobolev function as follows: $u = 0$ on ∂G iff $u \in W_0^{1,p}(G)$, where $W_0^{1,p}(G)$ for $p < \infty$ denotes the closure of $C_0^\infty(G)$ in $W^{1,p}(G)$. We will develop a rigorous definition of boundary values and show that it coincides with the former one. In order to do so we need functions which are integrable over the boundary of G. For $G \subset \mathbb{R}^d$ open let $L^p(\partial G)$ be equal to the set of all $\mathcal{H}^{d-1}$-measurable functions with

$$\|u\|_{L^p(\partial\Omega)} := \left(\int_{\partial G} |u|^p \, d\mathcal{H}^{d-1} \right)^{\frac{1}{p}} < \infty.$$

Here $\mathcal{H}^{d-1}$ denotes the $(d-1)$-dimensional Hausdorff measure. For C^1-functions we define the operator $\mathrm{tr} : C^1(\overline{G}) \to L^p(\partial G)$ by $\mathrm{tr}\, u := u|_{\partial G}$.

Lemma 1.1.1. *Let $u \in C^1(\overline{G})$ with $G \subset \mathbb{R}^d$ open and bounded with Lipschitz–boundary. Then we have*

$$\|\mathrm{tr}\, u\|_{L^p(\partial G)} \leq c(d, p, G) \|u\|_{W^{1,p}(G)}.$$

For $u \in W^{1,p}(G)$ we consider the approximation sequence $u_m \in C^\infty(\overline{G})$ with $\|u_m - u\|_{1,p} \to 0$. Its existence follows from Theorem 1.1.1. Lemma 1.1.1 shows that $(\mathrm{tr}\, u_m)$ is a Cauchy-sequence in $L^p(\partial G)$. We define its limit (which does not depend on the special choice of the sequence) as the trace of u. The result is a linear operator $\mathrm{tr} : W^{1,p}(G) \to L^p(\partial G)$ which coincides with the classical trace operator on $C^1(\overline{G})$.

Theorem 1.1.4. *Let $\Omega \subset \mathbb{R}^d$ be open and bounded with Lipschitz boundary. Then we have*

$$W_0^{1,p}(G) = \left\{ u \in W^{1,p}(G);\ \mathrm{tr}\, u = 0 \right\}$$

for $1 \leq p < \infty$.

All results of this section generalize in a straightforward manner to spaces of vector-valued functions. In order to keep the notation simple we do not use target spaces in the notation for our function spaces. It will follow from the context (and the bold-symbol) when we are dealing with these. In fluid mechanics the velocity field is a function from $\mathbb{R}^d \supset G \to \mathbb{R}^d$. In this setting we need function spaces of solenoidal (that is, divergence-free) functions. We will use the following notation for $1 \le p < \infty$

$$W^{1,p}_{\mathrm{div}}(G) := \{\boldsymbol{\psi} \in W^{1,p}(G) : \operatorname{div} \boldsymbol{\psi} = 0\},$$
$$C^{\infty}_{0,\mathrm{div}}(\Omega) := \{\boldsymbol{\psi} \in C^{\infty}_0(G) : \operatorname{div} \boldsymbol{\psi} = 0\},$$
$$L^p_{\mathrm{div}}(G) := \overline{C^{\infty}_{0,\mathrm{div}}(G)}^{L^2(G)}, \quad W^{1,p}_{0,\mathrm{div}}(G) := \overline{C^{\infty}_{0,\mathrm{div}}(G)}^{W^{1,p}(G)}.$$

Finally we write $W^{-1,p'}_{\mathrm{div}}(G)$ for the dual of $W^{1,p}_{0,\mathrm{div}}(G)$.

1.2 ORLICZ SPACES

In this section we present some important properties of Orlicz spaces (see [125] and [5]).

A function $A : [0,\infty) \to [0,\infty]$ is called a Young function if it is convex, left-continuous, vanishing at 0, and neither identically equal to 0 nor to ∞. Thus, with any such function, it is uniquely associated a (non-trivial) non-decreasing left-continuous function $a : [0,\infty) \to [0,\infty]$ such that

$$A(s) = \int_0^s a(r)\,\mathrm{d}r \quad \text{for} \quad s \ge 0. \tag{1.2.1}$$

The *Young conjugate* $\widetilde{A}$ of A is the Young function defined by

$$\widetilde{A}(s) = \sup\{rs - A(r) : r \ge 0\} \quad \text{for} \quad s \ge 0.$$

For $\widetilde{A}$ we have the representation formula

$$\widetilde{A}(s) = \int_0^s a^{-1}(r)\,\mathrm{d}r \qquad \text{for} \quad s \ge 0, \tag{1.2.2}$$

where a^{-1} denotes the (generalized) left-continuous inverse of a. Moreover, for every Young function A,

$$r \le A^{-1}(r)\widetilde{A}^{-1}(r) \le 2r \quad \text{for} \quad r \ge 0 \tag{1.2.3}$$

as well as

$$\widetilde{\widetilde{A}} = A. \tag{1.2.4}$$

Let A be a Young function of the form (1.2.1). Then the convexity of A and $A(0) = 0$ imply

$$A(\lambda s) \le \lambda A(s) \quad \text{for all} \quad \lambda \in [0,1] \tag{1.2.5}$$

and all $s \ge 0$. If $\lambda \ge 1$, then

$$\lambda A(s) \le A(\lambda s) \quad \text{for} \quad s \ge 0. \tag{1.2.6}$$

As a consequence, if $\lambda \geq 1$, then

$$A^{-1}(\lambda s) \leq \lambda A^{-1}(s) \quad \text{for} \quad s \geq 0, \tag{1.2.7}$$

where A^{-1} denotes the (generalized) right-continuous inverse of A.

A Young function A is said to satisfy the Δ_2-condition if there exists a positive constant K such that

$$A(2s) \leq KA(s) \quad \text{for} \quad s \geq 0. \tag{1.2.8}$$

If (1.2.8) just holds for $s \geq s_0$ for some $s_0 > 0$, then A is said to satisfy the Δ_2-condition near infinity. We say that A satisfies the ∇_2-condition [near infinity] if $\tilde{A}$ satisfies the Δ_2-condition [near infinity].

A Young function A is said to dominate another Young function B near infinity if there exist positive constants c and s_0 such that

$$B(s) \leq A(cs) \quad \text{for} \quad s \geq s_0\,. \tag{1.2.9}$$

The functions A and B are called equivalent near infinity if they dominate each other near infinity.

Let G be a measurable subset of $\mathbb{R}^d$, and let $u : G \to \mathbb{R}$ be a measurable function. Given a Young function A, the Luxemburg norm associated with A, of the function u is defined as

$$\|u\|_{L^A(G)} := \inf\left\{\lambda : \int_G A\left(\frac{|u|}{\lambda}\right) \mathrm{d}x \leq 1\right\}.$$

The collection of all measurable functions u for which this norm is finite is called the Orlicz space $L^A(G)$. It turns out to be Banach function space. The subspace of $L^A(G)$ of those functions u such that $\int_G u(x)\,\mathrm{d}x = 0$ will be denoted by $L^A_\perp(G)$. A Hölder-type inequality in Orlicz spaces takes the form

$$\|v\|_{L^{\tilde{A}}(G)} \leq \sup_{u \in L^A(G)} \frac{\int_G u(x)v(x)\,\mathrm{d}x}{\|u\|_{L^A(G)}} \leq 2\|v\|_{L^{\tilde{A}}(G)} \tag{1.2.10}$$

for every $v \in L^{\tilde{A}}(G)$. If $|G| < \infty$, then

$$L^A(G) \hookrightarrow L^B(G) \quad \text{if and only if } A \text{ dominates } B \text{ near infinity}. \tag{1.2.11}$$

The decreasing rearrangement $u^* : [0, \infty) \to [0, \infty]$ of a measurable function $u : G \to \mathbb{R}$ is the (unique) non-increasing, right-continuous function which is equimeasurable with u. Thus,

$$u^*(s) = \sup\{t \geq 0 : |\{x \in G : |u(x)| > t\}| > s\} \quad \text{for} \quad s \geq 0.$$

The equimeasurability of u and u^* implies that

$$\|u\|_{L^A(G)} = \|u^*\|_{L^A(0,|G|)} \tag{1.2.12}$$

for every $u \in L^A(G)$.

The Lebesgue spaces $L^p(G)$, corresponding to the choice $A_p(t) = t^p$, if $p \in [1, \infty)$, and $A_\infty(t) = \infty \cdot \chi_{(1,\infty)}(t)$, if $p = \infty$, are a basic example of Orlicz spaces. Other instances of Orlicz spaces are provided by the Zygmund spaces $L^p \log^\alpha L(G)$, and by the exponential spaces $\exp L^\beta(G)$. If either $p > 1$ and $\alpha \in \mathbb{R}$, or $p = 1$ and $\alpha \geq 0$, then $L^p \log^\alpha L(G)$ is the Orlicz space associated with a Young function equivalent to $t^p(\log t)^\alpha$ near infinity. Given $\beta > 0$, $\exp L^\beta(G)$ denotes the Orlicz space built upon a Young function equivalent to e^{t^β} near infinity.

An important tool will be the following characterization of Hardy type inequalities in Orlicz spaces [46, Lemma 1].

Lemma 1.2.1. *Let A and B be Young functions, and let $L \in (0, \infty]$.*

(i) *There exists a constant C such that*

$$\left\| \frac{1}{s} \int_0^s f(r)\, dr \right\|_{L^B(0,L)} \leq C\|f\|_{L^A(0,L)} \tag{1.2.13}$$

for every $f \in L^A(0, L)$ if and only if either $L < \infty$ and there exist constants $c > 0$ and $t_0 \geq 0$ such that

$$t\int_{t_0}^t \frac{B(s)}{s^2}\, ds \leq A(ct) \qquad \text{for } t \geq t_0, \tag{1.2.14}$$

or $L = \infty$ and (1.2.14) holds with $t_0 = 0$. In particular, in the latter case, the constant C in (1.2.13) depends only on the constant c appearing in (7.0.1).

(ii) *There exists a constant C such that*

$$\left\| \int_s^L f(r) \frac{dr}{r} \right\|_{L^B(0,L)} \leq C\|f\|_{L^A(0,L)} \tag{1.2.15}$$

for every $f \in L^A(0, L)$ if and only if either $L < \infty$ and there exist constants $c > 0$ and $t_0 \geq 0$ such that

$$t\int_{t_0}^t \frac{\widetilde{A}(s)}{s^2}\, ds \leq \widetilde{B}(ct) \qquad \text{for } t \geq t_0, \tag{1.2.16}$$

or $L = \infty$ and (1.2.16) holds with $t_0 = 0$. In particular, in the latter case, the constant C in (1.2.15) depends only on the constant c appearing in (1.2.16).

Assume now that G is an open set. The Orlicz–Sobolev space $W^{1,A}(G)$ is the set of all functions in $L^A(G)$ whose distributional gradient also belongs to $L^A(G)$. It is a Banach space endowed with the norm

$$\|u\|_{W^{1,A}(G)} := \|u\|_{L^A(G)} + \|\nabla u\|_{L^A(G)}.$$

We also define the subspace of $W^{1,A}(G)$ of those functions which vanish on ∂G as

$$W_0^{1,A}(G)$$
$$= \{u \in W^{1,A}(G) : \text{the continuation of } u \text{ by } 0 \text{ is weakly differentiable}\}.$$

In the case where $A(t) = t^p$ for some $p \geq 1$, and ∂G is regular enough, such a definition of $W_0^{1,A}(G)$ can be shown to reproduce the usual space $W_0^{1,p}(G)$ defined as the closure in $W^{1,p}(G)$ of the space $C_0^\infty(G)$ of smooth compactly supported functions in G. In general, the set of smooth bounded functions is dense in $L^A(G)$ only if A satisfies the Δ_2-condition (just near infinity when $|G| < \infty$), and hence, for arbitrary A, our definition of $W_0^{1,A}(G)$ yields a space which can be larger than the closure of $C_0^\infty(G)$ in $W_0^{1,A}(G)$ even for smooth domains. On the other hand, if G is a Lipschitz domain, namely a bounded open set in $\mathbb{R}^d$ which is locally the graph of a Lipschitz function of $d-1$ variables, then

$$W_0^{1,A}(G) = W^{1,A}(G) \cap W_0^{1,1}(G),$$

where $W_0^{1,1}(G)$ is defined as usual.

Lemma 1.2.2. *Let A be a Young-function satisfying the Δ_2-condition and $G \subset \mathbb{R}^d$ open.*

- *The following holds*

$$W_0^{1,A}(G) = \overline{C_0^\infty(G)}^{W^{1,A}(G)} = W^{1,A}(G) \cap W_0^{1,1}(G).$$

- *For every $u \in W^{1,A}$ there is a sequence $(u_k) \in C^\infty(G) \cap W^{1,A}(G)$ such that $u_k \to u$ in $W^{1,A}(G)$.*

Lemma 1.2.3. *Let A be a Young function satisfying the Δ_2- and the ∇_2-condition. Then $L^A(G)$ is reflexive with*

$$L^A(G)' \cong L^{\tilde{A}}(G).$$

1.3 BASICS ON LIPSCHITZ TRUNCATION

The purpose of the Lipschitz truncation technique is to approximate a Sobolev function $u \in W^{1,p}$ by λ-Lipschitz functions u_λ that coincide with u up to a set of small measure. The functions u_λ are constructed nonlinearly by modifying u on the level set of the Hardy–Littlewood maximal function of the gradient ∇u. This idea goes back to Acerbi and Fusco [1–3]. Lipschitz truncations are used in various areas of analysis: calculus of variations, in the existence theory of partial differential equations, and in regularity theory. We refer to [62] for a longer list of references.

The basic idea is to take a function $u \in W^{1,p}(\mathbb{R}^d)$, where $p \geq 1$, and cut values where the maximal function of its gradient is large. The Hardy–Littlewood maximal operator is defined by

$$M(v)(x) = \sup_{B:\, x \in B} \fint_B |v| \, dy$$

for $v \in L^1_{loc}(\mathbb{R}^d)$ which can be extended to vector- (or matrix-)valued functions by setting $M(\mathbf{v}) = M(|\mathbf{v}|)$. Basic properties of the maximal operator are summarized in the following lemma (see e.g. [134] and [141, Lemma 3.2] for d)).

Lemma 1.3.1. ***a)*** *Let $v \in L^1_{loc}(\mathbb{R}^d)$ and $\lambda > 0$. The level-set $\{x \in \mathbb{R}^d : |M(v)(x)| > \lambda\}$ is open.*

b) *The strong-type estimate*

$$\|M(v)\|_{L^p(\mathbb{R}^d)} \leq c_p \|v\|_{L^p(\mathbb{R}^d)} \quad \forall v \in L^p(\mathbb{R}^d)$$

holds for all $p \in (1, \infty]$.

c) *The weak-type estimate*

$$\mathcal{L}^d\Big(\{x \in \mathbb{R}^d : |M(v)(x)| > \lambda\}\Big) \leq c_p \frac{\|v\|_p^p}{\lambda^p} \quad \forall v \in L^p(\mathbb{R}^d)$$

holds for all $p \in [1, \infty)$ and all $\lambda > 0$.

d) *We have the estimate*

$$\int_{\{M(v)>\lambda\}} |v|^p \, dx \leq c_p \int_{\{|v|>\lambda/2\}} |v|^p \, dx \quad \forall p \in [1, \infty)$$

for all $\lambda > 0$.

The Lipschitz truncation will be defined via the maximal function of the gradient. For $u \in W^{1,p}(\mathbb{R}^d)$ the "bad set" is defined by

$$\mathcal{O}_\lambda := \{x \in \mathbb{R}^d : M(\nabla u)(x) > \lambda\}, \tag{1.3.17}$$

where $\lambda \geq 0$. If we have $u \in W^{1,p}(G)$ for $G \neq \mathbb{R}^d$ it has to be extended to $u \in W^{1,p}(\mathbb{R}^d)$. This can be done in an obvious way if $u \in W^{1,p}_0(G)$ where the extension is zero outside G. In the general case we may apply [5, Thm. 4.26]. Now, for $x, y \in \mathbb{R}^d \setminus \mathcal{O}_\lambda$ we have a.e.

$$|u(x) - u(y)| \leq c|x - y|\big(M(\nabla u)(x) + M(\nabla u)(y)\big) \leq 2c\lambda|x - y|,$$

see, e.g., [112]. Hence u is Lipschitz-continuous in $\mathbb{R}^d \setminus \mathcal{O}_\lambda$ with Lipschitz constant proportional to λ. By a standard extension theorem (see e.g. [71,

p. 201]) we can extend u (defined in $\mathbb{R}^d \setminus \mathcal{O}_\lambda$) to u_λ (defined in $\mathbb{R}^d$) such that the Lipschitz constant is preserved. (When dealing with this simple extension it is necessary to cut large values of $M(u)$ as well. We neglect this for brevity.) This means we have

$$|\nabla u_\lambda| \le c\lambda \quad \text{in} \quad \mathbb{R}^d. \tag{1.3.18}$$

Moreover, by construction we have

$$\{x \in \mathbb{R}^d : u \ne u_\lambda\} \subset \mathcal{O}_\lambda.$$

This and Lemma 1.3.1 c) imply

$$\mathcal{L}^d\Big(\{x \in \mathbb{R}^d : u \ne u_\lambda\}\Big) \le \mathcal{L}^d(\mathcal{O}_\lambda) \le \frac{c\|\nabla u\|_p^p}{\lambda^p}. \tag{1.3.19}$$

Combining (1.3.18) and (1.3.19) shows

$$\begin{aligned}\int_{\mathbb{R}^d} |\nabla u_\lambda|^p \,\mathrm{d}x &= \int_{\mathbb{R}^d\setminus\mathcal{O}_\lambda} |\nabla u_\lambda|^p \,\mathrm{d}x + \int_{\mathcal{O}_\lambda} |\nabla u_\lambda|^p \,\mathrm{d}x \\ &\le \int_{\mathbb{R}^d\setminus\mathcal{O}_\lambda} |\nabla u|^p \,\mathrm{d}x + c\lambda^p \mathcal{L}^d(\mathcal{O}_\lambda) \\ &\le c\int_{\mathbb{R}^d} |\nabla u|^p \,\mathrm{d}x.\end{aligned}$$

We obtain the following stability result

$$\|\nabla u_\lambda\|_p \le c\|\nabla u\|_p. \tag{1.3.20}$$

The basic properties (1.3.18)–(1.3.20) are already enough to make the Lipschitz truncation a powerful tool for numerous applications. We present two rather classical ones.

Lower semi-continuity in $W^{1,p}$.

Let $G \subset \mathbb{R}^d$ be an open and bounded with Lipschitz boundary. Suppose further that $F : G \times \mathbb{R}^{d\times D} \to [0, \infty)$ is a continuous function with p-growth ($p > 1$), i.e.,

$$F(x, \mathbf{Q}) \le c|\mathbf{Q}|^p + g(x) \quad \forall \mathbf{Q} \in \mathbb{R}^{d\times D} \tag{1.3.21}$$

with a constant $c \ge 0$ and a non-negative function $g \in L^1(G)$. We are interested in minimizing the functional

$$\mathcal{G}_F[\mathbf{w}] = \int_G F(x, \nabla \mathbf{w}) \,\mathrm{d}x$$

defined for functions $\mathbf{w} : G \to \mathbb{R}^D$. An important concept in showing the existence of minimizers is the lower semi-continuity of $\mathcal{G}_F$ with respect to an appropriate topology. The functional $\mathcal{G}_F$ is called $W^{1,p}$-weakly lower semi-continuous if

$$\mathcal{G}_F[\mathbf{v}] \le \liminf_{n\to\infty} \mathcal{G}_F[\mathbf{v}_n] \tag{1.3.22}$$

provided $\mathbf{v}_n \rightharpoonup \mathbf{v}$ in $W^{1,p}(G)$ for $m \to \infty$. The functional $\mathcal{G}_F$ is called $W^{1,\infty}$-weakly* lower semi-continuous if

$$\mathcal{G}_F[\mathbf{v}] \le \liminf_{n\to\infty} \mathcal{G}_F[\mathbf{v}_n]$$

provided $\mathbf{v}_n \rightharpoonup^* \mathbf{v}$ in $W^{1,\infty}(G)$ for $n \to \infty$.

Due to (1.3.21) the right concept for the functional $\mathcal{G}_F$ is $W^{1,p}$-weak lower semi-continuity. It can be deduced from $W^{1,\infty}$-weak* lower semi-continuity by the Lipschitz truncation, see Lemma 1.3.3 below. Using this idea Acerbi and Fusco [1] showed the $W^{1,p}$ lower semi-continuity of $\mathcal{G}_F$ in the case where F is only quasi-convex. Note that $W^{1,\infty}$-weak* lower semi-continuity is a consequence of the definition of quasi-convexity, see [1, Thm. 2.1]. For brevity we do not discuss the concept of quasi-convexity and refer instead to the fundamental papers [16] and [115].

Lower integrability for the p-Laplace system.

Consider the system

$$\operatorname{div}(|\nabla \mathbf{v}|^{p-2}\nabla \mathbf{v}) = \operatorname{div} \mathbf{F} \quad \text{in} \quad G, \tag{1.3.23}$$

$$\mathbf{v} = 0 \quad \text{on} \quad \partial G. \tag{1.3.24}$$

Here, G is an open set in $\mathbb{R}^d$, with $d \ge 2$, the exponent $p \in (1,\infty)$ and the function $\mathbf{F} : \Omega \to \mathbb{R}^{d\times D}$ is given. A weak solution to (1.3.23) is a function $\mathbf{v} \in W_0^{1,p}(G)$ such that

$$\int_G |\nabla \mathbf{v}|^{p-2}\nabla \mathbf{v} : \boldsymbol{\varphi}\, dx = \int_G \mathbf{F} : \nabla \boldsymbol{\varphi}\, dx$$

for all $\boldsymbol{\varphi} \in W_0^{1,p}(G)$. Its existence can be shown via standard methods provided $\mathbf{F} \in L^{p'}(G)$. We are concerned here with the question of how the regularity of $\mathbf{F}$ transfers to $\mathbf{v}$ (particularly to $|\nabla \mathbf{v}|^{p-2}\nabla \mathbf{v}$). In the linear case $p = 2$ this is answered by the classical theory of Calderón and Zygmund [43]. It says that $\mathbf{F} \in L^q(G)$ implies $\nabla \mathbf{v} \in L^q(G)$ for all $q \in (1,\infty)$. Note that the case $q < 2$, where q is below the duality exponent p', is included. In that situation existence of weak solutions is not clear a priori. There has been a great deal of effort in obtaining a corresponding result for the nonlinear case $p \neq 2$ such that

$$\mathbf{F} \in L^q(G) \quad \Rightarrow \quad |\nabla \mathbf{v}|^{p-2}\nabla \mathbf{v} \in L^q(G) \quad \forall q \in (1,\infty) \tag{1.3.25}$$

together with a corresponding estimate. This has been positively answered in the fundamental paper by Iwaniec [97] provided $q \geq p'$. An improvement to $q > p' - \delta$ for some small $\delta > 0$ has been carried out in [98] by different methods (for an overview and further references see [113]). We remark that the case $q \in (1, p' - \delta)$ is still open. Based on the Lipschitz truncation we can give a relatively easy proof for the estimate in the case $q \in (p' - \delta, p')$ using the approach in [141] (see also [40] for a more general setting and [101] for the parabolic problem), see Lemma 1.3.4 below.

Before we give proofs of these applications we present an important improvement of the Lipschitz truncation which firstly appeared in [62]. It concerns the smallness of the level-sets. Similar ideas have been used earlier for the L^∞-truncation in [78].

Lemma 1.3.2. *Let $v \in L^p(\mathbb{R}^d)$ with $p \in (1, \infty)$. Then there exist $j_0 \in \mathbb{N}$ and a sequence $\lambda_j \in \mathbb{R}$ with $2^{2^j} \leq \lambda_j \leq 2^{2^{j+1}-1}$ such that*

$$\lambda_j^p \mathcal{L}^d\Big(\{x \in \mathbb{R}^d : M(v) > \lambda_j\}\Big) \leq c^p 2^{-j} \|v\|_p^p$$

for all $j \geq j_0$ where $c = c_p$ is the constant in Lemma 1.3.1 b).

Proof. We have

$$\begin{aligned} \|M(v)\|_p^p &= \int_{\mathbb{R}^d} \int_0^\infty \vartheta^{p-1} \chi_{\{|v|>\vartheta\}} \, \mathrm{d}\vartheta \, \mathrm{d}x \\ &= \int_{\mathbb{R}^d} \sum_{m \in \mathbb{Z}} \int_{2^m}^{2^{m+1}} \vartheta^{p-1} \chi_{\{|M(v)|>\vartheta\}} \, \mathrm{d}\vartheta \, \mathrm{d}x \\ &\geq \int_{\mathbb{R}^d} \sum_{m \in \mathbb{Z}} (2^m)^p \chi_{\{|M(v)|>2^{m+1}\}} \, \mathrm{d}x \\ &\geq \sum_{j \in \mathbb{N}} \sum_{k=2^j}^{2^{j+1}-1} \int_{\mathbb{R}^d} (2^k)^p \chi_{\{|M(v)|>2 \cdot 2^k\}} \, \mathrm{d}x. \end{aligned} \tag{1.3.26}$$

The continuity of M on $L^p(\mathbb{R}^d)$, see Lemma 1.3.1 b), implies

$$\sum_{j \in \mathbb{N}} \sum_{k=2^j}^{2^{j+1}-1} \int_{\mathbb{R}^d} (2^k)^p \chi_{\{|M(v)|>2 \cdot 2^k\}} \, \mathrm{d}x \leq c^p \|v\|_p^p.$$

In particular, for all $j \in \mathbb{N}$

$$\sum_{k=2^j}^{2^{j+1}-1} \int_{\mathbb{R}^d} (2^k)^p \chi_{\{|M(v)|>2 \cdot 2^k\}} \, \mathrm{d}x \leq c^p \|v\|_p^p.$$

Since the sum contains 2^j summands, there is at least one index k_j such that

$$\int_{\mathbb{R}^d} (2^{k_j})^p \chi_{\{|M(v)|>2\cdot 2^{k_j}\}} \,\mathrm{d}x \le c^p \|v\|_p^p 2^{-j}. \tag{1.3.27}$$

Define $\lambda_j := 2^{k_j}$ and we conclude from (1.3.27) that

$$\int_{\mathbb{R}^d} (\lambda_j)^p \chi_{\{|M(v)|>2\,\lambda_j\}} \,\mathrm{d}x \le c^p \|v\|_p^p 2^{-j}.$$

This proves the claim. □

Lemma 1.3.2 shows that there is a particular sequence of levels (λ_j) such that

$$\lambda_j^p \mathcal{L}^d\left(\mathcal{O}_{\lambda_j}\right) \le \kappa_j \|\nabla \mathbf{v}\|_p^p \tag{1.3.28}$$

with $\kappa_j \to 0$ for $j \to \infty$. This improves the estimate (1.3.19) and does not follow from the original results by Acerbi and Fusco. It allows us to simplify the original proof of $W^{1,p}$-lower semi-continuity from [1].

Lemma 1.3.3. *Assume that the functionals $\mathcal{G}_F$ defined in* (1.3.22) *are $W^{1,\infty}$-weakly* lower semi-continuous for any choice of F satisfying* (1.3.21). *Then they are $W^{1,p}$-weakly lower semi-continuous.*

Proof. Let $(\mathbf{v}_n) \subset W^{1,p}(G)$ be a sequence with weak limit $\mathbf{v}$ such that

$$\mathbf{u}_n := \mathbf{v}_n - \mathbf{v} \rightharpoonup 0 \quad \text{in} \quad W^{1,p}(G).$$

We take the sequence (λ_j) in accordance with Lemma 1.3.2 for the level-sets of $\nabla \mathbf{u}$. We apply the Lipschitz truncation to the sequence $(\mathbf{u}_n)$ with level $\lambda = \lambda_j$, see the construction after (1.3.17), and obtain for the double sequence $(\mathbf{u}_{n,j} := \mathbf{u}_{n,\lambda_j})$

$$\|\nabla \mathbf{u}_{n,j}\|_\infty \le c\lambda_j \tag{1.3.29}$$

$$\mathbf{u}_{n,j} \rightharpoonup^* 0 \quad \text{in} \quad W^{1,\infty}(G) \tag{1.3.30}$$

$$\lambda_j^p \mathcal{L}^d\left(\{x \in \mathbb{R}^d : \mathbf{u}_n \neq \mathbf{u}_{n,j}\}\right) \le \kappa_j \tag{1.3.31}$$

due to (1.3.18)–(1.3.20), where $\kappa_j \to 0$ for $j \to \infty$. We obtain

$$\begin{aligned}
\liminf_{m\to\infty} \mathcal{G}[\mathbf{v}_n] &= \liminf_{n\to\infty} \int_G F(\cdot, \nabla \mathbf{v} + \nabla \mathbf{u}_n)\,\mathrm{d}x \\
&\ge \liminf_{n\to\infty} \int_{G\setminus\mathcal{O}_{\lambda_j}} F(\cdot, \nabla \mathbf{v} + \nabla \mathbf{u}_n)\,\mathrm{d}x = \liminf_{n\to\infty} \int_{G\setminus\mathcal{O}_{\lambda_j}} F(\nabla \mathbf{v} + \nabla \mathbf{u}_{n,j})\,\mathrm{d}x \\
&\ge \liminf_{n\to\infty} \int_G F(\cdot, \nabla \mathbf{v} + \nabla \mathbf{u}_{n,j})\,\mathrm{d}x - \limsup_{n\to\infty} \int_{\mathcal{O}_{\lambda_j}} F(\cdot, \nabla \mathbf{v} + \nabla \mathbf{u}_{n,j})\,\mathrm{d}x.
\end{aligned} \tag{1.3.32}$$

We can use the functional $\mathcal{G}_{\tilde{F}}$ with

$$\tilde{F}(x, \mathbf{Q}) = F(x, \nabla \mathbf{v} + \mathbf{Q}), \quad \mathbf{Q} \in \mathbb{R}^{d \times D}.$$

The function $\tilde{F}$ has p-growth as required in (1.3.21) (by $\nabla \mathbf{v} \in L^p(\Omega)$) such that

$$\liminf_{n\to\infty} \int_G F(\cdot, \nabla \mathbf{v} + \nabla \mathbf{u}_{n,j})\, dx = \liminf_{n\to\infty} \mathcal{G}_{\tilde{F}}[\mathbf{u}_{n,\lambda_j}] \geq \mathcal{G}_{\tilde{F}}[\mathbf{0}] = \mathcal{G}_F[\mathbf{v}]. \quad (1.3.33)$$

This is a consequence of the $W^{1,\infty}$-weak* lower semi-continuity of $\mathcal{G}_{\tilde{F}}$. Finally we have

$$\begin{aligned}\int_{\mathcal{O}_{\lambda_j}} F(\cdot, \nabla \mathbf{v} + \nabla \mathbf{u}_{n,j})\, dx &\leq c \int_{\mathcal{O}_{\lambda_j}} \big(|\nabla \mathbf{v}|^p + g\big)\, dx + c \int_{\mathcal{O}_{\lambda_j}} |\nabla \mathbf{u}_{n,j}|^p\, dx \\ &\leq c \int_{\mathcal{O}_{\lambda_j}} \big(|\nabla \mathbf{v}|^p + g\big)\, dx + c\lambda^d \mathcal{L}^d(\mathcal{O}_{\lambda_j})\end{aligned}$$

such that

$$\limsup_{\lambda\to\infty} \limsup_{n\to\infty} \int_{\mathcal{O}_{\lambda_j}} F(\cdot, \nabla \mathbf{v} + \nabla \mathbf{u}_{n,j})\, dx = 0 \quad (1.3.34)$$

by (1.3.31) and $|\nabla \mathbf{v}|^p + g \in L^1(G)$. Combining (1.3.32)–(1.3.34) shows that

$$\mathcal{G}_F[\mathbf{v}] \leq \liminf_{m\to\infty} \mathcal{G}_F[\mathbf{v}_n],$$

i.e. $\mathcal{G}_F$ is $W^{1,p}$-weakly lower semi-continuous. □

We now turn to the proof of the lower integrability for the p-Laplace system. In addition to the Lipschitz truncation crucial ingredients are the following integral identities. Let $0 < \varrho < \infty$, $0 \leq \underline{\delta} < \varrho < \overline{\delta}$ and $(\mathfrak{X}, \Sigma, \mu)$ be a measure space. There holds for every μ-measurable function f with $|f|^\varrho \in L^1(\mathfrak{X}, \Sigma, \mu)$

$$\int_0^\infty \vartheta^{\varrho-1-\underline{\delta}} \bigg(\int_{\{|f|>\vartheta\}} |f|^{\underline{\delta}}\, d\mu \bigg)\, d\vartheta = \frac{1}{\varrho - \underline{\delta}} \int_{\mathfrak{X}} |f|^\varrho\, d\mu, \quad (1.3.35)$$

$$\int_0^\infty \vartheta^{\varrho-1-\overline{\delta}} \bigg(\int_{\{|f|\leq\vartheta\}} |f|^{\overline{\delta}}\, d\mu \bigg)\, d\vartheta = \frac{1}{\overline{\delta} - \varrho} \int_{\mathfrak{X}} |f|^\varrho\, d\mu. \quad (1.3.36)$$

Both equalities are easy consequences of Fubini's Theorem. As we will apply (1.3.35) and (1.3.36) several times it is important that all estimates hold for any $\lambda > 0$. So Lemma 1.3.2 is no use.

Lemma 1.3.4. *There is a number $\delta > 0$ such that for all $q \in (p' - \delta)$ the following holds. Let $\mathbf{v} \in W_0^{1,q(p-1)}(G)$ be a weak solution to (1.3.23) with $\mathbf{F} \in L^q(G)$. Then we have*

$$\int_G \big||\nabla \mathbf{v}|^{p-2} \nabla \mathbf{v}\big|^q\, dx \leq c \int_G |\mathbf{F}|^q\, dx.$$

Proof. Take the solution $\mathbf{v}$ to (1.3.23) and use its Lipschitz truncation $\mathbf{v}_\lambda$ as a test-function, see the construction after (1.3.17). Note that the Lipschitz truncation can preserve zero boundary values at least in the case of a Lipschitz boundary, cf. [62, Thm. 3.2]. We obtain

$$\int_{\mathbb{R}^d\setminus\mathcal{O}_\lambda} |\nabla\mathbf{v}|^{p-2}\nabla\mathbf{v} : \nabla\mathbf{v}_\lambda \,\mathrm{d}x = -\int_{\mathcal{O}_\lambda} |\nabla\mathbf{v}|^{p-2}\nabla\mathbf{v} : \nabla\mathbf{v}_\lambda \,\mathrm{d}x + \int\int_{\mathbb{R}^d} \mathbf{F} : \nabla\mathbf{v}_\lambda \,\mathrm{d}x$$

which implies by (1.3.18)

$$\int_{\mathbb{R}^d\setminus\mathcal{O}_\lambda} |\nabla\mathbf{v}|^p \,\mathrm{d}x \le c\lambda \int_{\mathcal{O}_\lambda} |\nabla\mathbf{v}|^{p-1} \,\mathrm{d}x + \int_{\mathbb{R}^d} |\mathbf{F}||\nabla\mathbf{v}_\lambda| \,\mathrm{d}x. \tag{1.3.37}$$

Furthermore, the following holds

$$\begin{aligned}\int_{\{|\nabla\mathbf{v}|\le\lambda\}} |\nabla\mathbf{v}|^p \,\mathrm{d}x &= \int_{\{|\nabla\mathbf{v}|\le\lambda\}\cap\mathcal{O}_\lambda} |\nabla\mathbf{v}|^p \,\mathrm{d}x + \int_{\{|\nabla\mathbf{v}|\le\lambda\}\setminus\mathcal{O}_\lambda} |\nabla\mathbf{v}|^p \,\mathrm{d}x \\ &\le \lambda \int_{\mathcal{O}_\lambda} |\nabla\mathbf{v}|^{p-1} \,\mathrm{d}x + \int_{\mathbb{R}^d\setminus\mathcal{O}_\lambda} |\nabla\mathbf{v}|^p \,\mathrm{d}x.\end{aligned}$$

Inserting this into (1.3.37) yields

$$\int_{\{|\nabla\mathbf{v}|\le\lambda\}} |\nabla\mathbf{v}|^p \,\mathrm{d}x \le c\lambda \int_{\mathcal{O}_\lambda} |\nabla\mathbf{v}|^{p-1} \,\mathrm{d}x + \int_{\mathbb{R}^d} |\mathbf{F}||\nabla\mathbf{v}_\lambda| \,\mathrm{d}x.$$

As a consequence of Lemma 1.3.1 d) we deduce

$$\int_{\{|\nabla\mathbf{v}|\le\lambda\}} |\nabla\mathbf{v}|^p \,\mathrm{d}x \le c\lambda \int_{\{|\nabla\mathbf{v}|>\lambda/2\}} |\nabla\mathbf{v}|^{p-1} \,\mathrm{d}x + \int_{\mathbb{R}^d} |\mathbf{F}||\nabla\mathbf{v}_\lambda| \,\mathrm{d}x.$$

After multiplying with λ^{q-1-p} and integrating we have

$$\begin{aligned}\int_0^\infty \lambda^{q-1-p}\Big(\int_{\{|\nabla\mathbf{v}|\le\lambda\}} |\nabla\mathbf{v}|^p \,\mathrm{d}x\Big)\,\mathrm{d}\lambda &\le c\int_0^\infty \lambda^{q-p} \int_{\{|\nabla\mathbf{v}|>\lambda\}} |\nabla\mathbf{v}|^{p-1} \,\mathrm{d}x\,\mathrm{d}\lambda \\ &\quad + \int_0^\infty \lambda^{q-1-p} \int_{\mathbb{R}^d} |\mathbf{F}||\nabla\mathbf{v}_\lambda| \,\mathrm{d}x\,\mathrm{d}\lambda.\end{aligned}$$

Setting $\chi_\lambda := \int_{\mathbb{R}^d} |\mathbf{F}||\nabla\mathbf{v}_\lambda| \,\mathrm{d}x$ we obtain on account of (1.3.35) and (1.3.36)

$$\frac{1}{p-q}\int |\nabla\mathbf{v}|^q \,\mathrm{d}x \le \frac{c}{q-p+1}\int |\nabla\mathbf{v}|^q \,\mathrm{d}x + \int_0^\infty \lambda^{q-1-p}\chi(\lambda)\,\mathrm{d}\lambda.$$

If q is close enough to p, say $p-q<\tilde{\delta}$, we have

$$\frac{c}{q-p+1} < \frac{1}{p-q}$$

and hence

$$\int |\nabla\mathbf{v}|^q \,\mathrm{d}x \le c\int_0^\infty \lambda^{q-1-p}\chi(\lambda)\,\mathrm{d}\lambda.$$

We split $\chi(\lambda) = \chi_1(\lambda) + \chi_2(\lambda)$ where

$$\chi_1(\lambda) = \int_{\{M(\nabla \mathbf{v}) \leq \lambda\}} |\mathbf{F}||\nabla \mathbf{v}_\lambda| \, \mathrm{d}x, \quad \chi_1(\lambda) = \int_{\{M(\nabla \mathbf{v}) > \lambda\}} |\mathbf{F}||\nabla \mathbf{v}_\lambda| \, \mathrm{d}x.$$

Note that we have $|\nabla \mathbf{v}_\lambda| = |\nabla \mathbf{v}| \leq M(\nabla \mathbf{v})$ on $\{M(\nabla \mathbf{v}) \leq \lambda\}$. Setting $\mu = |\mathbf{F}|\mathcal{L}^d$ and using (1.3.36) (with $f = M(\nabla \mathbf{v})$, $\overline{\delta} = 1$ and $\varrho = q - p + 1$) as well as Hölder's inequality we obtain

$$\begin{aligned}
\int_0^\infty \lambda^{q-1-p} \chi_1(\lambda) \, \mathrm{d}\lambda &\leq \int_0^\infty \lambda^{q-1-p} \int_{\{M(\nabla \mathbf{v}) \leq \lambda\}} M(\nabla \mathbf{v}) \, \mathrm{d}\mu \, \mathrm{d}\lambda \\
&= \frac{1}{p-q} \int_{\mathbb{R}^d} M(\nabla \mathbf{v})^{q-p+1} \, \mathrm{d}\mu \\
&= \frac{1}{p-q} \int_{\mathbb{R}^d} M(\nabla \mathbf{v})^{q-p+1} |\mathbf{F}| \, \mathrm{d}x \\
&\leq \frac{1}{p-q} \|\mathbf{F}\|_{q/(p-1)} \|M(\nabla \mathbf{v})\|_q^{q-p+1} \\
&\leq c \|\mathbf{F}\|_{q/(p-1)} \|\nabla \mathbf{v}\|_q^{q-p+1}
\end{aligned}$$

and similarly by (1.3.35) (with $f = 1$, $\underline{\delta} = 0$ and $\varrho = q - p + 1$)

$$\begin{aligned}
\int_0^\infty \lambda^{q-1-p} \chi_2(\lambda) \, \mathrm{d}\lambda &\leq c \int_0^\infty \lambda^{q-p} \int_{\{M(\nabla \mathbf{v}) > \lambda\}} \mathrm{d}\mu \, \mathrm{d}\lambda \\
&= \frac{c}{q-p+1} \int_{\mathbb{R}^d} M(\nabla \mathbf{v})^{q-p+1} \, \mathrm{d}\mu \\
&\leq c \|\mathbf{F}\|_{q/(p-1)} \|\nabla \mathbf{v}\|_q^{q-p+1}.
\end{aligned}$$

Combining the estimates above implies

$$\int_G |\nabla \mathbf{v}|^q \, \mathrm{d}x \leq c \int_G |\mathbf{F}|^{\frac{q}{p-1}} \, \mathrm{d}x \quad \forall q \in (p - \tilde{\delta}, p)$$

or equivalently on setting $\delta = \frac{\tilde{\delta}}{p-1}$

$$\int_G ||\nabla \mathbf{v}|^{p-2} \nabla \mathbf{v}|^q \, \mathrm{d}x \leq c \int_G |\mathbf{F}|^q \, \mathrm{d}x \quad \forall q \in (p' - \delta, p'). \qquad \square$$

We now turn to an alternative approach for the extension of $u|_{\mathbb{R}^d \setminus \mathcal{O}_\lambda}$ into the "bad set" which has been used in [33] and [60]. Instead of using classical extension theorems as in the definition after (1.3.17) one can work with a Whitney covering of the "bad set" and local approximations. This is much more flexible and allows for instants to cut only parts of the gradient (in particular the symmetric gradient) or to work with higher derivatives, cf. Chapter 3. In fact, this is so far the only successful method for parabolic problems, cf. Section 5.2. The following lemma shows how to decompose an open set. It has been proved in [33] and [65] by slightly modifying the family of closed dyadic cubes given in [93].

Lemma 1.3.5. *Let $\mathcal{O} \subset \mathbb{R}^d$ be open. There is a Whitney covering $\{Q_i\}$ of $\mathcal{O}$ with the following properties.*

(W1) $\bigcup_j \overline{Q_j} = \mathcal{O}$ *and* $Q_j \cap Q_k = \emptyset$ *for* $j \neq k$.

(W2) $8\sqrt{d}\ell(Q_j) \leq \operatorname{dist}(Q_j, \partial\mathcal{O}) \leq 32\sqrt{d}\ell(Q_j)$. *In particular, if* $c_d := 2 + 32\sqrt{d}$, *then* $(c_d Q_j) \cap (\mathbb{R}^d \setminus \mathcal{O}) \neq \emptyset$.

(W3) *If the boundaries of the two cubes* Q_j *and* Q_k *touch, then*

$$\frac{1}{2} \leq \frac{\ell(Q_j)}{\ell(Q_k)} \leq 2.$$

(W4) *For a given* Q_j *there exists at most* $(3^d - 1)2^d$ *cubes* Q_k *that touch* Q_j.

On setting $Q_j^* := \frac{9}{8} Q_j$ and $r_j := \ell(Q_j^*)$ we have the following properties.

Corollary 1.3.1. *Under the assumptions of Lemma 1.3.5 the following holds*

(W5) $\bigcup_j \overline{Q_j^*} = \mathcal{O}$.

(W6) *If* Q_j^* *and* Q_k^* *intersect, then the boundaries of* Q_j *and* Q_k *touch and* $Q_j^* \subset 5Q_k^*$, *moreover* $r_j \sim r_k$ *and* $|Q_j^* \cap Q_k^*| \sim |Q_j^*| \sim |Q_k^*|$.

(W7) *The family* Q_j^* *is locally* 6^d *finite.*

(W8) $\sum_j \mathcal{L}^d(Q_j^*) \leq c(d)\mathcal{L}^d(\mathcal{O})$.

Lemma 1.3.6. *Let $\mathcal{O} \subset \mathbb{R}^d$ be open, $\{Q_j\}$ its Whitney covering from Lemma 1.3.5 and $Q_j^* = \frac{9}{8}Q_j$. Then there is a partition of unity $\{\varphi_j\}$ having the following properties.*

(U1) $\varphi_j \in C_0^\infty(\mathbb{R}^d)$ *and* $\operatorname{supp}\varphi_j = Q_j^*$.

(U2) $\chi_{\frac{7}{9}Q_j^*} = \chi_{\frac{7}{8}Q_j} \leq \varphi_j \leq \chi_{\frac{9}{8}Q_j} = \chi_{Q_j^*}$.

(U3) $|\nabla\varphi_j| \leq \dfrac{c\,\chi_{Q_j^*}}{r_j}$ *and* $|\nabla^2\varphi_j| \leq \dfrac{c\,\chi_{Q_j^*}}{r_j^2}$.

Proof. Let $\tilde{\varphi}_j \in C_0^\infty(\mathbb{R}^d)$ be such that $\operatorname{supp}\tilde{\varphi}_j = Q_j^*$ and

$$\chi_{\frac{7}{9}Q_j^*} = \chi_{\frac{7}{8}Q_j} \leq \tilde{\varphi}_j \leq \chi_{\frac{9}{8}Q_j} = \chi_{Q_j^*}.$$

Moreover, we assume that all $\widetilde{\varphi}_j$ the same function are up to translation and dyadic scaling. We define $\gamma := \sum_j \widetilde{\varphi}_j$ and $\varphi_j := \frac{\widetilde{\varphi}_j}{\gamma}$ such that $1 \leq \gamma \leq 6^d$ as well as

$$|\nabla\gamma|\,\chi_{Q_j^*} \leq c\frac{1}{r_j} \quad \forall j \in \mathbb{N}.$$

Thus φ_j defines a partition of unity with the required properties. □

For $\mathbf{u} \in W_0^{1,p}(G)$ (extended by zero to $\mathbb{R}^d$) we define as in (1.3.17) "the bad" set by

$$\mathcal{O}_\lambda := \{x \in \mathbb{R}^d : M(\nabla u) > \lambda\}.$$

We apply Corollary 1.3.1 and Lemma 1.3.6 to $\mathcal{O}_\lambda$ to obtain a covering $\{Q_j^*\}$ and functions $\{\varphi_j\}$. Now we define

$$\mathbf{u}_\lambda := \mathbf{u} - \sum_{i \in \mathcal{I}} \varphi_i(\mathbf{u} - \mathbf{u}_i), \tag{1.3.38}$$

where $\mathbf{u}_i := \mathbf{u}_{Q_i^*} := \fint_{Q_i^*} \mathbf{u}\, dx\, dt$. (In order to obtain a truncation with zero boundary values one has to involve cut-off function, see Chapter 3, or set $\mathbf{u}_i = 0$ close to the boundary, see [60].) We show first that the sum in (1.3.38) converges absolutely in $L^1(\mathbb{R}^d)$:

$$\int_{\mathbb{R}^d} |\mathbf{u} - \mathbf{u}_\lambda|\, dx \le c \sum_i \int_{Q_i^*} |\mathbf{u} - \mathbf{u}_i|\, dx \le c \sum_i \int_{Q_i^*} |\mathbf{u}|\, dx \le c \int_{\mathbb{R}^d} |\mathbf{u}|\, dx,$$

where we used (U2) and the finite intersection property of Q_i^*, cf. (W7). We proceed by showing the estimate for the gradient

$$\begin{aligned}\int_{\mathbb{R}^d} |\nabla(\mathbf{u} - \mathbf{u}_\lambda)|\, dx &\le c \sum_i \int_{Q_i^*} \big|\nabla(\varphi_i(\mathbf{u} - \mathbf{u}_i))\big|\, dx \\ &\le c \sum_i \int_{Q_i^*} |\nabla \mathbf{u}| + \left|\frac{\mathbf{u} - \mathbf{u}_i}{r_i}\right| dx\, dt \le c \sum_i \int_{Q_i^*} |\nabla \mathbf{u}|\, dx \\ &\le c \int_{\mathbb{R}^d} |\nabla \mathbf{u}|\, dx,\end{aligned}$$

where we used Poincaré's inequality. This shows that the definition in (1.3.38) makes sense. In particular we have

$$\mathbf{u}_\lambda = \begin{cases} \mathbf{u} & \text{in } \mathbb{R}^d \setminus \mathcal{O}_\lambda, \\ \sum_i \varphi_i \mathbf{u}_i & \text{in } \mathcal{O}_\lambda. \end{cases} \tag{1.3.39}$$

In the following we show that $\mathbf{u}_\lambda$ is indeed Lipschitz continuous with Lipschitz constant bounded by λ.

Lemma 1.3.7. *The following holds*

$$\|\nabla \mathbf{u}_\lambda\|_{L^\infty(\mathbb{R}^d)} \le c\lambda.$$

Proof. Let $x \in Q_i^*$ and $A_i := \{j : Q_j^* \cap Q_i^* \ne \emptyset\}$, then

$$\begin{aligned}|\nabla \mathbf{u}_\lambda(x)| &= \left|\sum_{j \in A_i} \nabla(\varphi_j \mathbf{u}_j)(x)\right| \le \sum_{j \in A_i} |\nabla(\varphi_j(\mathbf{u}_j - \mathbf{u}_i))(x)| \\ &\le c \sum_{j \in A_i} \left|\frac{\mathbf{u}_j - \mathbf{u}_i}{r_i}\right| \le c \fint_{5Q_i^*} \left|\frac{\mathbf{u} - \mathbf{u}_i}{r_i}\right| dx\end{aligned}$$

because $\{\varphi_j\}$ is a partition of unity, $r_i \sim r_j$ and $\mathbf{u}_i$ is constant. We also used (W6), (U3) as well as $\#A_j \le c$. By Poincaré's inequality, (W2) and the definition of $\mathcal{O}_\lambda$ we have

$$|\nabla \mathbf{u}_\lambda(x)| \le c \fint_{5Q_i^*} |\nabla \mathbf{u}| \, \mathrm{d}x \le c \fint_{c_d Q_i} |\nabla \mathbf{u}| \, \mathrm{d}x \le c\lambda.$$

As the $\{Q_i^*\}$ cover $\mathcal{O}_\lambda$ and $|\nabla \mathbf{u}_\lambda| = |\nabla \mathbf{u}| \le \lambda$ outside $\mathcal{O}_\lambda$ the claim follows. □

1.4 EXISTENCE RESULTS FOR POWER LAW FLUIDS

The stationary flow of a homogeneous incompressible fluid in a bounded body $G \subset \mathbb{R}^d$ $(d = 2, 3)$ is described by the equations

$$\begin{cases} \operatorname{div} \mathbf{S}(\boldsymbol{\varepsilon}(\mathbf{v})) = \rho(\nabla \mathbf{v})\mathbf{v} + \nabla \pi - \rho \mathbf{f} & \text{in } G \\ \operatorname{div} \mathbf{v} = 0 & \text{in } G, \\ \mathbf{v} = 0 & \text{on } \partial G. \end{cases} \tag{1.4.40}$$

See for instance [23]. In physical terms this means that the fluid reached a steady state – a situation of balance. The unknown quantities are the velocity field $\mathbf{v} : G \to \mathbb{R}^d$ and the pressure $\pi : G \to \mathbb{R}$. The function $\mathbf{f} : G \to \mathbb{R}^d$ represents a system of volume forces, while $\mathbf{S} : G \to \mathbb{R}^{d\times d}_{\mathrm{sym}}$ is the viscous stress tensor and $\rho > 0$ is the density of the fluid. In order to describe a specific fluid one needs a constitutive law relating the viscous stress tensor $\mathbf{S}$ to the symmetric gradient $\boldsymbol{\varepsilon}(\mathbf{v}) := \frac{1}{2}\big(\nabla \mathbf{v} + \nabla \mathbf{v}^T\big)$ of the velocity $\mathbf{v}$. In the simplest case this relation is linear, i.e.,

$$\mathbf{S} = \mathbf{S}(\boldsymbol{\varepsilon}(\mathbf{v})) = 2\nu \boldsymbol{\varepsilon}(\mathbf{v}), \tag{1.4.41}$$

where $\nu > 0$ is the viscosity of the fluid. In this case we have $\operatorname{div} \mathbf{S} = \nu \Delta \mathbf{v}$ and (1.4.40) are the stationary Navier–Stokes equations (for a recent approach see [85,86]). The existence of a weak solution (where derivatives are to be understood in a distributional sense) can be established by arguments which are nowadays standard. In the case of the constitutive relation (1.4.41) the system (1.4.40) can be analysed like a linear system – the arguments used to handle the perturbation caused by $(\nabla \mathbf{v})\mathbf{v}$ are of a technical nature (note that this is quite different from the parabolic situation), and standard techniques lead to smooth solutions (see for instance [86]).

Only fluids with simple molecular structure e.g. water, oil and certain gases satisfy a linear relation such as (1.4.41). Those which do not are called non-Newtonian fluids (see [13]). A special class among these are generalized Newtonian fluids. Here, the viscosity is assumed to be a function of the shear rate $|\boldsymbol{\varepsilon}(\mathbf{v})|$ and the constitutive relation is

$$\mathbf{S}(\boldsymbol{\varepsilon}(\mathbf{v})) = \nu(|\boldsymbol{\varepsilon}(\mathbf{v})|)\boldsymbol{\varepsilon}(\mathbf{v}). \tag{1.4.42}$$

An external force can produce two different reactions:

- The fluid becomes thicker (for example batter): the viscosity of a shear thickening fluid is an increasing function of the shear rate.
- The fluid becomes thinner (for example ketchup): the viscosity of a shear thinning fluid is a decreasing function of the shear rate.

The power law model for non-Newtonian/generalized Newtonian fluids

$$\mathbf{S}(\boldsymbol{\varepsilon}(\mathbf{v})) = \nu_0\big(1 + |\boldsymbol{\varepsilon}(\mathbf{v})|\big)^{p-2}\boldsymbol{\varepsilon}(\mathbf{v}) \tag{1.4.43}$$

is very popular among rheologists. Here $\nu_0 > 0$ and $p \in (1, \infty)$ is specified by physical experiments. An extensive list of specific p-values for different fluids can be found in [23]. It becomes clear that many interesting p-values lie in the interval $[\frac{3}{2}, 2]$. In the following we give a historical overview concerning the theory of weak solutions to (1.4.40) and sketch the proofs, cf. [29].

Monotone operator theory (1969).

The mathematical discussion of power law models started in the late sixties with the work of Lions and Ladyshenskaya (see [106–108] and [109]). Due to the appearance of the convective term $\mathrm{div}(\mathbf{v} \otimes \mathbf{v})$ the equations for power law fluids (the constitutive law is given by (1.4.43)) depend significantly on the value of p. In the stationary case, the existence of a weak solution to (1.4.44), (1.4.43) can be shown by monotone operator theory for $p \geq \frac{3d}{d+2}$. To be precise, there is a function $\mathbf{v} \in W^{1,p}_{0,\mathrm{div}}(G)$ such that

$$\int_G \mathbf{S}(\boldsymbol{\varepsilon}(\mathbf{v})) : \boldsymbol{\varepsilon}(\boldsymbol{\varphi})\,\mathrm{d}x = -\rho \int_G (\nabla\mathbf{v})\mathbf{v} \cdot \boldsymbol{\varphi}\,\mathrm{d}x + \rho \int_G \mathbf{f} \cdot \boldsymbol{\varphi}\,\mathrm{d}x \tag{1.4.44}$$

for all $\boldsymbol{\varphi} \in C^{\infty}_{0,\mathrm{div}}(G)$. Note that this formulation has the advantage that the pressure does not appear but can easily be recovered later by De Rahm theory (this was first used in [109]). For the recovery of the pressure see Theorem 2.2.10. Also note that the divergence-free constraint and homogeneous boundary conditions are incorporated in the definition of the space $W^{1,p}_{0,\mathrm{div}}(G)$. The condition

$$p > \frac{3d}{d+2} \tag{1.4.45}$$

ensures that the solution itself is a test-function and the convective term is a compact perturbation. We begin with the approach based on monotone operator theory (see [109]). It does not yet contain truncations, but it is the basis of the existence theory and everything is build upon it. Let us assume that (1.4.45) holds and that we have a sequence of approximate solutions, i.e. $(\mathbf{v}_n) \subset W^{1,p}_{0,\mathrm{div}}(G)$ solving (1.4.44). We want to pass to the limit. By (1.4.45), Sobolev's embedding Theorem and smooth approximation, (1.4.44) holds also for all $\boldsymbol{\varphi} \in W^{1,p}_{0,\mathrm{div}}(G)$. So $\mathbf{v}_n$ is an admissible test-function. Since $\int_G (\nabla \mathbf{v}_n)\mathbf{v}_n \cdot \mathbf{v}_n \, \mathrm{d}x = 0$ we obtain a uniform a priori estimate in $W^{1,p}(G)$ and (after choosing an appropriate subsequence)

$$\mathbf{v}_n \rightharpoonup \mathbf{v} \quad \text{in} \quad W^{1,p}_{0,\mathrm{div}}(G). \tag{1.4.46}$$

Note that we also used the coercivity from (1.4.43) and Korn's inequality. Using (1.4.43) again yields

$$\mathbf{S}(\boldsymbol{\varepsilon}(\mathbf{v}_n)) \rightharpoonup \tilde{\mathbf{S}} \quad \text{in} \quad L^{p'}(G). \tag{1.4.47}$$

The nonlinearity in the convective term $(\nabla \mathbf{v}_n)\mathbf{v}_n$ can be overcome by compactness arguments. Kondrachov's Theorem and (1.4.45) imply

$$\mathbf{v}_n \to \mathbf{v} \quad \text{in} \quad L^{2p'}(G) \tag{1.4.48}$$

and so

$$(\nabla \mathbf{v}_n)\mathbf{v}_n \rightharpoonup (\nabla \mathbf{v})\mathbf{v} \quad \text{in} \quad L^{\frac{2p}{p+1}}(G). \tag{1.4.49}$$

Using (1.4.46)–(1.4.49) we can pass to the limit in the equation and obtain

$$\int_G \tilde{\mathbf{S}} : \boldsymbol{\varepsilon}(\boldsymbol{\varphi}) \, \mathrm{d}x = - \int_G (\nabla \mathbf{v})\mathbf{v} \cdot \boldsymbol{\varphi} \, \mathrm{d}x + \int_G \mathbf{f} \cdot \boldsymbol{\varphi} \, \mathrm{d}x \tag{1.4.50}$$

for all $\boldsymbol{\varphi} \in W^{1,p}_{0,\mathrm{div}}(G)$. It remains to be shown

$$\tilde{\mathbf{S}} = \mathbf{S}(\boldsymbol{\varepsilon}(\mathbf{v})). \tag{1.4.51}$$

As $\mathbf{S}$ is nonlinear the weak convergence in (1.4.46) is not enough for this limit procedure. We have to apply methods from monotone operator theory. Let us consider the integral

$$\begin{aligned} &\int_G \big(\mathbf{S}(\boldsymbol{\varepsilon}(\mathbf{v}_n)) - \mathbf{S}(\boldsymbol{\varepsilon}(\mathbf{v}))\big) : \big(\boldsymbol{\varepsilon}(\mathbf{v}_n) - \boldsymbol{\varepsilon}(\mathbf{v})\big) \, \mathrm{d}x \\ &\quad = \int_G \mathbf{S}(\boldsymbol{\varepsilon}(\mathbf{v}_n)) : \big(\boldsymbol{\varepsilon}(\mathbf{v}_n) - \boldsymbol{\varepsilon}(\mathbf{v})\big) \, \mathrm{d}x - \int_G \mathbf{S}(\boldsymbol{\varepsilon}(\mathbf{v})) : \big(\boldsymbol{\varepsilon}(\mathbf{v}_n) - \boldsymbol{\varepsilon}(\mathbf{v})\big) \, \mathrm{d}x. \end{aligned}$$

The second term on the right-hand-side vanishes for $n \to \infty$ as a consequence of (1.4.46) and $\mathbf{S}(\boldsymbol{\varepsilon}(\mathbf{v})) \in L^{p'}(G)$. For the first term one we use the equation for $\mathbf{v}^n$ and obtain

$$\begin{aligned} &\int_G \mathbf{S}(\boldsymbol{\varepsilon}(\mathbf{v}_n)) : \big(\boldsymbol{\varepsilon}(\mathbf{v}_n) - \boldsymbol{\varepsilon}(\mathbf{v})\big) \, \mathrm{d}x \\ &\quad = - \int_G (\nabla \mathbf{v}_n)\mathbf{v}_n \cdot (\mathbf{v}_n - \mathbf{v}) \, \mathrm{d}x + \int_G \mathbf{f} \cdot (\mathbf{v}_n - \mathbf{v}) \, \mathrm{d}x \longrightarrow 0, \quad n \to \infty. \end{aligned}$$

This is a consequence of (1.4.46) and (1.4.49). Plugging all together we have shown

$$\int_G \big(\mathbf{S}(\boldsymbol{\varepsilon}(\mathbf{v}_n)) - \mathbf{S}(\boldsymbol{\varepsilon}(\mathbf{v}))\big) : \big(\boldsymbol{\varepsilon}(\mathbf{v}_n) - \boldsymbol{\varepsilon}(\mathbf{v})\big)\,\mathrm{d}x \longrightarrow 0, \quad n \to \infty.$$

The strict monotonicity of $\mathbf{S}$ implies $\boldsymbol{\varepsilon}(\mathbf{v}_n) \to \boldsymbol{\varepsilon}(\mathbf{v})$ a.e. and hence (1.4.51).

L^∞-truncation (1997).

Examining the three-dimensional situation we see that the bound $p > \frac{9}{5}$ is very restrictive since many interesting liquids lie beyond it. For example polyethylene oxide (polyethylene is the most common plastic) has lower flow behaviour indices: the experiments presented in [23] (table 4.1-2, p. 175) suggest values between 1.53 and 1.6 depending on the temperature. The first attempt to lower the bound for p was an approach via L^∞-truncation by Frehse, Málek and Steinhauer (see [78], see also [129]). The term

$$\int_G (\nabla\mathbf{v})\mathbf{v} \cdot \boldsymbol{\varphi}\,\mathrm{d}x$$

is defined for all $\boldsymbol{\varphi} \in L^\infty(G)$ if

$$p > \frac{2d}{d+1}. \tag{1.4.52}$$

Instead of testing the equation by $\mathbf{v}$ (which is not permitted) they used the function $\mathbf{v}_L \in L^\infty(G)$, $L \gg 1$, whose L^∞-norm is bounded by L and which equals $\mathbf{v}$ on a large set.

In order to give an overview of this method we assume that (1.4.52) holds and that we have a sequence of approximated solutions to (1.4.44) with uniform a priori estimates in $W^{1,p}_{0,\mathrm{div}}(G)$. Note that test-functions have to be bounded as $(\nabla\mathbf{v})\mathbf{v}$ is only an integrable function. We will demonstrate how to obtain a weak solution combining ideas of [78] and [140].

Again we have (1.4.46) and (1.4.47) but instead of (1.4.48) and (1.4.49) only the following hold

$$\mathbf{v}_n \to \mathbf{v} \quad \text{in} \quad L^{p'}(G), \tag{1.4.53}$$

$$(\nabla\mathbf{v}_n)\mathbf{v}_n \rightharpoonup (\nabla\mathbf{v})\mathbf{v} \quad \text{in} \quad L^{\sigma}(G), \tag{1.4.54}$$

where $\sigma := \frac{pd}{p(d+1)-2d} \in (1,\infty)$, cf. (1.4.52). We still obtain (1.4.50) for all $\boldsymbol{\varphi} \in W^{1,p}_{0,\mathrm{div}} \cap L^\infty(G)$ and the goal is to show (1.4.51). We are faced with the problem that the solution is not an admissible test-function any more. So an approach via monotone operator theory as described before will fail. Instead of testing with $\mathbf{u}_n := \mathbf{v}_n - \mathbf{v}$ we use a truncated function. As functions from the class $W^{1,p}_{0,\mathrm{div}} \cap L^\infty(G)$ are admissible we cut values of $\mathbf{u}_n$ which are too large and obtain a bounded function. For $L \in \mathbb{N}$ we define

$$\Psi_L := \sum_{\ell=1}^{L} \psi_{2^{-\ell}}, \qquad \psi_\delta(s) := \psi(\delta s),$$

where $\psi \in C_0^\infty([0,2])$, $0 \le \psi \le 1$, $\psi \equiv 1$ on $[0,1]$ and $0 \le -\psi' \le 2$. Now we use the test-function $\mathbf{u}_{n,L} := \Psi_L(|\mathbf{u}_n|)\mathbf{u}_n$ and neglect for a moment the fact that it is not divergence-free. For fixed L the function $\mathbf{u}_{n,L}$ is essentially bounded (in terms of L) and we obtain for $n \to \infty$

$$\mathbf{u}_{n,L} \to 0 \quad \text{in} \quad L^q(G) \quad \text{for all} \quad q < \infty. \tag{1.4.55}$$

Now we test with $\mathbf{u}_{n,L}$ which implies (using (1.4.54) and (1.4.55) for the integral $\int_G (\nabla \mathbf{v}_n)\mathbf{v}_n \cdot \mathbf{u}_{n,L}\,\mathrm{d}x$)

$$\begin{aligned}
&\limsup_n \int_G \Psi_L(|\mathbf{u}_n|)\big(\mathbf{S}(\boldsymbol{\varepsilon}(\mathbf{v}_n)) - \mathbf{S}(\boldsymbol{\varepsilon}(\mathbf{v}))\big) : \boldsymbol{\varepsilon}(\mathbf{u}_n)\,\mathrm{d}x \qquad (1.4.56)\\
&\le \limsup_n \int_G \Psi_L(|\mathbf{u}_n|)\big(\mathbf{S}(\boldsymbol{\varepsilon}(\mathbf{v}_n)) - \mathbf{S}(\boldsymbol{\varepsilon}(\mathbf{v}))\big) : \nabla \Psi_L(|\mathbf{u}_n|) \otimes \mathbf{u}_n\,\mathrm{d}x.
\end{aligned}$$

Now one needs that

$$\nabla \Psi_L(|\mathbf{u}_n|) \otimes \mathbf{u}_n \in L^p(G)$$

uniformly in L and n which follows from the definition of Ψ_L. This allows us to show that the left-hand-side of (1.4.56) is bounded in L and hence there is a subsequence (in fact one has to take a diagonal sequence) such that for $n \to \infty$

$$\sigma_{\ell,n} := \int_G \big(\mathbf{S}(\boldsymbol{\varepsilon}(\mathbf{v}_n)) - \tilde{\mathbf{S}})\big) : \psi_{2^{-\ell}}(|\mathbf{u}_n|)\boldsymbol{\varepsilon}(\mathbf{u}_n)\,\mathrm{d}x \longrightarrow \sigma_\ell, \qquad \forall \ell \in \mathbb{N}_0.$$

One can show easily that σ_ℓ is increasing in ℓ and so $\sigma_0 = 0$, i.e.,

$$\int_G \big(\mathbf{S}(\boldsymbol{\varepsilon}(\mathbf{v}_n)) - \mathbf{S}(\boldsymbol{\varepsilon}(\mathbf{v}))\big) : \psi_1(|\mathbf{u}_n|)\boldsymbol{\varepsilon}(\mathbf{u}_n)\,\mathrm{d}x \longrightarrow 0, \qquad n \to 0. \tag{1.4.57}$$

As $\psi_1(t) = 1$ for $t \le 1$ and $\mathbf{u}_n \to 0$ in $L^2(G)$ this yields

$$\int_G \Big(\big(\mathbf{S}(\boldsymbol{\varepsilon}(\mathbf{v}_n)) - \mathbf{S}(\boldsymbol{\varepsilon}(\mathbf{v}))\big) : \boldsymbol{\varepsilon}(\mathbf{u}_n)\Big)^\Theta \mathrm{d}x \longrightarrow 0, \qquad n \to 0, \tag{1.4.58}$$

for all $\Theta < 1$. Due to the monotonicity of $\mathbf{S}$ we deduce (1.4.51).

As $\operatorname{div}\mathbf{u}_{n,L} \neq 0$ we have to correct the divergence by means of the Bogovskiĭ-operator. It is solution operator to the divergence equation with respect to zero boundary conditions. See Section 2.1. Additional terms appear which can be handled similarly.

Remark 1.4.5. In [78] the limit case $p = \frac{2d}{d+1}$ is also included based on the fact that $(\nabla \mathbf{v})\mathbf{v}$ has div – curl structure and hence belongs to the Hardy space $\mathcal{H}^1(\mathbb{R}^d)$.

Lipschitz truncation (2003).

Although we can now cover a wide range of power law fluids there remain several with lower values of p. The experiments presented in [23] (table 4.1-2, p. 175) suggest values for 2% hydroxyethylcellulose (hydroxyethylcellulose is a gelling and thickening agent derived from cellulose, used in cosmetics, cleaning solutions, and other household products) between 1.19 and 1.25 depending on the temperature.

Since $\operatorname{div} \mathbf{v} = 0$ we can rewrite

$$\int_G (\nabla \mathbf{v})\mathbf{v} \cdot \boldsymbol{\varphi} \, dx = -\int_G \mathbf{v} \otimes \mathbf{v} : \boldsymbol{\varepsilon}(\boldsymbol{\varphi}) \, dx,$$

so that appropriate test-functions have to be Lipschitz continuous provided $\mathbf{v} \otimes \mathbf{v} \in L^1(G)$. This condition is satisfied for $p \geq \frac{2d}{d+2}$ by Sobolev's embedding. Otherwise one cannot define the convective term (at least in the stationary case). This bound therefore seems to be optimal.

In the case

$$p > \frac{2d}{d+2} \tag{1.4.59}$$

the existence of a weak solution to (1.4.44), (1.4.43) was first established in [79]. This is the first paper where the Lipschitz truncation was used in the context of fluid mechanics. Here one approximates the function $\mathbf{v}$ by a Lipschitz continuous function $\mathbf{v}_\lambda$ with $\|\nabla \mathbf{v}_\lambda\|_\infty \leq c\lambda$ instead of a bounded function as in the approach via L^∞-truncation.

Assume that (1.4.59) holds and that we have a sequence of solutions $(\mathbf{v}_n) \subset W^{1,p}_{0,\operatorname{div}}(G)$ to

$$\int_G \mathbf{S}(\boldsymbol{\varepsilon}(\mathbf{v}_n)) : \boldsymbol{\varepsilon}(\boldsymbol{\varphi}) \, dx = \int_G \mathbf{v}_n \otimes \mathbf{v}_n : \nabla \boldsymbol{\varphi} \, dx + \int_G \mathbf{f} \cdot \boldsymbol{\varphi} \, dx \tag{1.4.60}$$

for all $\boldsymbol{\varphi} \in W^{1,\infty}_{0,\operatorname{div}}(G)$ which is uniformly bounded. Again we have (1.4.46) and (1.4.47) and by Kondrachov's Theorem and (1.4.59)

$$\mathbf{v}_n \to \mathbf{v} \quad \text{in} \quad L^{2\sigma}(G), \qquad \mathbf{v}_n \otimes \mathbf{v}_n \rightharpoonup \mathbf{v} \otimes \mathbf{v} \quad \text{in} \quad L^{\sigma}(G), \tag{1.4.61}$$

where $\sigma \in \left(1, \frac{1}{2}\frac{pd}{d-p}\right)$, cf. (1.4.59). So we can pass to the limit in (1.4.60) and obtain

$$\int_G \tilde{\mathbf{S}} : \boldsymbol{\varepsilon}(\boldsymbol{\varphi}) \, dx = \int_G \mathbf{v} \otimes \mathbf{v} : \nabla \boldsymbol{\varphi} \, dx + \int_G \mathbf{f} \cdot \boldsymbol{\varphi} \, dx. \tag{1.4.62}$$

In order to show $\tilde{\mathbf{S}} = \mathbf{S}(\boldsymbol{\varepsilon}(\mathbf{v}))$ it is enough to have (1.4.58). Introduce the Lipschitz truncation $\mathbf{u}_{n,\lambda}$ of $\mathbf{u}_n := \mathbf{v}_n - \mathbf{v}$, cf. Section 1.3. Then (1.4.58) follows from

$$\int_G \big(\mathbf{S}(\boldsymbol{\varepsilon}(\mathbf{v}_n)) - \mathbf{S}(\boldsymbol{\varepsilon}(\mathbf{v}))\big) : \boldsymbol{\varepsilon}(\mathbf{u}_{n,\lambda})\,\mathrm{d}x \longrightarrow 0, \quad n \to 0, \tag{1.4.63}$$

and (1.3.31). As a consequence of $\|\nabla \mathbf{u}_{n,\lambda}\|_\infty \leq c\lambda$ the Lipschitz truncation features much better convergence properties than the original function. In particular, we have

$$\mathbf{u}_{n,\lambda} \to 0 \quad \text{in} \quad L^\infty(G), \quad \nabla \mathbf{u}_{n,\lambda} \rightharpoonup^* 0 \quad \text{in} \quad L^\infty(G),$$

recall (1.3.29) and (1.3.30). Taking this into account, (1.4.63) follows from (1.4.60) and (1.4.61).

We again neglected the fact that $\operatorname{div} \mathbf{u}_{n,\lambda} \neq 0$. There are two options for overcoming this. In [79] the authors introduce the pressure π_n and decompose it with respect to the terms appearing in the equation. This requires some technical effort but all terms can be handled. An easier way is presented in [62] where the divergence is corrected using the Bogovskiĭ operator as indicated in the approach via L^∞-truncation.

CHAPTER 2

Fluid mechanics & Orlicz spaces

Contents

Abstract

We extend some classical tools from fluid mechanics – Korn's inequality, the Bogovskiĭ operator and the pressure recovery – to the setting of Orlicz spaces. As a special case the known L^p-theory is included as well as the case of Orlicz spaces generated by a nice Young function (i.e., under Δ_2 and ∇_2 condition). In the general case there is some loss of integrability, for instance in the limit cases $L\log L \to L^1$ and $L^\infty \to \mathrm{Exp}(L)$. The results are shown to be optimal in the sense of Orlicz spaces.

A crucial tool in the mathematical approach to the behaviour of Newtonian fluids is Korn's inequality: given a bounded open domain $G \subset \mathbb{R}^d$, $d \geq 2$, with Lipschitz boundary ∂G we have

$$\int_G |\nabla \mathbf{v}|^2 \,\mathrm{d}x \leq 2 \int_G |\boldsymbol{\varepsilon}(\mathbf{v})|^2 \,\mathrm{d}x \tag{2.0.1}$$

for all $\mathbf{v} \in W_0^{1,2}(G)$. For smooth functions with compact support (2.0.1) can be shown by integration by parts. The general case is treated by approximation. A first proof was given by Korn in [104]. We note that variants of Korn's inequality in L^2 have been established by Courant and Hilbert [53], Friedrichs [84], Èidus [70] and Mihlin [114]. Many problems in the mathematical theory of generalized Newtonian fluids and mechanics of solids lead to the following question (compare for example the monographs of Málek, Nečas, Rokyta and Růžička [111], of Duvaut and Lions [66] and of Zeidler [143]): is it possible to bound a suitable energy depending on $\nabla \mathbf{v}$ by the corresponding functional of $\boldsymbol{\varepsilon}(\mathbf{v})$, that is

$$\int_G |\nabla \mathbf{v}|^p \,\mathrm{d}x \leq c(p, G) \int_G |\boldsymbol{\varepsilon}(\mathbf{v})|^p \,\mathrm{d}x \tag{2.0.2}$$

for functions $\mathbf{v} \in W_0^{1,p}(G)$? As shown by Gobert [91,92], Nečas [119], Mosolov and Mjasnikov [116], Temam [138] and later by Fuchs [74] this is

Existence Theory for Generalized Newtonian Fluids.
DOI: http://dx.doi.org/10.1016/B978-0-12-811044-7.00003-3

true for all $1 < p < \infty$ (we remark that the inequality fails in the case $p = 1$, see [120] and [52]).

A first step in the generalization of (2.0.2) is mentioned in [4]: Acerbi and Mingione prove a variant for the Young function

$$A(t) = (1 + t^2)^{\frac{p-2}{2}} t^2.$$

More precisely, they show that

$$\|\nabla \mathbf{v}\|_{L^A(G)} \le c(\varphi, G)\|\boldsymbol{\varepsilon}(\mathbf{v})\|_{L^A(G)} \tag{2.0.3}$$

for all functions $\mathbf{v} \in W_0^{1,A}(G)$. Although they only consider a special case they provide tools for much more general situations. Note that they only obtain inequalities in the Luxembourg-norm which is not appropriate in many situations (for example in regularity theory, see [35]). A general theorem is proved in [64], namely that

$$\int_G A(|\nabla \mathbf{v} - (\nabla \mathbf{v})_G|)\, dx \le c(A, G) \int_G A(|\boldsymbol{\varepsilon}(\mathbf{v}) - (\boldsymbol{\varepsilon}(\mathbf{v}))_G|)\, dx \tag{2.0.4}$$

for all $\mathbf{v} \in W^{1,A}(G)$, where A is a Young function satisfying the Δ_2- and ∇_2-condition. Furthermore, Fuchs [75] obtains (2.0.4) for functions with zero traces and the same class of Young functions by a different approach. It is shown in [32] that the Δ_2- and ∇_2-condition are also necessary for the inequality (2.0.4). We remark that the constitutive law

$$\mathbf{S} = \frac{A'(|\boldsymbol{\varepsilon}(\mathbf{v})|)}{|\boldsymbol{\varepsilon}(\mathbf{v})|}\boldsymbol{\varepsilon}(\mathbf{v})$$

for a Young function A is a quite general model to describe the motion of generalized Newtonian fluids (see, i.e., [35], [25] and [59]).

In order to characterize the behaviour of Prandtl–Eyring fluids (see Chapter 4) Eyring [69] suggested the constitutive law

$$\mathbf{S} = DW(\boldsymbol{\varepsilon}(\mathbf{u})), \qquad W(\boldsymbol{\varepsilon}) = h(|\boldsymbol{\varepsilon}|) = |\boldsymbol{\varepsilon}|\log(1 + |\boldsymbol{\varepsilon}|). \tag{2.0.5}$$

This leads in a natural way to the question about Korn's inequality in the space $L^h(G)$. Since we have

$$\tilde{h}(t) \approx t(\exp(t) - 1),$$

the ∇_2-condition fails in this case, hence the results mentioned above do not apply. So the following question remains: *given some integrability of the symmetric gradient – in the sense of Orlicz spaces – what is the best integrability for the full gradient we can expect?*

A second fundamental question in fluid mechanics is the recovery of the pressure. It is common (and very useful) to study pressure-free formulations of (generalized) Navier–Stokes equations. So one starts by finding a velocity field which solves the corresponding system in the sense of distributions on divergence-free test-functions. Afterwards there is the question about the existence of the pressure function in order to have a weak solution in the sense of distributions.

Let us be a bit more precise and consider the equation

$$\int_G \mathbf{H} : \nabla\boldsymbol{\varphi}\,\mathrm{d}x = 0 \quad \text{for all} \quad \boldsymbol{\varphi} \in C^{\infty}_{0,\mathrm{div}}(G), \tag{2.0.6}$$

where $\mathbf{H}$ is an integrable function (in case of a stationary generalized Navier–Stokes equation we have $\mathbf{H} = \mathbf{S}(\boldsymbol{\varepsilon}(\mathbf{v})) + \nabla\Delta^{-1}\mathbf{f} - \mathbf{v}\otimes\mathbf{v}$). Secondly, the pressure π is reconstructed in the sense that

$$\int_G \mathbf{H} : \nabla\boldsymbol{\varphi}\,\mathrm{d}x = \int_\Omega \pi \,\mathrm{div}\,\boldsymbol{\varphi}\,\mathrm{d}x \quad \text{for all} \quad \boldsymbol{\varphi} \in C^{\infty}_{0}(G). \tag{2.0.7}$$

The existence of a pressure in the sense of distributions is a consequence of the classical theorem by De Rahm (see [131] for an appropriate version). It is also well-known that – if $1 < p < \infty$ – then $\mathbf{H} \in L^p(G)$ implies $\pi \in L^p(G)$. This result breaks down in the limit cases. Again motivated by the Prandtl–Eyring model the following question remains: *given a function* $\mathbf{H}$ *solving* (2.0.6) *– located in some Orlicz space – what is the optimal integrability of the pressure in* (2.0.7)*?*

In classical L^p-spaces, both questions raised above can be answered by Nečas' negative norm theorem [119]. The negative Sobolev norm of the distributional gradient of a function $u \in L^1(G)$ can be defined as

$$\|\nabla u\|_{W^{-1,p}(G)} = \sup_{\boldsymbol{\varphi}\in C_0^\infty(G)} \frac{\int_G u \,\mathrm{div}\,\boldsymbol{\varphi}\,\mathrm{d}x}{\|\nabla\boldsymbol{\varphi}\|_{L^{p'}(G)}}\,\mathrm{d}x, \tag{2.0.8}$$

where $1 \le p \le \infty$. In (2.0.8), and in similar occurrences throughout this chapter, we tacitly assume that the supremum is extended over all functions $\mathbf{v}$ which do not vanish identically. We remark that the quantity on the right-hand side of (2.0.8) agrees with the norm of ∇u, when regarded as an element of the dual of $W^{1,p'}_0(G)$. Nečas showed that, if G is regular enough – a bounded Lipschitz domain, say – and $1 < p < \infty$, then the $L^p(G)$ norm of a function is equivalent to the $W^{-1,p}(G)$ norm of its gradient. Namely, there exist positive constants $C_1 = C_1(G,p)$ and $C_2 = C_2(d)$, such that

$$C_1\|u - u_G\|_{L^p(G)} \le \|\nabla u\|_{W^{-1,p}(G)} \le C_2\|u - u_G\|_{L^p(G)} \tag{2.0.9}$$

for every $u \in L^1(G)$.

Using the formula

$$\Delta \mathbf{u} = \operatorname{div} \mathbf{V}(\mathbf{u}), \qquad V_{ij}(\mathbf{u}) = 2\boldsymbol{\varepsilon}^D(\mathbf{u}) - \Big(\frac{1}{2} - \frac{1}{d}\Big)(\operatorname{div} \mathbf{u})I,$$

where $\boldsymbol{\varepsilon}^D = \boldsymbol{\varepsilon} - \frac{1}{d}\operatorname{tr}\boldsymbol{\varepsilon} I$, the proof of Korn's inequality based on (2.0.9) is elementary. Moreover, a combination of De Rahm's Theorem and (2.0.9) shows that if $\mathbf{H} \in L^p(G)$ satisfies (2.0.6) then there is $\pi \in L^p(G)$ such that (2.0.7) holds.

In order to understand how Korn's inequality and the pressure recovery work in Orlicz spaces, we have to understand Orlicz versions of (2.0.9). Let A be a Young function, and let G be a bounded domain in $\mathbb{R}^d$. We define the negative Orlicz–Sobolev norm associated with A of the distributional gradient of a function $u \in L^1(G)$ as

$$\|\nabla u\|_{W^{-1,A}(G)} = \sup_{\varphi \in C_0^\infty(G,\mathbb{R}^n)} \frac{\int_G u \operatorname{div}\boldsymbol{\varphi} \, dx}{\|\nabla \boldsymbol{\varphi}\|_{L^{\tilde{A}}(G)}}. \tag{2.0.10}$$

The alternative notation $W^{-1}L^A(G)$ will also occasionally be employed to denote the negative Orlicz–Sobolev norm $W^{-1,A}(G)$ associated with the Orlicz space $L^A(G)$. As (2.0.9) is known to break down in the limit cases, an Orlicz-version with the same Young function on both sides cannot hold in general. In fact, our Orlicz–Sobolev space version of the negative norm theorem involves pairs of Young functions A and B which obey the following balance conditions:

$$t\int_0^t \frac{B(s)}{s^2}\, ds \le A(ct) \quad \text{for} \quad t \ge 0, \tag{2.0.11}$$

and

$$t\int_0^t \frac{\widetilde{A}(s)}{s^2}\, ds \le \widetilde{B}(ct) \quad \text{for} \quad t \ge 0, \tag{2.0.12}$$

for some positive constant c. Note that the same conditions come into play in the study of singular integral operators in Orlicz spaces [47].

If either (2.0.11) or (2.0.12) holds, then A dominates B globally [48, Proposition 3.5]. In a sense, the assumptions (2.0.11) and (2.0.12) provide us with a quantitative information about how much weaker the norm $\|\cdot\|_{L^B(G)}$ is than $\|\cdot\|_{L^A(G)}$. Under these assumptions a version of the negative norm theorem can be restored in Orlicz–Sobolev spaces.

Theorem 2.0.5. *Let A and B be Young functions, fulfilling* (2.0.11) *and* (2.0.12)*. Assume that G is a bounded domain with the cone property in $\mathbb{R}^d$,*

$d \geq 2$. *Then there exist constants* $C_1 = C_1(G, c)$ *and* $C_2 = C_2(d)$ *such that*

$$C_1 \|u - u_G\|_{L^B(G)} \leq \|\nabla u\|_{W^{-1,A}(G)} \leq C_2 \|u - u_G\|_{L^A(G)} \tag{2.0.13}$$

for every $u \in L^1(G)$. *Here,* c *denotes the constant appearing in* (2.0.11) *and* (2.0.12).

Remark 2.0.6. Inequality (2.0.13) continues to hold even if conditions (2.0.11) and (2.0.12) are just fulfilled for $t \geq t_0$ for some $t_0 > 0$, but with constants C_1 and C_2 depending also on A, B, t_0 and $|G|$. Indeed, the Young functions A and B can be replaced, if necessary, with Young functions which are equivalent near infinity and fulfil (2.0.11) and (2.0.12) for every $t > 0$. Due to (1.2.11), such replacement leaves the quantities $\|\cdot\|_{L^A(G)}$, $\|\cdot\|_{L^B(G)}$ and $\|\nabla \cdot\|_{W^{-1,A}(G)}$ unchanged, up to multiplicative constants depending on A, B, t_0 and $|G|$.

The situations when (2.0.11), or (2.0.12), holds with $B = A$ can be precisely characterized. Membership of A to Δ_2 is a necessary and sufficient condition for (2.0.12) to hold with $B = A$ [103, Theorem 1.2.1]. Therefore, under this condition, assumption (2.0.12) can be dropped in Theorem 2.0.5. On the other hand, $A \in \nabla_2$ if and only if $\widetilde{A} \in \Delta_2$. Hence membership of A to ∇_2 is a necessary and sufficient condition for (2.0.11) to hold with $B = A$. Thus, under this condition, assumption (2.0.11) can be dropped in Theorem 2.0.5. Particularly, if $A \in \Delta_2 \cap \nabla_2$, then both conditions (2.0.11) and (2.0.12) are fulfilled with $B = A$. Hence, we have the following corollary which also follows from the results of [64].

Corollary 2.0.1. *Assume that* G *is a bounded domain with the cone property in* $\mathbb{R}^d$, $d \geq 2$. *Let* A *be a Young function in* $\Delta_2 \cap \nabla_2$. *Then there are two constants* $C = C(G, A)$ *and* $C_2 = C_2(d)$ *such that*

$$C_1 \|u - u_G\|_{L^A(G)} \leq \|\nabla u\|_{W^{-1,A}(G)} \leq C_2 \|u - u_G\|_{L^A(G)} \tag{2.0.14}$$

for every $u \in L^1(G)$.

A typical situation where condition (2.0.11) does not hold with $B = A$ is when A grows linearly, or "almost linearly", near infinity. In this case, $A \notin \nabla_2$. In fact, as already mentioned, the standard negative norm theorem expressed by (2.0.9) breaks down in the borderline case $p = 1$. On the other hand, condition (2.0.12) fails, with $B = A$, if, for example, A has a very fast – faster than any power – growth. In this case, $A \notin \Delta_2$. Loosely speaking, the norm $\|\cdot\|_{L^A(G)}$ is now "close" to $\|\cdot\|_{L^\infty(G)}$, and, as a matter of fact, equation (2.0.9) is not true with $p = \infty$.

These, however, are not the only situations when (2.0.11), or (2.0.12), fail with $B = A$. For instance, there are functions A which neither satisfy the Δ_2 condition, nor the ∇_2 condition. Therefore, neither (2.0.12) nor (2.0.11) can hold with $B = A$. In those cases $A(t)$ "oscillates" between two different powers t^p and t^q, with $1 < p < q < \infty$. Functions of this kind are referred to as (p, q)-growth in the literature. Partial differential equations, and associated variational problems, whose nonlinearity is governed by this growth, have been extensively studied. In the framework of non-Newtonian fluids, they have been analysed in [22].

All the circumstances described above can be handled via Theorem 2.0.5. A few examples involving customary families of Young functions are presented hereafter.

Example 1. Assume that $A(t)$ is a Young function equivalent to $t^p \log^\alpha(1+t)$ near infinity, where either $p > 1$ and $\alpha \in \mathbb{R}$, or $p = 1$ and $\alpha \geq 1$. Hence, if $|G| < \infty$, then

$$L^A(G) = L^p \log^\alpha L(G).$$

Assume that G is a bounded domain with the cone property in $\mathbb{R}^d$. If $p > 1$, then $A \in \Delta_2 \cap \nabla_2$, and hence Corollary 2.0.1 tells us that

$$C_1 \|u - u_G\|_{L^p \log^\alpha L(G)} \leq \|\nabla u\|_{W^{-1} L^p \log^\alpha L(G)} \leq C_2 \|u - u_G\|_{L^p \log^\alpha L(G)} \tag{2.0.15}$$

for every $u \in L^1(G)$. However, if $p = 1$, then $A \in \Delta_2$, but $A \notin \nabla_2$. An application of Theorem 2.0.5 now implies

$$C_1 \|u - u_G\|_{L \log^{\alpha-1} L(G)} \leq \|\nabla u\|_{W^{-1} L \log^\alpha L(G)} \leq C_2 \|u - u_G\|_{L \log^\alpha L(G)} \tag{2.0.16}$$

for every $u \in L^1(G)$. In particular,

$$C_1 \|u - u_G\|_{L^1(G)} \leq \|\nabla u\|_{W^{-1} L \log L(G)} \leq C_2 \|u - u_G\|_{L \log L(G)} \tag{2.0.17}$$

for every $u \in L^1(G)$.

Example 2. Let $\beta > 0$, and let $A(t)$ be a Young function equivalent to $\exp(t^\beta)$ near infinity. Then

$$L^A(G) = \exp L^\beta(G)$$

if $|G| < \infty$. One has that $A \in \nabla_2$, but $A \notin \Delta_2$. Theorem 2.0.5 ensures that, if G is a bounded domain with the cone property in $\mathbb{R}^d$, then

$$C_1 \|u - u_G\|_{\exp L^{\frac{\beta}{\beta+1}}(G)} \leq \|\nabla u\|_{W^{-1} \exp L^\beta(G)} \leq C_2 \|u - u_G\|_{\exp L^\beta(G)} \tag{2.0.18}$$

for every $u \in L^1(G)$. Moreover,

$$C_1 \|u - u_G\|_{\exp L(G)} \leq \|\nabla u\|_{W^{-1}L^\infty(G)} \leq C_2 \|u - u_G\|_{L^\infty(G)} \tag{2.0.19}$$

for every $u \in L^1(G)$.

Our approach is based on a study of the Bogovskiĭ operator [24] in Orlicz spaces in Theorem 2.1.7 which is already interesting itself. The Bogovskiĭ operator is a solution operator to the divergence equation with respect to zero boundary conditions. The continuity of the Bogovskiĭ operator implies the negative norm Theorem from which we can deduce both, the pressure recovery and Korn's inequality.

In Theorem 2.2.10 we give the precise statement of the pressure recovery in Orlicz spaces. In fact, $\mathbf{H} \in L^A(G)$ implies $\pi \in L^B(G)$ where A and B are linked through (2.0.11) and (2.0.12). Moreover, the following inequality holds

$$\int_G B(|\pi|)\,\mathrm{d}x \leq \int_G A(C|\mathbf{H} - \mathbf{H}_G|)\,\mathrm{d}x.$$

Theorem 2.3.12 contains a version of Korn's inequality in general Orlicz spaces which says

$$\int_G B(|\nabla \mathbf{u} - (\nabla \mathbf{u})_G|)\,\mathrm{d}x \leq \int_G A(C|\boldsymbol{\varepsilon}(\mathbf{u}) - (\boldsymbol{\varepsilon}(\mathbf{u}))_G|)\,\mathrm{d}x.$$

The final question which remains is the sharpness of the mentioned results. In Section 2.3 we are going to show that the balance conditions (2.0.11) and (2.0.12) are also necessary for a Korn's inequality. This implies that also the results about negative norms in Theorem 2.0.5 and the Bogovskiĭ operator in Theorem 2.1.6 are optimal.

2.1 BOGOVSKII OPERATOR

Our proof of Theorem 2.0.5 relies upon an analysis of the divergence equation

$$\begin{cases} \operatorname{div} \mathbf{u} = f & \text{in} \quad G, \\ \mathbf{u} = 0 & \text{on} \quad \partial G, \end{cases} \tag{2.1.20}$$

in Orlicz spaces which we analyse in the following. Subsequently, we set

$$C_{0,\perp}^\infty(G) = \{u \in C_0^\infty(G) : u_G = 0\},$$
$$L_\perp^A(G) = \{u \in L^A(G) : u_G = 0\},$$

where $u_G = \fint_G u\,\mathrm{d}x$ denotes the mean value of the function u.

Theorem 2.1.6. *Assume that G is a bounded domain with the cone property in $\mathbb{R}^d$, $d \geq 2$. Let A and B be Young functions fulfilling* (2.0.11) *and* (2.0.12). *Then there exists a bounded linear operator*

$$\mathrm{Bog}_G : L^A_\perp(G) \to W^{1,B}_0(G) \tag{2.1.21}$$

such that

$$\mathrm{Bog}_G : C^\infty_{0,\perp}(G) \to C^\infty_0(G) \tag{2.1.22}$$

and

$$\mathrm{div}\,(\mathrm{Bog}_G f) = f \quad \text{in } G \tag{2.1.23}$$

for every $f \in L^A_\perp(G)$. In particular, there exists a constant $C = C(G, c)$ such that

$$\|\nabla(\mathrm{Bog}_G f)\|_{L^B(G)} \leq C\|f\|_{L^A(G)} \tag{2.1.24}$$

and

$$\int_G B(|\nabla(\mathrm{Bog}_G f)|)\,\mathrm{d}x \leq \int_G A(C|f|)\,\mathrm{d}x \tag{2.1.25}$$

for every $f \in L^A_\perp(G)$. Here, c denotes the constant appearing in (2.0.11) *and* (2.0.12).

Although it will not be used for our main purposes, we state in Theorem 2.1.7 below a result parallel to Theorem 2.1.6, dealing with a version of problem (2.1.20) in the case when the right-hand side of the equation is in divergence form. Namely,

$$\begin{cases} \mathrm{div}\,\mathbf{u} = \mathrm{div}\,\mathbf{g} & \text{in } G, \\ \mathbf{u} = 0 & \text{on } \partial G, \end{cases} \tag{2.1.26}$$

where $\mathbf{g} : G \to \mathbb{R}^d$ is a given function satisfying the compatibility condition (in a weak sense) that its normal component on ∂G vanishes. As a precise formulation of this condition we consider the space $H^A(G)$ of those vector-valued functions $\mathbf{u} : G \to \mathbb{R}^n$ for which the norm

$$\|\mathbf{u}\|_{H^A(G)} = \|\mathbf{u}\|_{L^A(G,\mathbb{R}^n)} + \|\,\mathrm{div}\,\mathbf{u}\|_{L^A(G)} \tag{2.1.27}$$

is finite. We denote by $H^A_0(G)$ its subspace of those functions $\mathbf{u} \in H^A(G)$ whose normal component on ∂G vanishes, in the sense that

$$\int_G \varphi\,\mathrm{div}\,\mathbf{u}\,\mathrm{d}x = -\int_G \mathbf{u} \cdot \nabla\varphi\,\mathrm{d}x \tag{2.1.28}$$

for every $\varphi \in C^\infty(\overline{G})$. It is easy to see that both $H^A(G)$ and $H^A_0(G)$ are Banach spaces.

Theorem 2.1.7. *Assume that G is a bounded Lipschitz domain in $\mathbb{R}^d$, $d \geq 2$. Let A and B be Young functions fulfilling* (2.0.11) *and* (2.0.12). *Then there exists a bounded linear operator*

$$\mathcal{E}_G : H_0^A(G) \to W_0^{1,B}(G) \tag{2.1.29}$$

such that

$$\operatorname{div}(\mathcal{E}_G \mathbf{g}) = \operatorname{div} \mathbf{g} \quad \textit{in} \quad G \tag{2.1.30}$$

for every $\mathbf{g} \in H_0^A(G)$. In particular, there exists a constant $C = C(G, c)$ such that

$$\|\nabla(\mathcal{E}_G \mathbf{g})\|_{L^B(G)} \leq C \|\operatorname{div} \mathbf{g}\|_{L^A(G)} \tag{2.1.31}$$

and

$$\|\mathcal{E}_G \mathbf{g}\|_{L^B(G)} \leq C \|\mathbf{g}\|_{L^A(G)} \tag{2.1.32}$$

for every $\mathbf{g} \in H_0^A(G)$. Here, c denotes the constant appearing in (2.0.11) *and* (2.0.12).

The proofs of Theorems 2.1.6 and 2.1.7 make use of a rearrangement estimate, which extends those of [18, Theorem 16.12] and [15], for a class of singular integral operators of the form

$$Tf(x) = \lim_{\varepsilon \to 0^+} \int_{\{y : |y - x| > \varepsilon\}} K(x, y) f(y)\, \mathrm{d}y \quad \text{for} \quad x \in \mathbb{R}^d, \tag{2.1.33}$$

for an integrable function $f : \mathbb{R}^d \to \mathbb{R}$. Here $K(x, y) = N(x, x - y)$, where the kernel $N : \mathbb{R}^d \times \mathbb{R}^d \to \mathbb{R}$ fulfills the following properties:

$$N(x, \lambda z) = \lambda^{-d} N(x, z) \quad \text{for} \quad x, z \in \mathbb{R}^d; \tag{2.1.34}$$

$$\int_{\mathbb{S}^{d-1}} N(x, z)\, \mathrm{d}\mathcal{H}^{d-1}(z) = 0 \quad \text{for} \quad x \in \mathbb{R}^d; \tag{2.1.35}$$

For every $\sigma \in [1, \infty)$, there exists a constant C_1 such that

$$\left(\int_{\mathbb{S}^{d-1}} |N(x, z)|^\sigma\, \mathrm{d}\mathcal{H}^{d-1}(y) \right)^{\frac{1}{\sigma}} \leq C_1 (1 + |x|)^d \quad \text{for} \quad x \in \mathbb{R}^d, \tag{2.1.36}$$

where $\mathbb{S}^{s-1}$ denotes the unit sphere, centered at 0 in $\mathbb{R}^d$, and $\mathcal{H}^{s-1}$ stands for the $(s-1)$-dimensional Hausdorff measure.

There exists a constant C_2 such that

$$|K(x, y)| \leq C_2 \frac{(1 + |x|)^d}{|x - y|^s} \quad \text{for} \quad x, y \in \mathbb{R}^d, x \neq y, \tag{2.1.37}$$

and, if $2|x-z|<|x-y|$, then

$$|K(x,y)-K(z,y)| \leq C_2(1+|y|)^d \frac{|x-z|}{|x-y|^{d+1}}, \tag{2.1.38}$$

$$|K(y,x)-K(y,z)| \leq C_2(1+|y|)^d \frac{|x-z|}{|x-y|^{d+1}}. \tag{2.1.39}$$

Theorem 2.1.8. *Let G be a bounded open set in $\mathbb{R}^d$, and let K be a kernel satisfying (2.1.34)–(2.1.39). If $f \in L^1(\mathbb{R}^d)$ and $f=0$ in $\mathbb{R}^d \setminus G$, then the singular integral operator T given by (2.1.33) is well defined for a.e. $x \in \mathbb{R}^d$, and there exists a constant $C = C(C_1, C_2, d, \operatorname{diam}(G))$ for which*

$$(Tf)^*(s) \leq C\left(\frac{1}{s}\int_0^s f^*(r)\,\mathrm{d}r + \int_s^{|G|} f^*(r)\frac{dr}{r}\right) \quad \textit{for} \quad s \in (0,|G|). \tag{2.1.40}$$

As a consequence of Theorem 2.1.8, the boundedness of singular integral operators given by (2.1.33) between Orlicz spaces associated with Young functions A and B fulfilling (2.0.11) and (2.0.12) can be established.

Theorem 2.1.9. *Let G, K and T be as in Theorem 2.1.8. Assume that A and B are Young functions satisfying (2.0.11) and (2.0.12). Then there exists a constant $C = C(C_1, C_2, d, \operatorname{diam}(G), c)$ such that*

$$\|Tf\|_{L^B(G)} \leq C\|f\|_{L^A(G)}, \tag{2.1.41}$$

and

$$\int_G B(|Tf|)\,\mathrm{d}x \leq \int_G A(C|f|)\,\mathrm{d}x \tag{2.1.42}$$

for every $f \in L^A(G)$. Here, c denotes the constant appearing in (7.0.1) and (5.3.12).

Proof. According to Lemma 1.2.1, if A and B are Young functions satisfying (2.0.11), then there exists a constant $C = C(c)$ such that

$$\left\|\frac{1}{s}\int_0^s \varphi(r)\,\mathrm{d}r\right\|_{L^B(0,\infty)} \leq C\|\varphi\|_{L^A(0,\infty)} \tag{2.1.43}$$

for every $\varphi \in L^A(0,\infty)$. Moreover, if A and B fulfil (2.0.12), then there exists a constant $C = C(c)$ such that

$$\left\|\int_s^\infty \varphi(r)\frac{\mathrm{d}r}{r}\right\|_{L^B(0,\infty)} \leq C\|\varphi\|_{L^A(0,\infty)} \tag{2.1.44}$$

for every $\varphi \in L^A(0,\infty)$. Combining (2.1.40), (2.1.43) and (2.1.44), and making use of property (1.2.12) implies inequality (2.1.41).

As far as (2.1.42) is concerned, observe that, inequalities (2.0.11) and (2.0.12) continue to hold, with the same constant c, if A and B are replaced with kA and kB, where k is any positive constant. Thus, inequality (2.1.41) continues to hold, with the same constant C, after this replacement, whatever k is, namely

$$\|Tf\|_{L^{kB}(G)} \leq C\|f\|_{L^{kA}(G)} \tag{2.1.45}$$

for every $f \in L^A(G)$. Now, given any such f, choose $k = \frac{1}{\int_G A(|f|)\,dx}$. The very definition of Luxemburg norm tells us that $\|f\|_{L^{kA}(G)} \leq 1$. Hence, by (2.1.45), $\|Tf\|_{L^{kB}(G)} \leq C$. The definition of Luxemburg norm again implies that $\int_G kB\left(\frac{|Tf|}{C}\right) dx \leq 1$, namely (2.1.42). □

Proof of Theorem 2.1.8. Let $R > 0$ be such that $G \subset B_R(0)$, the ball centered at 0, with radius R. Fix a smooth function $\eta : [0, \infty) \to [0, \infty)$ for which $\eta = 1$ in $[0, 3R]$ and $\eta = 0$ in $[4R, \infty)$. Define

$$\widehat{N}(x, z) = \eta(|x|)N(x, z) \quad \text{for} \quad x, z \in \mathbb{R}^d,$$
$$\widehat{K}(x, y) = \widehat{N}(x, x - y) \quad \text{for} \quad x, y \in \mathbb{R}^d.$$

By properties (2.1.34)–(2.1.39), one has that:

$$\widehat{N}(x, \lambda y) = \lambda^{-d}\widehat{N}(x, z) \quad \text{for} \quad x, z \in \mathbb{R}^d; \tag{2.1.46}$$

$$\int_{\mathbb{S}^{d-1}} \widehat{N}(x, z)\, d\mathcal{H}^{d-1}(z) = 0 \quad \text{for} \quad x \in \mathbb{R}^d; \tag{2.1.47}$$

for every $\sigma \in [1, \infty)$, there exists a constant $\widehat{C}_1 = \widehat{C}_1(C_1, \sigma, R, d)$ such that

$$\left(\int_{\mathbb{S}^{d-1}} |\widehat{N}(x, z)|^\sigma\, d\mathcal{H}^{d-1}(z)\right)^{\frac{1}{\sigma}} \leq \widehat{C}_1 \quad \text{for} \quad x \in \mathbb{R}^d, \tag{2.1.48}$$

where C_1 is the constant appearing in (2.1.36); there exists a constant $\widehat{C}_2 = \widehat{C}_2(C_2, R, d)$ for which

$$|\widehat{K}(x, y)| \leq \frac{\widehat{C}_2}{|x - y|^d} \quad \text{for} \quad x, y \in \mathbb{R}^d, x \neq y, \tag{2.1.49}$$

and, if $x \in \mathbb{R}^d$, $y \in G$ and $2|x - z| < |x - y|$, then

$$|\widehat{K}(x, y) - \widehat{K}(z, y)| \leq \widehat{C}_2 \frac{|x - z|}{|x - y|^{d+1}}, \tag{2.1.50}$$

$$|\widehat{K}(y, x) - \widehat{K}(y, z)| \leq \widehat{C}_2 \frac{|x - z|}{|x - y|^{d+1}}, \tag{2.1.51}$$

where C_2 is the constant appearing in (2.1.37)–(2.1.39). Define

$$\widehat{T}_\varepsilon f(x) = \int_{\{y:|y-x|>\varepsilon\}} \widehat{K}(x,y)f(y)\,\mathrm{d}y,$$
$$\widehat{T}_S f(x) = \sup_{\varepsilon>0} |\widehat{T}_\varepsilon(f)(x)|.$$

Inequality (2.1.40) will follow if we prove that

$$(\widehat{T}_S f)^*(s) \le C\left(\frac{1}{s}\int_0^s f^*(r)\,\mathrm{d}r + \int_s^{|G|} f^*(r)\,\frac{\mathrm{d}r}{r}\right) \quad \text{for} \quad s \in (0,\infty) \qquad (2.1.52)$$

for some constant $C = C(C_1, C_2, d, R)$, and for every $f \in L^1(\mathbb{R}^d)$ for which $f = 0$ in $\mathbb{R}^d \setminus B_R(0)$. A proof of inequality (2.1.52) can be accomplished along the same lines as that of Theorem 1 of [15], which in turn relies upon similar techniques as in [51]. For completeness, we give the details of the proof hereafter.

The key step in the derivation of (2.1.52) consists in showing that, for every $\gamma \in (0,1)$, there exists a constant $C = C(C_1, C_2, \gamma, d, R)$ such that

$$(\widehat{T}_S f)^*(s) \le C(Mf)^*(\gamma s) + (\widehat{T}_S f)^*(2s) \quad \text{for} \quad s \in (0,\infty) \qquad (2.1.53)$$

for every $f \in L^1(\mathbb{R}^d)$ with $f = 0$ in $\mathbb{R}^d \setminus B_R(0)$. Fix $s > 0$, and define

$$E = \{x \in \mathbb{R}^d : \widehat{T}_S f(x) > (\widehat{T}_S f)^*(2s)\}.$$

Then, there exists an open set $U \supset E$ for which $|U| \le 3s$. By Whitney's covering theorem, there exist a family of disjoint cubes $\{Q_k\}$ such that $U = \cup_{k=1}^\infty Q_k$, $\sum_{k=1}^\infty |Q_k| = |U| \le 3s$, and

$$\mathrm{diam}(Q_k) \le \mathrm{dist}(Q_k, \mathbb{R}^d \setminus U) \le 4\,\mathrm{diam}(Q_k) \quad \text{for} \quad k \in \mathbb{N}.$$

The operator $\widehat{T}_S$ is of weak type (1, 1), namely, there exists a constant C' such that

$$|\{x \in \mathbb{R}^d : \widehat{T}_S f(x) > \lambda\}| \le \frac{C'}{\lambda}\|f\|_{L^1(\mathbb{R}^d)} \qquad (2.1.54)$$

for $f \in L^1(\mathbb{R}^d)$. The proof of (2.1.54) follows from classical arguments: By (2.1.51) we have for all $r > 0$, $y \in \mathbb{R}^d$ and all $x \in B_r(z)$ that

$$\int_{|y-x|\ge 2r} |\widehat{K}(y,x) - \widehat{K}(y,z)|\,\mathrm{d}y \le C.$$

So $\widehat{K}$ satisfies condition (10) in [134, p. 33]. By [134, Cor. 1, p. 33] we obtain

$$|\{x \in \mathbb{R}^d : \widehat{T}_\varepsilon f(x) > \lambda\}| \le \frac{C}{\lambda}\|f\|_{L^1(\mathbb{R}^d)}, \qquad (2.1.55)$$

where C does not depend on ε. Now (2.1.55) implies (2.1.54) by taking the supremum in ε.

We now show that there exists a constant $\overline{C}$ such that

$$|\{x \in Q_k : \widehat{T}_S f(x) > \overline{C} Mf(x) + (\widehat{T}_S f)^*(2s)\}| \leq \frac{1-\gamma}{3} |Q_k| \quad \text{for} \quad k \in \mathbb{N}. \tag{2.1.56}$$

Fix any $k \in \mathbb{N}$, choose $x_k \in \mathbb{R}^d \setminus U$ such that $\operatorname{dist}(x_k, Q_k) \leq 4\operatorname{diam}(Q_k)$, and denote by Q the cube, centered at x_k, with $\operatorname{diam}(Q) = 20\operatorname{diam}(Q_k)$. Define

$$g = f\chi_Q, \quad h = f\chi_{\mathbb{R}^d \setminus Q},$$

so that $f = g + h$. If we prove that there exist constants $\overline{C}_1$ and $\overline{C}_2$ such that

$$\widehat{T}_S h(x) \leq \overline{C}_1 Mf(x) + (\widehat{T}_S f)^*(2s) \quad \text{for} \quad x \in Q_k, \tag{2.1.57}$$

and

$$|\{x \in Q_k : \widehat{T}_S g(x) > \overline{C}_2 Mf(x)\}| \leq \frac{1-\gamma}{3} |Q_k|, \tag{2.1.58}$$

then (2.1.56) follows with $\overline{C} = \overline{C}_1 + \overline{C}_2$. Consider (2.1.58) first. Let $\overline{C}_2$ be a constant for which $\frac{C'|Q|}{\overline{C}_2} \leq \frac{1-\gamma}{3}|Q_k|$. Let $\lambda = \frac{\overline{C}_2}{|Q|} \int_Q |g| \, dx$. Since $\overline{C}_2 Mf(x) \geq \lambda$ for $x \in Q_k$, an application of (2.1.54) with this choice of λ tells us that

$$\begin{aligned} |\{x \in Q_k : \widehat{T}_S g(x) > \overline{C}_2 Mf(x)\}| &\leq |\{\widehat{T}_S g(x) > \lambda\}| \\ &\leq \frac{C'}{\lambda} \int_Q |g| \, dx \leq \frac{C'|Q|}{\overline{C}_2} \leq \frac{1-\gamma}{3} |Q_k|, \end{aligned}$$

namely (2.1.58). In order to establish (2.1.57), it suffices to prove that, for every $\varepsilon > 0$,

$$|\widehat{T}_\varepsilon h(x)| \leq \overline{C}_1 Mf(x) + \widehat{T}_S f(x_k) \quad \text{for} \quad x \in Q_k. \tag{2.1.59}$$

Indeed, since $x_k \notin U$, we have that $\widehat{T}_S f(x_k) \leq (\widehat{T}_S f)^*(2s)$, and hence (2.1.59) implies (2.1.57). We may thus focus on (2.1.59). Fix $\varepsilon > 0$, and set $r = \max\{\varepsilon, \operatorname{dist}(x_k, \mathbb{R}^d \setminus Q)\}$. Observe that $r > 10\operatorname{diam}(Q_k)$. Given any $x \in Q_k$, define $V = B_\varepsilon(x) \triangle B_\varepsilon(x_k)$. One has that

$$\begin{aligned} |\widehat{T}_\varepsilon h(x)| &= \left| \int_{\{y : |y-x| > \varepsilon\}} \widehat{K}(x, y) h(y) \, dy \right| \\ &\leq \left| \int_{\{y : |y-x_k| > \varepsilon\}} \widehat{K}(x, y) h(y) \, dy \right| + \int_V |\widehat{K}(x, y) h(y)| \, dy. \end{aligned} \tag{2.1.60}$$

Observe that, if $y \in \operatorname{supp} h$, then $|x-y| > \frac{r}{2}$ and hence $\frac{1}{|x-y|^d} < \frac{2^d}{r^d}$. Thus, due to (2.1.49),

$$|\widehat{K}(x,y)| \le \frac{\widehat{C}_2}{r^d}.$$

Moreover, $V \subset B_{3r}(x)$. Therefore, there exists a constant $\widehat{C}$ such that

$$\int_V |\widehat{K}(x,y)h(y)|\,dy \le \widehat{C} \fint_{B_{3r}(x)} |h(y)|\,dy \le \widehat{C}Mh(x) \le \widehat{C}Mf(x). \tag{2.1.61}$$

On the other hand,

$$\begin{aligned} \left| \int_{\{y:|y-x_k|>\varepsilon\}} \widehat{K}(x,y)h(y)\,dy \right| &\le \left| \int_{\{y:|y-x_k|>r\}} \widehat{K}(x,y)h(y)\,dy \right| \qquad (2.1.62) \\ &\le \left| \int_{\{y:|y-x_k|>r\}} \widehat{K}(x_k,y)f(y)\,dy \right| + \int_{\{y:|y-x_k|>r\}} |\widehat{K}(x_k,y) - \widehat{K}(x,y)|\,|f(y)|\,dy \\ &\le \widehat{T}_S(x_k) + \int_{\{y:|y-x_k|>r\}} |\widehat{K}(x_k,y) - \widehat{K}(x,y)|\,|f(y)|\,dy, \end{aligned}$$

where the first inequality holds since $h(y) = 0$ in $\{y : |y-x_k| \le r\}$ if $r = \operatorname{dist}(x_k, \mathbb{R}^d \setminus Q)$, and trivially holds (with equality) if $r = \varepsilon$. Since $2|x-x_k| \le |x-y|$ in the last integral in (2.1.62), and f vanishes in $\mathbb{R}^d \setminus B_R(0)$, by (2.1.50)

$$|\widehat{K}(x_k,y) - \widehat{K}(x,y)| \le \widehat{C}_2 \frac{|x_k - x|}{|x-y|^{d+1}} \le \widehat{C}_2 \frac{\operatorname{diam}(Q_k)}{|x-y|^{d+1}}.$$

Hence,

$$\begin{aligned} &\int_{\{y:|y-x_k|>r\}} |\widehat{K}(x_k,y) - \widehat{K}(x,y)|\,|f(y)|\,dy \\ &\quad \le \int_{\{y:|y-x|>\operatorname{diam}(Q_k)\}} |f(y)| \frac{\operatorname{diam}(Q_k)}{|x-y|^{d+1}}\,dy \le \widetilde{C}Mf(x) \end{aligned} \tag{2.1.63}$$

for some constant $\widetilde{C}$. Note that, in the first inequality, we made use of the inclusion $\{y : |y-x_k| > r\} \subset \{y : |y-x| > \operatorname{diam}(Q_k)\}$, which holds since $|x-x_k| < 5\operatorname{diam}(Q_k)$, and $10\operatorname{diam}(Q_k) < r$.

Combining inequalities (2.1.60)–(2.1.63) implies (2.1.59). Inequality (2.1.56) is fully established. Via summation in $k \in Q_k$, we obtain from (2.1.56) that

$$|\{x \in \mathbb{R}^d : \widehat{T}_S f(x) > \widehat{C}Mf(x) + (\widehat{T}_S f)^*(2s)\}| \le (1-\gamma)s. \tag{2.1.64}$$

Combining (2.1.64) with the inequality

$$|\{x \in \mathbb{R}^d : Mf(x) > (Mf)^*(\gamma s)\}| \le \gamma s \tag{2.1.65}$$

tells us that

$$\begin{aligned}&|\{x\in\mathbb{R}^d : \widehat{T}_S f(x) > \widehat{C}(Mf)^*(\gamma s) + (\widehat{T}_S f)^*(2s)\}|\\ &\quad\le |\{\widehat{T}_S f(x) > \widehat{C}Mf(x) + (\widehat{T}_S f)^*(2s)\}| + |\{Mf(x) > (Mf)^*(\gamma s)\}| \le s.\end{aligned}$$

Hence (2.1.53) follows, by the very definition of decreasing rearrangement.

Starting from inequality (2.1.53) we apply the iteration argument from [15, Lemma 3.2] and obtain for $\gamma = 1/2$

$$(\widehat{T}_S f)^*(s) \le C\sum_{k=0}^{\infty}(Mf)^*(2^{k-1}s) + \lim_{s\to\infty}(\widehat{T}_S f)^*(s).$$

Therefore, we have for all f satisfying $\lim_{s\to\infty}(\widehat{T}_S f)^*(s) = 0$ that

$$(\widehat{T}_S f)^*(s) \le C\sum_{k=2}^{\infty}(Mf)^*(2^{k-1}s) + 2C(Mf)^*\Big(\frac{s}{2}\Big).$$

Since

$$(Mf)^*(2^{k-1}s) \le \int_{\{2^{k-2}s\le\sigma\le 2^{k-1}s\}}(Mf)^*(\sigma)\frac{\mathrm{d}\sigma}{\sigma}$$

we conclude

$$(\widehat{T}_S f)^*(s) \le C\int_s^{\infty}(Mf)^*(\sigma)\frac{\mathrm{d}\sigma}{\sigma} + 2C(Mf)^*\Big(\frac{s}{2}\Big). \tag{2.1.66}$$

Now we fix $s > 0$ and assume that $\int_s^{\infty}(Mf)^*(\sigma)\frac{\mathrm{d}\sigma}{\sigma}$ is finite. Since each f has compact support, $\widehat{T}_S f(x)$ converges to zero for all k as $|x|\to\infty$ (recall (2.1.49) and the definition of $\widehat{T}_S$). Therefore, we have that $\lim_{s\to\infty}(\widehat{T}_S f)^*(s) = 0$ for every k. Hence (2.1.66) implies

$$\begin{aligned}(\widehat{T}_S f)^*(s) &\le C\int_s^{\infty}(Mf)^*(\sigma)\frac{\mathrm{d}\sigma}{\sigma} + 2C(Mf)^*\Big(\frac{s}{2}\Big)\\ &\le C\bigg(\int_s^{\infty} f^*(r)\,\frac{dr}{r} + f^*\Big(\frac{s}{2}\Big)\bigg)\\ &\le C\bigg(\int_s^{\infty} f^*(r)\,\frac{dr}{r} + \frac{2}{s}\int_{\frac{s}{2}}^{s} f^*(r)\,\mathrm{d}r\bigg)\\ &\le C\bigg(\int_s^{\infty} f^*(r)\,\frac{dr}{r} + \frac{1}{s}\int_0^{s} f^*(r)\,\mathrm{d}r\bigg)\end{aligned}$$

which shows (2.1.40). □

Lemma 2.1.1. *Let G be a bounded domain with the cone property in $\mathbb{R}^d$, with $n\ge 2$. Then there exist $N\in\mathbb{N}$ and a finite family $\{G_i\}_{i=0,\ldots,N}$ of domains which are* *starshaped with respect to balls, for which* $G = \cup_{i=0}^{N} G_i$. *Moreover, given* $f \in L_{\perp}^{A}(G)$, *there exist* $f_i \in L_{\perp}^{A}(G)$, $i = 0, \dots N$, *such that* $f_i = 0$ *in* $G \setminus G_i$,

$$f = \sum_{i=0}^{N} f_i$$

and

$$\|f_i\|_{L^A(G)} \le C\|f\|_{L^A(G)} \quad \textit{for} \quad i = 0, \dots, N, \tag{2.1.67}$$

for some constant $C = C(G)$.

Proof, sketched. Any bounded open set with the cone property can be decomposed into a finite union of Lipschitz domains [5, Lemma 4.22]. On the other hand, any Lipschitz domain can be decomposed into a finite union of open sets which are starshaped with respect to balls [85, Lemma 3.4, Chapter 3]. This proves the existence of the domains $\{G_i\}_{i=0,\dots N}$ as in the statement. The same argument as in the proof of [85, Lemma 3.2, Chapter 3] then enables us to construct the desired family of functions f_i on G, $i = 1, \dots, N$, according to the following iteration scheme. We set $D_i = \cup_{j=i+1}^{N} G_j$, $g_0 = f$, and, for $i = 1, \dots, N-1$,

$$g_i(x) = \begin{cases} (1 - \chi_{G_i \cap D_i}(x))g_{i-1}(x) - \frac{\chi_{G_i \cap D_i}(x)}{|G_i \cap D_i|} \int_{D_i \setminus G_i} g_{i-1}(y)\, dy & \text{if } x \in D_i, \\ 0 & \text{otherwise,} \end{cases} \tag{2.1.68}$$

and

$$f_i(x) = \begin{cases} g_{i-1}(x) - \frac{\chi_{G_i \cap D_i}(x)}{|G_i \cap D_i|} \int_{G_i} g_{i-1}(y)\, dy & \text{if } x \in G_i, \\ 0 & \text{otherwise.} \end{cases} \tag{2.1.69}$$

Observe that, since G is connected, we can always relabel the sets $G_i \cap D_i$ in such a way that $|G_i \cap D_i| > 0$ for $i = 1, \dots, N-1$. Finally, we define

$$f_N = g_{N-1}. \tag{2.1.70}$$

The family $\{f_i\}$ satisfies the required properties. The only nontrivial property is (2.1.67). To verify it, fix i, and observe that, by (2.1.69), the second inequality in (1.2.10), inequality (1.2.3), and inequality (1.2.7)

$$\begin{aligned} \|f_i\|_{L^A(G)} &\le \|g_{i-1}\|_{L^A(G)} \left(1 + \frac{2}{|G_i \cap D_i|} \|1\|_{L^A(G_i \cap D_i)} \|1\|_{L^{\tilde{A}}(G_i)}\right) \\ &= \|g_{i-1}\|_{L^A(G)} \left(1 + \frac{2}{|G_i \cap D_i| A^{-1}(1/|G_i \cap D_i|)} \frac{1}{\tilde{A}^{-1}(1/|G_i|)}\right) \end{aligned}$$

$$\leq \|g_{i-1}\|_{L^A(G)}\left(1+4\frac{\widetilde{A}^{-1}(1/|G_i\cap D_i|)}{\widetilde{A}^{-1}(1/|G_i|)}\right)$$

$$\leq \|g_{i-1}\|_{L^A(G)}\left(1+4\frac{|G_i|}{|G_i\cap D_i|}\right). \tag{2.1.71}$$

On the other hand, by (2.1.68) and a chain similar to (2.1.71), one has that

$$\|g_{i-1}\|_{L^A(G)} \leq \|g_{i-2}\|_{L^A(G)}\left(1+\frac{2}{|G_{i-1}\cap D_{i-1}|}\|1\|_{L^A(G_{i-1}\cap D_{i-1})}\|1\|_{L^{\tilde{A}}(D_{i-1})}\right)$$

$$\leq \|g_{i-2}\|_{L^A(G)}\left(1+4\frac{\widetilde{A}^{-1}(|1/G_{i-1}\cap D_{i-1}|)}{\widetilde{A}^{-1}(|1/D_{i-1}|)}\right)$$

$$\leq \|g_{i-2}\|_{L^A(G)}\left(1+4\frac{|D_{i-1}|}{|G_{i-1}\cap D_{i-1}|}\right). \tag{2.1.72}$$

From (2.1.71), and an iteration of (2.1.72), one infers that

$$\|f_i\|_{L^A(G)} \leq \left(1+4\frac{|G_i|}{|G_i\cap D_i|}\right)\prod_{j=1}^{i-1}\left(1+4\frac{|D_j|}{|G_j\cap D_j|}\right)\|f\|_{L^A(G)},$$

and (2.1.67) follows. □

Proof of Theorem 2.1.6. By Lemma 2.1.1, it suffices to prove the statement in the case when G is a domain starshaped with respect to a ball $\mathcal{B}$, which, without loss of generality, can be assumed to be centered at the origin and with radius 1. In this case, we are going to show that the (gradient of the) Bogovskiĭ operator Bog_G, defined at a function $f \in L^A_\perp(G)$ is

$$\mathrm{Bog}_G f(x) = \int_G f(y)\left(\frac{x-y}{|x-y|^d}\int_{|x-y|}^{\infty}\omega\left(y+\zeta\frac{x-y}{|x-y|}\right)\zeta^{d-1}\,\mathrm{d}\zeta\right)\mathrm{d}y \tag{2.1.73}$$

for $x \in G$. Here ω is any (nonnegative) function in $C_0^\infty(\mathcal{B})$ with $\int_{\mathcal{B}}\omega\,\mathrm{d}x = 1$, agrees with a singular integral operator, whose kernel fulfills (2.1.34)–(2.1.39), plus two operators enjoying stronger boundedness properties. If $f \in C_0^\infty(G)$ it is easy to see that the same is true for $\mathrm{Bog}_G f$ using the representations

$$\mathrm{Bog}_G f(x)$$
$$= \int_G f(y)(x-y)\left(\frac{x-y}{|x-y|^d}\int_1^{\infty}\omega\Big(y+\zeta(x-y)\Big)\zeta^{d-1}\,\mathrm{d}\zeta\right)\mathrm{d}y$$
$$= \int_G f(y)(x-y)\left(\frac{x-y}{|x-y|^d}\int_0^{\infty}\omega\left(y+\zeta\frac{(x-y)}{|x-y|}\right)(|x-y|\zeta)^{d-1}\,\mathrm{d}\zeta\right)\mathrm{d}y.$$

Setting $\mathbf{u} = \mathrm{Bog}_G f$, we have

$$\mathbf{u}(x) = \int_G f(y)\mathbf{N}(x,y)\,\mathrm{d}y \quad \text{for} \quad x \in G,$$

where

$$\mathbf{N}(x,y) = \frac{x-y}{|x-y|^d} \int_{|x-y|}^{\infty} \omega\Big(y + \zeta \frac{x-y}{|x-y|}\Big) \zeta^{d-1}\, d\zeta \quad \text{for} \quad x, y \in G.$$

By standard arguments we obtain (see [85, Proof of Lemma III.3.1] for more details)

$$\partial_j u_i(x) = \int_G f(y) \partial_j \mathbf{N}_i(x,y)\, dy + f(x) \int_G \frac{(x-y)_i (x-y)_j}{|x-y|^2} \omega(y)\, dy.$$

Computing $\partial_j \mathbf{N}$ we see that

$$\partial_j \mathbf{N} = K_{ij}(x,y) + G_{ij}(x,y)$$

with

$$\begin{aligned} K_{ij}(x,y) = {} & \frac{\delta_{ij}}{|x-y|^d} \int_0^{\infty} \omega\Big(x + \zeta \frac{x-y}{|x-y|}\Big) \zeta^{d-1}\, d\zeta \\ & + \frac{x_i - y_i}{|x-y|^{d+1}} \int_0^{\infty} \partial_j \omega\Big(x + \zeta \frac{x-y}{|x-y|}\Big) \zeta^d\, d\zeta, \end{aligned} \tag{2.1.74}$$

$$G_{ij}(x,y) = \frac{x_i - y_i}{|x-y|^{d+1}} \int_0^{\infty} \partial_j \omega\Big(x + \zeta \frac{x-y}{|x-y|}\Big) \sum_{k=0}^{d-1} \binom{d}{k} |x-y|^k \zeta^{d-k}\, d\zeta. \tag{2.1.75}$$

Now we want to justify the formula for a non-smooth function f. We claim that $\mathbf{u} \in W_0^{1,1}(G)$, and

$$\frac{\partial u_i}{\partial x_j} = H_{ij} f \quad \text{for a.e.} \quad x \in G, \tag{2.1.76}$$

where H_{ij} is the linear operator defined at f as

$$\begin{aligned} (H_{ij} f)(x) = {} & \int_G K_{ij}(x,y) f(y)\, dy + \int_G G_{ij}(x,y) f(y)\, dy \\ & + f(x) \int_G \frac{(x-y)_i (x-y)_j}{|x-y|^2} \omega(y)\, dy \quad \text{for} \quad x \in G, \end{aligned} \tag{2.1.77}$$

for $i, j = 1, \ldots d$. Here, K_{ij} is the kernel of a singular integral operator satisfying the same assumptions as the kernel K in Theorem 2.1.8. Moreover, the following holds

$$|G_{ij}(x,y)| \le \frac{c}{|x-y|^{d-1}} \quad \text{for} \quad x, y \in \mathbb{R}^d,\ x \neq y. \tag{2.1.78}$$

Computing the divergence based on (2.1.76) and (2.1.77) we see that

$$\operatorname{div} \mathbf{u} = \int_G d f(y) \int_1^{\infty} \omega(y + r(x-y)) r^{d-1}\, dr\, dy$$

$$
\begin{aligned}
&+\int_G f(y)\sum_{i=1}^{d}\int_1^{\infty}(x_i-y_i)\partial_i\omega(y+r(x-y))r^d\,\mathrm{d}r\,\mathrm{d}y\\
&+\sum_{i=1}^{d}f(x)\int_G\frac{(x_i-y_i)^2}{|x-y|^2}\omega(y)\,\mathrm{d}y\\
&=\int_G df(y)\int_1^{\infty}\omega(y+r(x-y))r^{d-1}\,\mathrm{d}r\,\mathrm{d}y\\
&+\int_G f(y)\int_1^{\infty}\frac{d}{dr}\omega(y+r(x-y))r^d\,\mathrm{d}r\,\mathrm{d}y+f(x)\\
&=-\omega(x)\int_G f(y)\,\mathrm{d}y+f(x)
\end{aligned}
$$

using (2.1.74), (2.1.75) and $\int_G\omega\,\mathrm{d}y=1$. As $\int_G f\,\mathrm{d}y=0$ we obtain $\operatorname{div}\mathbf{u}=f$.

Now we pass to general functions $f\in L^A_\perp(G)$. Recall that, if $f\in C^\infty_{0,\perp}(G)$, then $\mathbf{u}\in C^\infty_0(G)$, and moreover the equations (2.1.76) and (2.1.23) hold for every $x\in G$. Due to (2.0.11), $L^A_\perp(G)\to L\mathrm{Log}L_\perp(G)$, since $B(t)$ grows at least linearly near infinity, and hence $A(t)$ dominates the function $t\log(1+t)$ near infinity. Since the space $C^\infty_{0,\perp}(G)$ is dense in $L\log L_\perp(G)$, there exists a sequence of functions $\{f_k\}\subset C^\infty_{0,\perp}(G)$ such that $f_k\to f$ in $L\log L(G)$. Hence

$$\mathrm{Bog}_G: L\log L(G)\to L^1(G)$$

(in fact, Bog_G is also bounded into $L\log L(G)$). Furthermore,

$$H_{ij}: L\log L(G)\to L^1(G),$$

as a consequence of (2.1.78) and of a special case of Theorem 2.1.9, with $L^A(G)=L\log L(G)$ and $L^B(G)=L^1(G)$. Thus, $\mathrm{Bog}f_k\to\mathrm{Bog}_G f$ in $L^1(G)$ and $H_{ij}f_k\to H_{ij}f$ in $L^1(G)$. This implies that $\mathbf{u}\in W^{1,1}_0(G)$, and (2.1.76) and (2.1.23) hold.

By Theorem 2.1.9, the singular integral operator defined by the first term on the right-hand side of (2.1.77) is bounded from $L^A(G)$ into $L^B(G)$. By inequality (2.1.78), the operator defined by the second term on the right-hand-side of (2.1.77) has (at least) the same boundedness properties as a Riesz potential operator with kernel $\frac{1}{|x-y|^{d-1}}$. Such an operator is bounded in $L^1(G)$ and in $L^\infty(G)$, with norms depending only on $|G|$ and on d. An interpolation theorem by Calderon [19, Theorem 2.12, Chap. 3] then ensures that it is also bounded from $L^A(G)$ into $L^A(G)$, and hence from $L^A(G)$ into $L^B(G)$, with norm depending on d and $|G|$. Finally, the operator given by the last term on the right-hand-side of (2.1.77) is pointwise bounded (in absolute value) by $|f(x)|$. Thus, it is bounded from $L^A(G)$ into

$L^A(G)$, and hence from $L^A(G)$ into $L^B(G)$. Equations (2.1.21) and (2.1.24) are thus established.

Inequality (2.1.25) follows from (2.1.24) via a scaling argument analogous to that which leads to (2.1.42) from (2.1.41) – see the Proof of Theorem 2.1.9. □

Proof of Theorem 2.1.7. By an argument as in the proofs of [85, Lemmas 3.4 and 3.5, and Theorem 3.3], it suffices to show that, if G and G are bounded Lipschitz domains, for which the domain $G_0 = G \cap D$ is star-shaped with respect to a ball $B \Subset G_0$, and f has the form

$$f = \zeta \operatorname{div} \mathbf{g} + \theta \int_G \varphi \operatorname{div} \mathbf{g} \, dy,$$

for some functions $\zeta \in C_0^\infty(G)$, $\theta \in C_0^\infty(G_0)$ and $\varphi \in C^\infty(\overline{G})$, and fulfills

$$\int_{G_0} f(x) \, dx = 0,$$

then there exists a function $\mathbf{w} \in W_0^{1,B}(G)$ such that

$$\operatorname{div} \mathbf{w} = f \quad \text{in} \quad G_0, \tag{2.1.79}$$

$$\|\nabla \mathbf{w}\|_{L^B(G_0)} \le C \| \operatorname{div} \mathbf{g}\|_{L^A(G)}, \tag{2.1.80}$$

and

$$\|\mathbf{w}\|_{L^B(G_0)} \le C \|\mathbf{g}\|_{L^A(G)}, \tag{2.1.81}$$

for some constant $C = C(\varphi, \theta, \zeta, c, B, G, G)$, where c is the constant appearing in (2.0.11) and (2.0.12).

Since $f \in L_\perp^A(G_0)$, an inspection of the proof of Theorem 2.1.6 then reveals that the function $\mathbf{w}$, given by

$$\mathbf{w}(x) = \int_{G_0} f(y) \mathbf{N}(x, y) \, dy \quad \text{for} \quad x \in G_0, \tag{2.1.82}$$

where

$$\mathbf{N}(x, y) = \frac{x - y}{|x - y|^d} \int_{|x-y|}^\infty \omega\left(y + \zeta \frac{x - y}{|x - y|}\right) \zeta^{d-1} \, d\zeta \quad \text{for} \quad x, y \in G, \tag{2.1.83}$$

and ω is any (nonnegative) function in $C_0^\infty(B)$ with $\int_B \omega \, dx = 1$, satisfies (2.1.79), and

$$\|\nabla \mathbf{w}\|_{L^B(G_0)} \le C \|f\|_{L^A(G_0)} \tag{2.1.84}$$

for some constant C. Since

$$\|f\|_{L^A(G_0)} \le C \| \operatorname{div} \mathbf{g}\|_{L^A(G)} \tag{2.1.85}$$

for some constant C, inequality (2.1.80) follows. It remains to prove (2.1.81). To this purpose, assume, for the time being, that $\operatorname{div} \mathbf{g} \in C_0^\infty(G)$. Then, by [85, Equation 3.35],

$$\begin{aligned} w_i(x) = &- \int_{G_0} N_i(x, y)\mathbf{g}(y) \cdot \nabla\zeta(y)\, dy \\ &- \int_{G_0} K_{ij}(x, x-y)\zeta(y)g_j(y)\, dy \\ &- \int_{G_0} \sum_{j=1}^{n} G_{ij}(x, y)\zeta(y)g_j(y)\, dy \\ &- \zeta(x) \sum_{j=1}^{d} g_j(x) \int_{G_0} \frac{(x_i - y_i)(x_j - y_j)}{|x-y|^2}\omega(y)\, dy \\ &- \left(\int_G \mathbf{g} \cdot \nabla\varphi\, dy\right) \int_{G_0} N_i(x, y)\theta(y)\, dy \quad \text{for a.e.} \quad x \in G_0, \end{aligned} \tag{2.1.86}$$

where the kernels K_{ij} and G_{ij} satisfy the same assumptions as the kernels in (2.1.77), and $|\mathbf{N}(x, y)| \le C|x-y|^{1-d}$ for some constant C. Note that condition (2.1.28) has been used in writing the last term on the right-hand side of equation (2.1.86).

We now drop the assumption that $\operatorname{div} \mathbf{g} \in C_0^\infty(G)$. Condition (2.0.11) entails that $L^A(G) \to L \log L(G)$, and hence $H_0^A(G) \to H_0^{L \log L}(G)$, where the latter space denotes $H_0^A(G)$ with $A(t)$ equivalent to $t \log(1+t)$ near infinity. Thus, $\mathbf{g} \in H_0^{L \log L}(G)$, and hence it can be approximated by a sequence of functions $\{\mathbf{g}_k\} \subset C_0^\infty(G)$ in such a way that $g_k \to g$ in $H^{L \log L}(G)$, cf. Theorem A.1.46. The first and third term on the right-hand side of (2.1.86) are integral operators applied to $\mathbf{g}$ whose kernel is bounded by a multiple of $|x-y|^{1-d}$. The fourth term is just bounded by a constant multiple of $|\mathbf{g}|$. The last term is a constant multiple of an integral of $\mathbf{g}$ against a bounded vector-valued function $\nabla\varphi$. Hence, all these operators are bounded from $L^A(G)$ into $L^A(G_0)$. The second term is a singular integral operator enjoying the same properties as the singular integral operator K in Theorems 2.1.8 and 2.1.9. Thus, since all operators appearing on the right-hand side of (2.1.86) are bounded from $L \log L(G)$ into $L^1(G_0)$, the right-hand side of (2.1.86), evaluated with $\mathbf{g}$ replaced by $\mathbf{g}_k$, converges in $L^1(G_0)$ to the right-hand-side of (2.1.86). On the other hand, equation (2.1.82) and the properties of the Bogovskiĭ operator tell us that the left-hand-side of (2.1.86), with w_i corresponding to $\mathbf{g}_k$, converges in $L^1(G_0)$ to the left-hand-side of (2.1.86). Altogether, we conclude that (2.1.86) actually holds even if $\mathbf{g}$ is just in $H_0^A(G)$.

The properties of the operators on the right-hand-side of (2.1.86) mentioned above ensure that they are bounded from $L^A(G)$ into $L^B(G_0)$. Inequality (2.1.81) thus follows from (2.1.86). □

2.2 NEGATIVE NORMS & THE PRESSURE

We need a last preliminary result in preparation of the proof of Theorem 2.0.5.

Proposition 2.2.1. *Let G be an open subset in $\mathbb{R}^d$ such that $|G| < \infty$, and let A be a Young function. Assume that $u \in L^A(G)$. Then we have*

$$\sup_{v \in L^{\widetilde{A}}(G)} \frac{\int_G uv\,dx}{\|v\|_{L^{\widetilde{A}}(G)}} = \sup_{\varphi \in C_0^\infty(G)} \frac{\int_G u\varphi\,dx}{\|\varphi\|_{L^{\widetilde{A}}(G)}}, \tag{2.2.87}$$

and

$$\sup_{v \in L_\perp^{\widetilde{A}}(G)} \frac{\int_G uv\,dx}{\|v\|_{L^{\widetilde{A}}(G)}} = \sup_{\varphi \in C_{0,\perp}^\infty(G)} \frac{\int_G u\varphi\,dx}{\|\varphi\|_{L^{\widetilde{A}}(G)}}. \tag{2.2.88}$$

Note that equation (2.2.87) is well known under the assumption that $A \in \nabla_2$ near infinity, namely $\widetilde{A} \in \Delta_2$ near infinity. This is because $C_0^\infty(G)$ is dense in $L^{\widetilde{A}}(G)$ in this case. Equation (2.2.88) also easily follows from this property when $A \in \nabla_2$ near infinity. The novelty of Proposition 2.2.1 is in the arbitrariness of A.

Proof of Proposition 2.2.1. Consider first (2.2.87). It clearly suffices to show that

$$\sup_{v \in L^{\widetilde{A}}(G)} \frac{\int_G uv\,dx}{\|v\|_{L^{\widetilde{A}}(G)}} = \sup_{v \in L^\infty(G)} \frac{\int_G uv\,dx}{\|v\|_{L^{\widetilde{A}}(G)}}, \tag{2.2.89}$$

and

$$\sup_{v \in L^\infty(G)} \frac{\int_G uv\,dx}{\|v\|_{L^{\widetilde{A}}(G)}} = \sup_{\varphi \in C_0^\infty(G)} \frac{\int_G u\varphi\,dx}{\|\varphi\|_{L^{\widetilde{A}}(G)}}. \tag{2.2.90}$$

Given any $v \in L^{\widetilde{A}}(G)$, define, for $k \in \mathbb{N}$, the function $v_k : G \to \mathbb{R}$ as

$$v_k = \operatorname{sign}(v) \min\{|v|, k\}. \tag{2.2.91}$$

Clearly, $v_k \in L^\infty(G)$, and $0 \le |v_k| \nearrow |v|$ a.e. in G as $k \to \infty$. Hence,

$$\int_G |uv_k|\,dx \nearrow \int_G |uv|\,dx \quad \text{as} \quad k \to \infty,$$

by the monotone convergence theorem for integrals, and, by the Fatou property of the Luxemburg norm,

$$\|v_k\|_{L^{\widetilde{A}}(G)} \nearrow \|v\|_{L^{\widetilde{A}}(G)} \quad \text{as} \quad k \to \infty.$$

Thus, since

$$\sup_{v \in L^{\widetilde{A}}(G)} \frac{\int_G uv \, \mathrm{d}x}{\|v\|_{L^{\widetilde{A}}(G)}} = \sup_{v \in L^{\widetilde{A}}(G)} \frac{\int_G |uv| \, \mathrm{d}x}{\|v\|_{L^{\widetilde{A}}(G)}},$$

equation (2.2.89) follows.

As far as (2.2.90) is concerned, consider an increasing sequence of compact sets E_k such that $\mathrm{dist}(E_k, \mathbb{R}^d \setminus G) \geq \frac{2}{k}$, $E_k \subset E_{k+1} \subset G$ for $k \in \mathbb{N}$, and $\cup_k E_k = G$. Moreover, let $\{\varrho_k\}$ be a family of (nonnegative) smooth mollifiers in $\mathbb{R}^d$, such that $\mathrm{supp}\varrho_k \subset B_{\frac{1}{k}}(0)$ and $\int_{\mathbb{R}^d} \varrho_k \, \mathrm{d}x = 1$ for $k \in \mathbb{N}$. Given $v \in L^\infty(G)$, define $w_k : \mathbb{R}^d \to \mathbb{R}$ as

$$w_k = \begin{cases} v & \text{in } E_k, \\ 0 & \text{elsewhere,} \end{cases}$$

and $\varphi_k : \mathbb{R}^d \to \mathbb{R}$ as

$$\varphi_k(x) = \int_{\mathbb{R}^d} w_k(y) \varrho_k(x - y) \, \mathrm{d}y \quad \text{for} \quad x \in \mathbb{R}^d. \tag{2.2.92}$$

Classical properties of mollifiers ensure that

$$\varphi_k \in C_0^\infty(G), \quad \varphi_k \to v \quad \text{a.e. in } G \text{ as} \quad k \to \infty,$$
$$\|\varphi_k\|_{L^\infty(G)} \leq \|v\|_{L^\infty(G)} \quad \text{for} \quad k \in \mathbb{N}.$$

Thus, if $u \in L^A(G)$, then

$$\int_G u\varphi_k \, \mathrm{d}x \to \int_G uv \, \mathrm{d}x \quad \text{as} \quad k \to \infty, \tag{2.2.93}$$

by the dominated convergence theorem for integrals. Moreover,

$$\|\varphi_k\|_{L^{\widetilde{A}}(G)} \to \|v\|_{L^{\widetilde{A}}(G)} \quad \text{as} \quad k \to \infty. \tag{2.2.94}$$

Indeed, by dominated convergence and the definition of Luxemburg norm,

$$\int_G \widetilde{A}\left(\frac{|\varphi_k|}{\|v\|_{L^{\widetilde{A}}(G)}}\right) \mathrm{d}x \to \int_G \widetilde{A}\left(\frac{|v|}{\|v\|_{L^{\widetilde{A}}(G)}}\right) \mathrm{d}x \leq 1 \quad \text{as} \quad k \to \infty.$$

In particular, for every $\varepsilon > 0$, there exists k_ε such that

$$\int_G \widetilde{A}\left(\frac{|\varphi_k|}{\|v\|_{L^{\widetilde{A}}(G)}}\right) \mathrm{d}x < 1 + \varepsilon \quad \text{if} \quad k > k_\varepsilon.$$

Hence, by the arbitrariness of ε and the definition of Luxemburg norm,

$$\liminf_{k\to\infty} \|\varphi_k\|_{L^{\tilde{A}}(G)} \geq \|v\|_{L^{\tilde{A}}(G)}. \tag{2.2.95}$$

We also have that

$$\limsup_{k\to\infty} \|\varphi_k\|_{L^{\tilde{A}}(G)} \leq \|v\|_{L^{\tilde{A}}(G)}. \tag{2.2.96}$$

Indeed, assume that (2.2.96) fails. Then, there exists $\sigma > 0$ and a subsequence of $\{\varphi_k\}$, still denoted by $\{\varphi_k\}$, such that

$$1 < \int_G \tilde{A}\left(\frac{|\varphi_k|}{\|v\|_{L^{\tilde{A}}(G)} + \sigma}\right) dx \to \int_G \tilde{A}\left(\frac{|v|}{\|v\|_{L^{\tilde{A}}(G)} + \sigma}\right) dx \leq 1,$$

which is a contradiction. Equation (2.2.94) follows from (2.2.95) and (2.2.96). Coupling (2.2.93) with (2.2.94) implies (2.2.90). The proof of (2.2.87) is complete.

The proof of (2.2.88) follows along the same lines, and, in particular, via the equations

$$\sup_{v\in L_\perp^{\tilde{A}}(G)} \frac{\int_G uv\, dx}{\|v\|_{L^{\tilde{A}}(G)}} = \sup_{v\in L_\perp^{\infty}(G)} \frac{\int_G uv\, dx}{\|v\|_{L^{\tilde{A}}(G)}}, \tag{2.2.97}$$

and

$$\sup_{v\in L_\perp^{\infty}(G)} \frac{\int_G uv\, dx}{\|v\|_{L^{\tilde{A}}(G)}} = \sup_{\varphi\in C_{0,\perp}^{\infty}(G)} \frac{\int_G u\varphi\, dx}{\|\varphi\|_{L^{\tilde{A}}(G)}}. \tag{2.2.98}$$

For any $v \in L_\perp^{\tilde{A}}(G)$, we define the sequence of functions $\{\overline{v}_k\} \subset L_\perp^\infty(G)$ by

$$\overline{v}_k = v_k - (v_k)_G.$$

Here we have $k \in \mathbb{N}$ and v_k is given by (2.2.91). We can prove equation (2.2.97) via a slight variant of the argument used for (2.2.89). Here, one has to use the fact that $(v_k)_G \to 0$ as $k \to \infty$.

Equation (2.2.98) can be established similarly to (2.2.90). Let $v \in L_\perp^\infty(G)$ be given. We have to replace the sequence $\{\varphi_k\}$ defined by (2.2.92) with the sequence $\{\overline{\varphi}_k\} \subset C_{0,\perp}^\infty(G)$ defined by

$$\overline{\varphi}_k = \varphi_k - (\varphi_k)_G \psi \quad \text{for} \quad k \in \mathbb{N}.$$

Here ψ is any function in $C_0^\infty(G)$ such that $\int_G \psi\, dx = 1$. For every $\varepsilon > 0$, there exists $k_\varepsilon \in \mathbb{N}$ such that $\|\overline{\varphi}_k\|_{L^\infty(G)} \leq \|v\|_{L^\infty(G)} + \varepsilon$, provided that $k > k_\varepsilon$.

□

Proof of Theorem 2.0.5. Let $u \in L^1(G)$. Then

$$\begin{aligned}
\|u-u_G\|_{L^B(G)} &= \sup_{v\in L^{\widetilde{B}}(G)} \frac{\int_G (u-u_G)\,v\,\mathrm{d}x}{\|v\|_{L^{\widetilde{B}}(G)}} = \sup_{v\in L^{\widetilde{B}}(G)} \frac{\int_G (u-u_G)\,(v-v_G)\,\mathrm{d}x}{\|v\|_{L^{\widetilde{B}}(G)}} \\
&= \sup_{v\in L^{\widetilde{B}}(G)} \frac{\int_G u\,(v-v_G)\,\mathrm{d}x}{\|v\|_{L^{\widetilde{B}}(G)}} \le 3 \sup_{v\in L^{\widetilde{B}}(G)} \frac{\int_G u\,(v-v_G)\,\mathrm{d}x}{\|v-v_G\|_{L^{\widetilde{B}}(G)}} \\
&= 3 \sup_{v\in L^{\widetilde{B}}_{\perp}(G)} \frac{\int_G u\,v\,\mathrm{d}x}{\|v\|_{L^{\widetilde{B}}(G)}} = 3 \sup_{\varphi\in C^\infty_{0,\perp}(G)} \frac{\int_G u\varphi\,\mathrm{d}x}{\|\varphi\|_{L^{\widetilde{B}}(G)}}. \qquad (2.2.99)
\end{aligned}$$

Note that the inequality in (2.2.99) holds since, by the first inequality in (1.2.3),

$$\begin{aligned}
\|v-v_G\|_{L^{\widetilde{B}}(G)} &\le \|v\|_{L^{\widetilde{B}}(G)} + \|v_G\|_{L^{\widetilde{B}}(G)} \le \|v\|_{L^{\widetilde{B}}(G)} + |v_G|\,\|1\|_{L^{\widetilde{B}}(G)} \\
&\le \|v\|_{L^{\widetilde{B}}(G)} + \tfrac{2}{|G|}\|v\|_{L^{\widetilde{B}}(G)}\|1\|_{L^{B}(G)}\|1\|_{L^{\widetilde{B}}(G)} \\
&= \|v\|_{L^{\widetilde{B}}(G)} + \tfrac{2}{|G|}\|v\|_{L^{\widetilde{B}}(G)}\frac{1}{B^{-1}(|G|)}\,\frac{1}{\widetilde{B}^{-1}(|G|)} \le 3\|v\|_{L^{\widetilde{B}}(G)}.
\end{aligned}$$

The last equality in (2.2.99) follows from (2.2.88). By Theorem 2.1.6, applied with A and B replaced with $\widetilde{B}$ and $\widetilde{A}$, respectively, there exists a constant $C = C(G, c)$ such that

$$\begin{aligned}
\sup_{\varphi\in C^\infty_{0,\perp}(G)} \frac{\int_G u\varphi\,\mathrm{d}x}{\|\varphi\|_{L^{\widetilde{B}}(G)}} &= \sup_{\varphi\in C^\infty_{0,\perp}(G)} \frac{\int_G u\,\mathrm{div}\,(\mathrm{Bog}_G\varphi)\,\mathrm{d}x}{\|\varphi\|_{L^{\widetilde{B}}(G)}} \\
&\le C \sup_{\varphi\in C^\infty_{0,\perp}(G)} \frac{\int_G u\,\mathrm{div}\,(\mathrm{Bog}_G\varphi)\,\mathrm{d}x}{\|\nabla \mathrm{Bog}_G\varphi\|_{L^{\widetilde{A}}(G)}} \\
&\le C \sup_{\boldsymbol{\varphi}\in C^\infty_0(G)} \frac{\int_G u\,\mathrm{div}\,\boldsymbol{\varphi}\,\mathrm{d}x}{\|\nabla\boldsymbol{\varphi}\|_{L^{\widetilde{A}}(G)}}\,\mathrm{d}x = C\,\|u\|_{W^{-1,A}(G)}. \qquad (2.2.100)
\end{aligned}$$

The first inequality in (2.0.13) follows from (2.2.99) and (2.2.100). The second inequality is trivial, since

$$\begin{aligned}
&\|u\|_{W^{-1,A}(G)} \\
&= \sup_{\boldsymbol{\varphi}\in C^\infty_0(G)} \int_G \frac{u\,\mathrm{div}\,\boldsymbol{\varphi}}{\|\nabla\boldsymbol{\varphi}\|_{L^{\widetilde{A}}(G)}}\,\mathrm{d}x = \sup_{\boldsymbol{\varphi}\in C^\infty_0(G)} \frac{\int_G \left(u-u_G\right)\,\mathrm{div}\,\boldsymbol{\varphi}\,\mathrm{d}x}{\|\nabla\boldsymbol{\varphi}\|_{L^{\widetilde{A}}(G)}} \\
&\le C \sup_{\boldsymbol{\varphi}\in C^\infty_0(G)} \frac{\int_G \left(u-u_G\right)\,\mathrm{div}\,\boldsymbol{\varphi}\,\mathrm{d}x}{\|\,\mathrm{div}\,\boldsymbol{\varphi}\|_{L^{\widetilde{A}}(G)}}\,\mathrm{d}x \le C \sup_{\varphi\in C^\infty_{0,\perp}(G)} \frac{\int_G \left(u-u_G\right)\varphi\,\mathrm{d}x}{\|\varphi\|_{L^{\widetilde{A}}(G)}}\,\mathrm{d}x \\
&\le 2C\|u-u_G\|_{L^A(G)},
\end{aligned}$$

for some constant $C = C(d)$. □

In the remaining part of this section, we focus on the second question raised in the introduction of this chapter, namely the reconstruction of the pressure π in a correct Orlicz space. In case of fluids governed by a general constitutive law, the function $\mathbf{H}$ belongs to some Orlicz space $L^A(G)$. If $A \in \Delta_2 \cap \nabla_2$, then $\pi \in L^A(G)$ as well. However, in general, one can only expect that π belongs to some larger Orlicz space $L^B(G)$. The balance between the Young functions A and B is determined by conditions (2.0.11) and (2.0.12), as stated in the following result.

Theorem 2.2.10. *Let A and B be Young functions fulfilling* (2.0.11) *and* (2.0.12). *Let G be a bounded domain with the cone property in $\mathbb{R}^d$, $d \geq 2$. Assume that $\mathbf{H} \in L^A(G)$ satisfies*

$$\int_G \mathbf{H} : \nabla \boldsymbol{\varphi} \, dx = 0$$

for every $\boldsymbol{\varphi} \in C^\infty_{0,\mathrm{div}}(G)$. Then there exists a unique function $\pi \in L^B_\perp(G)$ such that

$$\int_G \mathbf{H} : \nabla \boldsymbol{\varphi} \, dx = \int_G \pi \,\mathrm{div}\, \boldsymbol{\varphi} \, dx \tag{2.2.101}$$

for every $\boldsymbol{\varphi} \in C^\infty_0(G)$. Moreover, there exists a constant $C = C(G, c)$ such that

$$\|\pi\|_{L^B(G)} \leq C \|\mathbf{H} - \mathbf{H}_G\|_{L^A(G)}, \tag{2.2.102}$$

and

$$\int_G B(|\pi|) \, dx \leq \int_G A(C|\mathbf{H} - \mathbf{H}_G|) \, dx. \tag{2.2.103}$$

Here, c denotes the constant appearing in (2.0.11) *and* (2.0.12).

In particular, Theorem 2.2.10 reproduces, within a unified framework, various results appearing in the literature. For instance, when the power law model is in force, the function $A(t)$ is just a power t^q for some $q > 1$. So $L^A(G)$ agrees with the Lebesgue space $L^q(G)$, and Theorem 2.2.10 recovers the fact that π belongs to the same Lebesgue space $L^q(G)$.

As far as the simplified system (without convective term) for the Eyring–Prandtl model (see Chapter 4) is concerned, under appropriate assumptions on the function $\mathbf{f}$ one has that $\mathbf{H} = \mathbf{S}(\boldsymbol{\varepsilon}(\mathbf{v})) + \nabla\Delta^{-1}\mathbf{f} \in \exp L(G)$. Hence, via Theorem 2.2.10, we obtain the existence of a pressure $\pi \in \exp L^{\frac{1}{2}}(G)$. More generally, if $\mathbf{H} \in \exp L^\beta(G)$ for some $\beta > 0$, one has that $\pi \in \exp L^{\beta/(\beta+1)}(G)$. The complete system for the Eyring–Prandtl model, in the 2-dimensional case, admits a weak solution $\mathbf{v}$ such that $\mathbf{v} \otimes \mathbf{v} \in L\log L^2(G)$ and hence $\mathbf{H} = \mathbf{S}(\boldsymbol{\varepsilon}(\mathbf{v})) + \nabla\Delta^{-1}\mathbf{f} - \mathbf{v} \otimes \mathbf{v} \in L\log L^2(G)$, see Chapter 4. Again, one cannot expect that the pressure π belongs to the same space. In fact, Theorem 2.2.10 implies the existence of a pressure $\pi \in L\log L(G)$. This reproduces a result from [33]. In general, if $\mathbf{H} \in L\log L^\alpha(G)$ for some $\alpha \geq 1$, then we obtain that $\pi \in L\log L^{\alpha-1}(G)$.

Proof of Theorem 2.2.10. By De Rahms Theorem, in the version of [131], there exists a distribution Ξ such that

$$\int_G \mathbf{H} : \nabla\boldsymbol{\varphi}\,\mathrm{d}x = \Xi(\operatorname{div}\boldsymbol{\varphi}) \tag{2.2.104}$$

for every $\boldsymbol{\varphi} \in C_0^\infty(G)$. Replacing $\boldsymbol{\varphi}$ with $\mathrm{Bog}_G(\varphi - \varphi_G)$ in (2.2.104), where $\varphi \in C_0^\infty(G)$, implies

$$\int_G \mathbf{H} : \nabla\mathrm{Bog}_G(\varphi - \varphi_G)\,\mathrm{d}x = \Xi(\varphi - \varphi_G)$$

for every $\varphi \in C_0^\infty(G)$. We claim that the linear functional $C_0^\infty(G) \ni \varphi \mapsto \Xi(\varphi - \varphi_G)$ is bounded on $C_0^\infty(G)$ equipped with the $L^\infty(G)$ norm. Indeed, by (2.0.11), one has that $L^A(G) \to L\log L(G)$. Moreover, by a special case of Theorem 2.1.6, $\nabla\mathrm{Bog}_G : L_\perp^\infty(G) \to \exp L(G)$. Thus, since $L\log L(G)$ and $\exp L(G)$ are Orlicz spaces generated by Young functions which are conjugate of each other,

$$\begin{aligned}\left|\int_G \mathbf{H} : \nabla\mathrm{Bog}_G(\varphi - \varphi_G)\,\mathrm{d}x\right| & \\ \leq C\|\mathbf{H}\|_{L\log L(G)}\|\nabla\mathrm{Bog}_G(\varphi - \varphi)_G)\|_{\exp L(G)} & \\ \leq C'\|\mathbf{H}\|_{L^A(G)}\|\varphi - \varphi_G\|_{L^\infty(G)} & \\ \leq C''\|\mathbf{H}\|_{L^A(G)}\|\varphi\|_{L^\infty(G)}\,, &\end{aligned} \tag{2.2.105}$$

for every $\varphi \in C_0^\infty(G)$, where $C = C(|G|, d)$ and $C' = C'(G, c)$. Hence, the relevant functional can be continued to a bounded linear functional on $\varphi \in C_0(G)$, with the same norm.

Now, as a consequence of Riesz's representation Theorem, there exists a Radon measure μ such that

$$\Xi(\varphi - \varphi_G) = \int_G \varphi\,\mathrm{d}\mu$$

for every $\varphi \in C_0(G)$. Fix any open set $E \subset G$. By Theorem 2.1.6 again, there exists a constant C such that

$$\begin{aligned}\mu(E) &= \sup_{\varphi\in C_0^0(E),\ \|\varphi\|_\infty=1} \Xi(\varphi - \varphi_G)\\ &= \sup_{\varphi\in C_0^0(E),\ \|\varphi\|_\infty=1} \int_G \mathbf{H} : \nabla\mathrm{Bog}_G(\varphi - \varphi_G)\,\mathrm{d}x\\ &\leq \sup_{\varphi\in C_0^0(E),\ \|\varphi\|_\infty=1} \|\mathbf{H}\|_{L\log L(E)}\|\nabla\mathrm{Bog}_G(\varphi - \varphi_G)\|_{\exp L(G)} \qquad (2.2.106)\\ &\leq C \sup_{\varphi\in C_0^0(E),\ \|\varphi\|_\infty=1} \|\mathbf{H}\|_{L\log L(E)}\|\varphi - (\varphi)_G\|_{L^\infty(G)} \leq C\,\|\mathbf{H}\|_{L\log L(E)}.\end{aligned}$$

One can verify that the norm $\|\cdot\|_{L\log L(E)}$ is absolutely continuous in the following sense. For every $\varepsilon > 0$ there exists $\delta > 0$ such that $\|\mathbf{H}\|_{L\log L(E)} < \varepsilon$ if $|E| < \delta$. Since any Lebesgue measurable set can be approximated from outside by open sets, inequality (2.2.106) implies that the measure μ is absolutely continuous with respect to the Lebesgue measure. Hence, μ has a density with respect to the Lebesgue measure. So Ξ can be represented by a function $\pi \in L^1(G)$ fulfilling (2.2.101) holds. The function π is uniquely determined if we assume that $\pi_G = 0$. By this assumption, Theorem 2.0.5, and equation (2.2.101) we have that

$$
\begin{aligned}
\|\pi\|_{L^B(G)} &\le C\|\nabla\pi\|_{W^{-1,A}(G)} = C \sup_{\varphi\in C_0^\infty(G)} \frac{\int_G \pi \operatorname{div}\boldsymbol{\varphi}\,\mathrm{d}x}{\|\nabla\boldsymbol{\varphi}\|_{L^{\tilde{A}}(G)}} \\
&= C \sup_{\varphi\in C_0^\infty(G)} \frac{\int_G \mathbf{H} : \nabla\boldsymbol{\varphi}\,\mathrm{d}x}{\|\nabla\boldsymbol{\varphi}\|_{L^{\tilde{A}}(G)}} = C \sup_{\varphi\in C_0^\infty(G)} \frac{\int_G (\mathbf{H}-\mathbf{H}_G) : \nabla\boldsymbol{\varphi}\,\mathrm{d}x}{\|\nabla\boldsymbol{\varphi}\|_{L^{\tilde{A}}(G)}} \\
&\le 2C\|\mathbf{H}-\mathbf{H}_G\|_{L^A(G)},
\end{aligned}
$$

where $C = C(G, c)$. This proves inequality (2.2.102). Inequality (2.2.103) follows from (2.2.102), by replacing A and B with kA and kB, respectively, with $k = \frac{1}{\int_G A(|\mathbf{H}-\mathbf{H}_G|)\,\mathrm{d}x}$, via an argument analogous to that of the proof of (2.1.42). □

2.3 SHARP CONDITIONS FOR KORN-TYPE INEQUALITIES

In order to formulate the main result of this section we need to introduce the Banach function space

$$
E^A(G) := \{\mathbf{u} \in L^A(G) : \boldsymbol{\varepsilon}(\mathbf{u}) \in L^A(G)\},
$$
$$
\|\mathbf{u}\|_{E^A(G)} := \|\mathbf{u}\|_{L^A(G)} + \|\boldsymbol{\varepsilon}(\mathbf{u})\|_{L^A(G)},
$$

and its subspace

$$
E_0^A(G) := \{\mathbf{u} \in E^A(G) : \text{ the continuation of } \mathbf{u} \text{ by zero belongs to } E^A(\mathbb{R}^d)\}.
$$

If A is of power growth (belongs to $\Delta_2 \cap \nabla_2$) then there is no need for this definition. In this case the space $E^A(G)$ coincides with the standard Lebesgue spaces (Orlicz spaces). However, this is not true in our general context.

Theorem 2.3.11. *Let G be any open bounded set in $\mathbb{R}^d$. Let A and B be Young functions such that*

$$
t\int_{t_0}^{t} \frac{B(s)}{s^2}\,\mathrm{d}s \le A(ct) \qquad \text{for} \quad t \ge t_0, \tag{2.3.107}
$$

and

$$t\int_{t_0}^{t}\frac{\widetilde{A}(s)}{s^2}\,\mathrm{d}s \le \widetilde{B}(ct) \qquad \text{for} \quad t \ge t_0, \tag{2.3.108}$$

for some constants $c > 0$ *and* $t_0 \ge 0$. *Then* $E_0^A(G) \subset W_0^{1,B}(G)$, *and*

$$\|\nabla\mathbf{u}\|_{L^B(G)} \le C\|\boldsymbol{\varepsilon}(\mathbf{u})\|_{L^A(G)} \tag{2.3.109}$$

for some constant $C = C(t_0, G, c, A, B)$ *and for every* $\mathbf{u} \in E_0^A(G)$. *Moreover,*

$$\int_G B(C|\nabla\mathbf{u}|)\,\mathrm{d}x \le C_1 + \int_G A(C|\boldsymbol{\varepsilon}(\mathbf{u})|)\,\mathrm{d}x$$

for every $\mathbf{u} \in E_0^A(G)$ *with* $C_1 = C_1(t_0, G, c, A, B)$. *If* $t_0 = 0$ *then* $C_1 = 0$ *and* $C = C(G, c)$.

Theorem 2.3.12. *Let* G *be an open bounded Lipschitz domain in* $\mathbb{R}^d$. *Assume that* A *and* B *are Young functions fulfilling conditions* (2.3.107) *and* (2.3.108). *Then* $E^A(G) \subset W^{1,B}(G)$, *and*

$$\|\nabla\mathbf{u} - (\nabla\mathbf{u})_G\|_{L^B(G)} \le C\|\boldsymbol{\varepsilon}(\mathbf{u}) - (\boldsymbol{\varepsilon}(\mathbf{u}))_G\|_{L^A(G)} \tag{2.3.110}$$

for some constant $C = C(t_0, G, c; A, B)$ *and for every* $\mathbf{u} \in E^A(G)$. *Moreover,*

$$\int_G B(C|\nabla\mathbf{u} - (\nabla\mathbf{u})_G|)\,\mathrm{d}x \le C_1 + \int_G A(C|\boldsymbol{\varepsilon}(\mathbf{u}) - (\boldsymbol{\varepsilon}(\mathbf{u}))_G|)\,\mathrm{d}x \tag{2.3.111}$$

for every $\mathbf{u} \in E^A(G)$ *with* $C_1 = C_1(t_0, G, c, A, B)$. *If* $t_0 = 0$ *then* $C_1 = 0$ *and* $C = C(G, c)$.

Instead of subtracting the mean value it is also useful to subtract an element from the kernel of the differential operator $\boldsymbol{\varepsilon}$ given by

$$\mathcal{R} = \left\{\mathbf{w} : \mathbb{R}^d \to \mathbb{R}^d : \mathbf{w}(x) = \mathbf{b} + \mathbf{Q}x : \ \mathbf{b} \in \mathbb{R}^d, \ \mathbf{Q} \in \mathbb{R}^{d\times d}, \ \mathbf{Q} = -\mathbf{Q}^T\right\}.$$

Corollary 2.3.1. *Let* G *be an open bounded Lipschitz domain in* $\mathbb{R}^d$. *Assume that* A *and* B *are Young functions fulfilling conditions* (2.3.107) *and* (2.3.108). *Then* $E^A(G) \subset W^{1,B}(G)$, *and there is* $\mathbf{w} \in \mathcal{R}$ *such that*

$$\|\nabla\mathbf{u} - \nabla\mathbf{w}\|_{L^B(G)} \le C\|\boldsymbol{\varepsilon}(\mathbf{u})\|_{L^A(G)} \tag{2.3.112}$$

for some constant $C = C(t_0, G, c; A, B)$ *and for every* $\mathbf{u} \in E^A(G)$. *Moreover,*

$$\int_G B(C|\nabla\mathbf{u} - \nabla\mathbf{w}|)\,\mathrm{d}x \le C_1 + \int_G A(C|\boldsymbol{\varepsilon}(\mathbf{u})|)\,\mathrm{d}x \tag{2.3.113}$$

for every $\mathbf{u} \in E^A(G)$ *with* $C_1 = C_1(t_0, G, c, A, B)$. *If* $t_0 = 0$ *then* $C_1 = 0$ *and* $C = C(G, c)$.

Theorem 2.3.13. *Let G be an open bounded Lipschitz domain in $\mathbb{R}^d$ and let A be any Young function. Then there is $\mathbf{w} \in \mathcal{R}$ such that*

$$\|\mathbf{u}-\mathbf{w}\|_{L^A(G)} \le C\|\boldsymbol{\varepsilon}(\mathbf{u})\|_{L^A(G)} \tag{2.3.114}$$

for some constant $C = C(G, A)$ and for every $\mathbf{u} \in E^A(G)$. Moreover,

$$\int_G A(C|\mathbf{u}-\mathbf{w}|)\,\mathrm{d}x \le \int_G A(C|\boldsymbol{\varepsilon}(\mathbf{u})|)\,\mathrm{d}x \tag{2.3.115}$$

for every $\mathbf{u} \in E^A(G)$.

Proof of Theorem 2.3.12. Let us introduce negative norms for single partial derivatives as follows. Given $u \in L^1(G)$, we set

$$\left\|\frac{\partial u}{\partial x_k}\right\|_{W^{-1,A}(G)} = \sup_{\varphi\in C_0^\infty(G)} \frac{\int_G u \frac{\partial \varphi}{\partial x_k}\,\mathrm{d}x}{\|\nabla\varphi\|_{L^{\tilde{A}}(G)}} \quad \text{for} \quad k=1,\dots,d.$$

Obviously, the following holds

$$\left\|\frac{\partial u}{\partial x_k}\right\|_{W^{-1,A}(G)} \le \|\nabla u\|_{W^{-1,A}(G)} \quad \text{for} \quad k=1,\dots d. \tag{2.3.116}$$

On the other hand,

$$\begin{aligned}
\|\nabla u\|_{W^{-1,A}(G)} &= \sup_{\boldsymbol{\varphi}\in C_0^\infty(G)} \frac{\int_G u \operatorname{div}\boldsymbol{\varphi}\,\mathrm{d}x}{\|\nabla\boldsymbol{\varphi}\|_{L^{\tilde{A}}(G)}} = \sup_{\boldsymbol{\varphi}\in C_0^\infty(G)} \sum_{k=1}^d \frac{\int_G u \frac{\partial \varphi_k}{\partial x_k}\,\mathrm{d}x}{\|\nabla\boldsymbol{\varphi}\|_{L^{\tilde{A}}(G)}} \\
&\le \sup_{\boldsymbol{\varphi}\in C_0^\infty(G)} \sum_{k=1}^d \frac{\int_G u \frac{\partial \varphi_k}{\partial x_k}\,\mathrm{d}x}{\|\nabla\varphi_k\|_{L^{\tilde{A}}(G)}} \le \sum_{k=1}^d \sup_{\varphi\in C_0^\infty(G)} \frac{\int_G u \frac{\partial \varphi}{\partial x_k}\,\mathrm{d}x}{\|\nabla\varphi\|_{L^{\tilde{A}}(G)}} \\
&= \sum_{k=1}^d \left\|\frac{\partial u}{\partial x_k}\right\|_{W^{-1,A}(G)},
\end{aligned} \tag{2.3.117}$$

where φ_k denotes the k-th component of $\boldsymbol{\varphi}$. Next, notice the identity

$$\frac{\partial^2 v_i}{\partial x_k \partial x_j} = \frac{\partial(\boldsymbol{\varepsilon}(\mathbf{v}))_{ij}}{\partial x_k} + \frac{\partial(\boldsymbol{\varepsilon}(\mathbf{v}))_{ik}}{\partial x_j} - \frac{\partial(\boldsymbol{\varepsilon}(\mathbf{v}))_{jk}}{\partial x_i} \tag{2.3.118}$$

for every weakly differentiable function $\mathbf{v} : G \to \mathbb{R}^d$. Thus, the following chain holds for every $\mathbf{u} \in W^{1,1}(G) \cap E^A(G)$

$$\begin{aligned}
\|\nabla\mathbf{u} - (\nabla\mathbf{u})_G\|_{L^B(G)} &\le C \sum_{i,j=1}^d \left\|\frac{\partial u_i}{\partial x_j} - \left(\frac{\partial u_i}{\partial x_j}\right)_G\right\|_{L^B(G)} \\
&\le C \sum_{i,j=1}^n \left\|\nabla\frac{\partial u_i}{\partial x_j}\right\|_{W^{-1,A}(G)} \le C \sum_{i,j,k=1}^d \left\|\frac{\partial^2 u_i}{\partial x_k \partial x_j}\right\|_{W^{-1,A}(G)}
\end{aligned}$$

$$\leq C \sum_{i,j,k=1}^{d} \left(\left\| \frac{\partial(\boldsymbol{\varepsilon}(\mathbf{u}))_{ij}}{\partial x_k} \right\|_{W^{-1,A}(G)} + \left\| \frac{\partial(\boldsymbol{\varepsilon}(\mathbf{u}))_{ik}}{\partial x_j} \right\|_{W^{-1,A}(G)} + \left\| \frac{\partial(\boldsymbol{\varepsilon}(\mathbf{u}))_{jk}}{\partial x_i} \right\|_{W^{-1,A}(G)} \right)$$

$$\leq C \sum_{i,j=1}^{d} \|\nabla(\boldsymbol{\varepsilon}(\mathbf{u}))_{ij}\|_{W^{-1,A}(G)} \leq C \sum_{i,j=1}^{n} \|(\boldsymbol{\varepsilon}(\mathbf{u}))_{ij} - ((\boldsymbol{\varepsilon}(\mathbf{u}))_{ij})_G\|_{L^A(G)}$$

$$\leq C \|\boldsymbol{\varepsilon}(\mathbf{u}) - (\boldsymbol{\varepsilon}(\mathbf{u}))_G\|_{L^A(G)}. \tag{2.3.119}$$

Note that the second inequality holds by Theorem 2.0.5 (see Remark 2.0.6 for the case $t_0 > 0$), the third by (2.3.117), the fourth by (2.3.118), the fifth by (2.3.116), and the sixth by Theorem 2.0.5 again.

Let us turn to the modular version. Suppose first that $t_0 = 0$ in (2.3.107) and (2.3.108). An inspection of the proof of Theorem (2.3.119) and of the statement of Theorem 2.0.5 tells us that the constant C in (2.3.119) depends only on G and on the constant c appearing in conditions (2.3.107) and (2.3.108). These conditions continue to hold if the functions A and B are replaced by the functions A_M and B_M given by $A_M(t) = A(t)/M$ and $B_M(t) = B(t)/M$ for some positive constant M. Given a function $\mathbf{u} \in E_0^A(G)$, set

$$M = \int_\Omega A(C|\boldsymbol{\varepsilon}(\mathbf{u}) - (\boldsymbol{\varepsilon}(\mathbf{u}))_G|)\, dx.$$

If $M = \infty$, then inequality (2.3.111) holds trivially. We may thus assume that

$$\|\nabla \mathbf{u} - (\nabla \mathbf{u})_G\|_{L^{A_M}(G)} \leq 1.$$

Hence, by inequality (2.3.110) applied with A and B replaced by A_M and B_M, we obtain

$$\int_\Omega B(|\nabla \mathbf{u} - (\nabla \mathbf{u})_G|)\, dx \leq \int_\Omega A(C|\boldsymbol{\varepsilon}(\mathbf{u}) - (\boldsymbol{\varepsilon}(\mathbf{u}))_G|)\, dx. \tag{2.3.120}$$

This is (2.3.111) with $C_1 = 0$.

Assume next that (2.3.107) and (2.3.108) just hold for some $t_0 > 0$. The functions A and B can be replaced by new Young functions $\overline{A}$ and $\overline{B}$, equivalent to A and B near infinity, and such that (2.3.107) and (2.3.108) hold for the new functions with $t_0 = 0$. The same argument as above implies (2.3.120) with A and B replaced by $\overline{A}$ and $\overline{B}$., i.e., we have

$$\int_\Omega \overline{B}(|\nabla \mathbf{u} - (\nabla \mathbf{u})_G|)\, dx \leq \int_\Omega \overline{A}(C|\boldsymbol{\varepsilon}(\mathbf{u}) - (\boldsymbol{\varepsilon}(\mathbf{u}))_G|)\, dx \tag{2.3.121}$$

for some constant C. Since $\overline{A}$ and $\overline{B}$ are equivalent to A and B near infinity, there exist constants $t_0 > 0$ and $c > 0$ such that

$$\overline{A}(t) \leq A(ct) \quad \text{if} \quad t \geq t_0, \qquad B(t) \leq \overline{B}(ct) \quad \text{if} \quad t \geq t_0. \tag{2.3.122}$$

From (2.3.121) and (2.3.122) we obtain

$$\begin{aligned}
&\int_G B(|\nabla \mathbf{u} - (\nabla \mathbf{u})_G|)\,\mathrm{d}x \\
&\quad = \int_{\{|\nabla \mathbf{u}-(\nabla \mathbf{u})_G|<t_0\}} B(|\nabla \mathbf{u} - (\nabla \mathbf{u})_G|)\,\mathrm{d}x \\
&\qquad + \int_{\{|\nabla \mathbf{u}-(\nabla \mathbf{u})_G|\geq t_0\}} B(|\nabla \mathbf{u} - (\nabla \mathbf{u})_G|)\,\mathrm{d}x \\
&\quad \leq B(t_0)|G| + \int_G \overline{B}(c|\nabla \mathbf{u} - (\nabla \mathbf{u})_G|)\,\mathrm{d}x \\
&\quad \leq B(t_0)|G| + \int_G \overline{A}(Cc|\boldsymbol{\varepsilon}(\mathbf{u}) - (\boldsymbol{\varepsilon}(\mathbf{u}))_G|)\,\mathrm{d}x \\
&\quad \leq B(t_0)|G| + \int_{\{Cc|\boldsymbol{\varepsilon}(\mathbf{u})-(\boldsymbol{\varepsilon}(\mathbf{u}))_G|<t_0\}} \overline{A}(Cc|\boldsymbol{\varepsilon}(\mathbf{u}) - (\boldsymbol{\varepsilon}(\mathbf{u}))_G|)\,\mathrm{d}x \\
&\qquad + \int_{\{Cc|\boldsymbol{\varepsilon}(\mathbf{u})-(\boldsymbol{\varepsilon}(\mathbf{u}))_G|\geq t_0\}} \overline{A}(Cc|\boldsymbol{\varepsilon}(\mathbf{u}) - (\boldsymbol{\varepsilon}(\mathbf{u}))_G|)\,\mathrm{d}x \\
&\quad \leq (B(t_0) + A(ct_0))|G| + \int_G A(Cc^2|\boldsymbol{\varepsilon}(\mathbf{u}) - (\boldsymbol{\varepsilon}(\mathbf{u}))_G|)\,\mathrm{d}x,
\end{aligned}$$

namely (2.3.111) with $C_1 = (B(t_0) + A(ct_0))|G|$. □

Proof of Theorem 2.3.11. If $\mathbf{u} \in W_0^{1,1}(G)$, then $(\nabla \mathbf{u})_G = (\boldsymbol{\varepsilon}(\mathbf{u}))_G = 0$, and Theorem 2.3.11 implies the claim. □

Proof of Corollary 2.3.1. Let us choose $\mathbf{w} \in \mathcal{R}$ through the condition $\nabla \mathbf{w} = (\nabla \mathbf{u})_G - (\boldsymbol{\varepsilon}(\mathbf{u}))_G$. Then the following holds by (2.3.110)

$$\|\nabla \mathbf{u} - \nabla \mathbf{w}\|_{L^B(G)} \leq \|\nabla \mathbf{u} - (\nabla \mathbf{u})_G\|_{L^B(G)} + \|(\boldsymbol{\varepsilon}(\mathbf{u}))_G\|_{L^B(G)} \leq c\,\|\boldsymbol{\varepsilon}(\mathbf{u})\|_{L^A(G)}.$$

The integral version follows form the norm version as in the proof of (2.3.111). □

Proof of Theorem 2.3.13.

Step 1: Star-shaped domains.

First, we assume that G is star-shaped with respect to some ball $\mathcal{B} \subset G$. Then, according to formula (2.39) in [126] each $\mathbf{v} \in C^\infty(G)$ can be represented as

$$\mathbf{u}(x) = P_G\mathbf{u}(x) + \mathscr{L}_{\boldsymbol{\varepsilon}}(\boldsymbol{\varepsilon}(\mathbf{u}))(x), \tag{2.3.123}$$

where $P_G : L^1(G) \to \mathcal{R}$ is a suitable linear projection operator into the space of rigid motions (compare (2.33)–(2.39) in [126]) defined even for

$L^1(G)$ functions. Furthermore, the operator $\mathscr{L}_\varepsilon$ is a weakly singular integral operator (compare (2.37) in [126]) given by

$$\mathscr{L}_\varepsilon(\boldsymbol{\varphi}) := \mathcal{S}_\varepsilon(\boldsymbol{\psi}) + \mathcal{T}_\varepsilon(\boldsymbol{\psi}), \quad \boldsymbol{\psi} \in C^\infty(G),$$
$$\mathcal{S}^i_\varepsilon(\boldsymbol{\varphi})(x) := \int_G \frac{\mathbf{G}^i(x,e)}{|x-z|^{d-1}} : \boldsymbol{\psi}(z)\,\mathrm{d}z, \quad i=1,\dots,d, \tag{2.3.124}$$
$$\mathcal{T}^i_\varepsilon(\boldsymbol{\varphi})(x) := \int_G \boldsymbol{\theta}^i(x,z) : \boldsymbol{\psi}(z)\,\mathrm{d}z, \quad i=1,\dots,d,$$

for $x \in G$. Here, $\mathbf{G}^i(x,e)$ are smooth functions ($e := (x-z)/|x-z|$) and $\boldsymbol{\theta}^i(x,z)$ are bounded continuous functions (both with values in $\mathbb{R}^{d\times d}$); see [126] after (2.38).

Since the kernel of $\mathcal{S}_\varepsilon$ is essentially homogeneous of degree $1-d$ the theory of Riesz potentials applies (see, e.g. [134] or [90]). Hence we have $\mathcal{S}_\varepsilon : L^1(G) \to L^1(G)$. Of course, the same is true for $\mathcal{T}_\varepsilon$ because the $\boldsymbol{\theta}^i(x,z)$ are bounded. So, we obtain

$$\mathscr{L}_\varepsilon : L^1(G) \to L^1(G) \tag{2.3.125}$$

continuously. We claim that equation (2.3.123) continues to hold even if $\mathbf{u} \in E^A(G)$. From [138, Proposition 1.3, Chapter 1], we deduce that $C^\infty(G)$ is dense in $E^1(G)$. Thus, there exists a sequence $\{\mathbf{u}_m\} \subset C^\infty(G)$ such that

$$\mathbf{u}_m \to \mathbf{u} \quad \text{in} \quad E^1(G).$$

We already know that formula (2.3.123) holds with $\mathbf{u}$ replaced by $\mathbf{u}_m$. Thus we can pass to the limit (passing to a subsequence if necessary) in the representation formula (2.3.123) applied to $\mathbf{u}_m$. This implies that it continues to hold also for $\mathbf{u}$. Note that this is a consequence of (2.3.125).

Now, we turn to the case of a Lipschitz domain. In order to do so, we consider a general linear projection operator $\Pi : L^1(G) \to \Sigma$ such that

$$\|\Pi\mathbf{u}\|_{L^1(G)} \le C\|\mathbf{u}\|_{L^1(G)}, \tag{2.3.126}$$

for some constant C, and every $\mathbf{u} \in L^1(G)$. We claim that there exists a constant C' such that

$$\inf_{\mathbf{w}\in\mathcal{R}} \|\mathbf{u}-\mathbf{w}\|_{L^A(G)} \le \|\mathbf{u}-\Pi\mathbf{u}\|_{L^A(G)} \le C(A) \inf_{\mathbf{w}\in\mathcal{R}} \|\mathbf{u}-\mathbf{w}\|_{L^A(G)}, \tag{2.3.127}$$

for every $\mathbf{u} \in L^A(G)$ and for every Young function A.

The left-wing inequality in (2.3.127) is trivial. As far as the right-wing inequality is concerned, given any $\mathbf{w} \in \mathcal{R}$, and any $\mathbf{u}$ in $L^A(G)$ we set

$$\mathbf{v} = \mathbf{w} + (\mathbf{u}-\mathbf{w})_G.$$

Since Π, restricted to $\mathcal{R}$, agrees with the identity map, we have that $\Pi\mathbf{v}=\mathbf{v}$. As a consequence,

$$\mathbf{u}-\Pi\mathbf{u}=(\mathbf{u}-\mathbf{v})-\Pi(\mathbf{u}-\mathbf{v}).$$

Thus,

$$\|\mathbf{u}-\Pi\mathbf{u}\|_{L^A(G}\le\|\mathbf{u}-\mathbf{v}\|_{L^A(G)}+\|\Pi(\mathbf{u}-\mathbf{v})\|_{L^A(G)}. \tag{2.3.128}$$

By the triangle inequality,

$$\|\mathbf{u}-\mathbf{v}\|_{L^A(G)}=\|\mathbf{u}-\mathbf{w}-(\mathbf{u}-\mathbf{w})_G\|_{L^A(G)}\le 2\|\mathbf{u}-\mathbf{w}\|_{L^A(G)}. \tag{2.3.129}$$

Since the range of Π is a finite dimensional space, where all norms are equivalent, there exists a constant C'' such that

$$\|\Pi(\mathbf{u}-\mathbf{v})\|_{L^A(G)}+\|\nabla\Pi(\mathbf{u}-\mathbf{v})\|_{L^A(G)}\le C''\,\|\Pi(\mathbf{u}-\mathbf{v})\|_{L^1(G)}. \tag{2.3.130}$$

Inequality (2.3.126) ensures that

$$\|\Pi(\mathbf{u}-\mathbf{v})\|_{L^1(G)}\le C\,\|\mathbf{u}-\mathbf{v}\|_{L^1(G)}=C\,\|\mathbf{u}-\mathbf{w}-\langle\mathbf{u}-\mathbf{w}\rangle_G\|_{L^1(G)}. \tag{2.3.131}$$

Now, by the triangle inequality,

$$\|\mathbf{u}-\mathbf{w}-\langle\mathbf{u}-\mathbf{w}\rangle_G\|_{L^1(G)}\le 2\|\mathbf{u}-\mathbf{w}\|_{L^1(G)}. \tag{2.3.132}$$

On the other hand, our assumptions on G ensure that a Poincaré-type inequality holds in $W^{1,1}(G)$. Hence there exists a constant C such that

$$\|\mathbf{u}-\mathbf{w}-\langle\mathbf{u}-\mathbf{w}\rangle_G\|_{L^1(G)}\le C\|\nabla(\mathbf{u}-\mathbf{w})\|_{L^1(G)}. \tag{2.3.133}$$

Altogether, inequality (2.3.127) follows.

Step 2: the union of two sets.

Let A and B be Young functions. An open set G in $\mathbb{R}^d$, $d\ge 2$, will be called admissible with respect to the couple (A, B) if there exists a constant C such that

$$\inf_{\mathbf{w}\in\mathcal{R}}\|\mathbf{u}-\mathbf{w}\|_{L^A(G)}\le C\|\boldsymbol{\varepsilon}(\mathbf{u})\|_{L^A(G)} \tag{2.3.134}$$

for every $\mathbf{u}\in E^A(G)$. We will show that, under certain assumptions, the union of two admissible sets will be admissible. So, assume that G_1 and G_2 are bounded connected open sets in $\mathbb{R}^d$ with Lipschitz boundary. Assume that each of them is admissible with respect to (A, B), and $G_1\cap G_2\neq\emptyset$. Then we have

$$\text{the set } G_1\cup G_2 \text{ is admissible with respect to } (A, B). \tag{2.3.135}$$

In order to prove (2.3.135) we consider a ball $\mathcal{B} \subset G_1 \cap G_2$. Fix $\omega \in C_0^\infty(\mathcal{B})$. Denote by $\mathcal{P}_1$ the space of polynomials of degree not exceeding 1, and by $\Pi_2\mathbf{u} \in \mathcal{P}_1$ the averaged Taylor polynomial of second-order with respect to ω of a function $\mathbf{u} \in L^1(G_1 \cup G_2)$ – see [37]. The operator $\Pi_2 : L^1(G_1 \cup G_2) \to \mathcal{P}_1$ is linear, and, by [37, Corollary 4.1.5], there exists a constant C such that

$$\|\Pi_2\mathbf{u}\|_{L^1(G_1\cup G_2)} \le C\,\|\mathbf{u}\|_{L^1(\mathcal{B})}$$

for every $\mathbf{u} \in L^1(G_1 \cup G_2)$. Let us denote by $\Pi_{\mathcal{R}}$ the L^2-orthogonal projection from $\mathcal{P}_1$ into $\mathcal{R}$. We have that

$$\|\Pi_{\mathcal{R}}\mathbf{p}\|_{L^1(G_1\cup G_2)} \le c\,\|\mathbf{p}\|_{L^1(G_1\cup G_2)}$$

for every $\mathbf{p} \in \mathcal{P}_1$. Thus, the linear operator $\Pi = \Pi_{\mathcal{R}} \circ \Pi_2$ maps $L^1(G_1 \cup G_2)$ into $\mathcal{R}$, and there exists a constant C such that

$$\|\Pi\mathbf{u}\|_{L^1(G_j)} \le \|\Pi\mathbf{u}\|_{L^1(G_1\cup G_2)} \le C\,\|\mathbf{u}\|_{L^1(\mathcal{B})} \le C\,\|\mathbf{u}\|_{L^1(G_j)} \quad j=1,2 \tag{2.3.136}$$

for every $\mathbf{u} \in L^1(G_1 \cup G_2)$. Due to inequality (2.3.136), inequality (2.3.127) ensures that there exists a constant C such that

$$\begin{aligned}\inf_{\mathbf{w}\in\mathcal{R}} \|\mathbf{u}-\mathbf{w}\|_{L^A(G_1\cup G_2)} &\le \|\mathbf{u}-\Pi\mathbf{u}\|_{L^A(G_1\cup G_2)} \le \sum_{j=1,2}\|\mathbf{u}-\Pi\mathbf{u}\|_{L^A(G_j)}\\ &\le C\sum_{j=1,2}\inf_{\mathbf{w}\in\mathcal{R}}\|\mathbf{u}-\mathbf{w}\|_{L^A(G_j)}\end{aligned} \tag{2.3.137}$$

for every $\mathbf{u} \in L^A(G)$. The conclusion (2.3.135) follows from (2.3.137) and (2.3.134) applied with $G = G_j$, for $j = 1, 2$.

Step 3: The general cases.

Any open Lipschitz domain G is the finite union of open sets G_i, $i = 1, \dots, k$, starshaped with respect to a ball. Since G is connected, after, possibly, relabelling, we may assume that the sets $\cup_{i=1}^{j-1} G_i$ and G_j have a non-empty intersection. The conclusion then follows from repeated use of (2.3.135). □

Remark 2.3.7. A proof of Corollary 2.3.1 can also be given based on the representation formula (2.3.123). After differentiating (2.3.123) the main part is a singular operator. It can be shown that it enjoys the properties (2.1.34)–(2.1.39). Hence it is continuous from $L^A(G)$ to $L^B(G)$ thanks to Theorem 2.1.9. We refer to [31] and [46] for details.

In the following we are concerned with the necessity of the conditions (2.3.107) and (2.3.108) for a Korn-type inequality.

Theorem 2.3.14. *Let G be an open bounded set in* $\mathbb{R}^d$. *Let A and B be Young functions such that*

$$\|\nabla \mathbf{u}\|_{L^B(G)} \leq C\|\boldsymbol{\varepsilon}(\mathbf{u})\|_{L^A(G)} \tag{2.3.138}$$

for some constant C and for every $\mathbf{u} \in W_0^{1,1}(G) \cap E_0^A(G)$. *Then conditions* (2.3.107) *and* (2.3.108) *hold.*

Our proof of inequality (2.3.107) is based on the technique of laminates as in [52]. In general a first order laminate is a probability measure ν on $\mathbb{R}^{d\times d}$ given by

$$\nu = \lambda\delta_{\mathbf{A}} + (1-\lambda)\delta_{\mathbf{B}}$$

where $\lambda \in (0,1)$ and $\mathbf{A}, \mathbf{B} \in \mathbb{R}^{d\times d}$ with $\operatorname{rank}(\mathbf{A}-\mathbf{B}) = 1$. Here $\delta_{\mathbf{X}}$ denotes the Dirac measure concentrated on the matrix $\mathbf{X}$. We say ν has average $\mathbf{C}$ if $\lambda\mathbf{A} + (1-\lambda)\mathbf{B} = \mathbf{C}$. We obtain a second order laminate if we replace $\delta_{\mathbf{A}}$ (resp. $\delta_{\mathbf{B}}$) by a first order laminate with average $\mathbf{A}$ (resp. $\mathbf{B}$). Iteratively we can define laminates of arbitrary order with a given average. For a detailed discussion we refer to [102] and [118].

Lemma 2.3.1 ([52], eq. (5)). *Let* ν *be a laminate with average C, then there is a sequence of uniformly Lipschitz continuous functions* $u_i : (0,r)^d \to \mathbb{R}^d$ *with boundary data Cx such that*

$$\int_{(0,r)^d} \Phi(|\nabla \mathbf{u}_i|)\,\mathrm{d}x \longrightarrow r^d \int_{\mathbb{R}^{d\times d}} \Phi(|\mathbf{X}|)\,\mathrm{d}\nu(\mathbf{X}), \tag{2.3.139}$$

for every continuous function Φ.

Proof of Theorem 2.3.14. Part 1: Inequality (2.3.108) holds.

Assume, without loss of generality, that the unit ball $\mathcal{B}_1$, centered at 0, is contained in G, and denote by ω_d its Lebesgue measure. Let us preliminarily observe that inequality (2.3.138) implies that

$$A \text{ dominates } B \text{ near infinity.} \tag{2.3.140}$$

Indeed, given any nonnegative function $h \in L^A(0,\omega_d)$, consider the function $\mathbf{v} : \mathcal{B}_1 \to \mathbb{R}^d$ given by

$$\mathbf{v}(x) = \left(\int_{|x|}^1 h(\omega_d r^d)\,\mathrm{d}r, 0, \ldots, 0\right) \quad \text{for} \quad x \in \mathcal{B}_1.$$

Then $\mathbf{v} \in L^A(\mathcal{B}_1)$, and

$$|\boldsymbol{\varepsilon}(\mathbf{v})(x)| \leq |\nabla \mathbf{v}(x)| = h(\omega_d |x|^d) \quad \text{for} \quad x \in \mathcal{B}_1.$$

An application of (2.3.138), with $\mathbf{u}$ replaced by $\mathbf{v}$, shows that

$$\|h\|_{L^B(0,\omega_d)} = \|\nabla \mathbf{v}\|_{L^B(\Omega)} \leq C\|\boldsymbol{\varepsilon}(\mathbf{v})\|_{L^A(\Omega)} \leq C\|\nabla \mathbf{v}\|_{L^A(\Omega)} = \|h\|_{L^A(0,\omega_d)}.$$

Thus $L^A(0,\omega_d) \to L^B(0,\omega_d)$, and (2.3.140) follows.

Now, given h as above, define the function $\rho : [0,1] \to [0,\infty]$ by

$$\rho(r) = \int_r^1 \frac{h(\omega_d t^d)}{t}\, \mathrm{d}t \quad \text{for} \quad r \in [0,1],$$

and the function $\mathbf{v} : \mathcal{B}_1 \to \mathbb{R}^d$ by

$$\mathbf{u}(x) = \mathbf{Q} x \rho(|x|) \quad \text{for} \quad x \in \mathcal{B}_1,$$

where $\mathbf{Q} \in \mathbb{R}^{d\times d}$ is any skew-symmetric matrix such that $|\mathbf{Q}| = 1$. We have that $\mathbf{u}$ is a weakly differentiable function, and

$$\boldsymbol{\varepsilon}(\mathbf{u})(x) = \frac{\mathbf{Q}x \odot x}{|x|^2}\rho'(|x|)|x|,$$
$$\nabla \mathbf{u}(x) = \mathbf{Q}\rho(|x|) + \frac{\mathbf{Q}x \otimes x}{|x|^2}\rho'(|x|)|x|$$

for a.e. $x \in \mathcal{B}_1$. Here, $\odot$ denotes the symmetric part of the tensor product of two vectors in $\mathbb{R}^d$. Hence,

$$|\boldsymbol{\varepsilon}(\mathbf{u})(x)| \leq |\rho'(|x|)||x| = h(\omega_d |x|),$$
$$\rho(|x|) \leq |\nabla \mathbf{u}(x)| + |\rho'(|x|)||x| = |\nabla \mathbf{u}(x)| + h(\omega_d |x|)$$

for a.e. $x \in \mathcal{B}_1$. Thus, due to (2.3.138) and (2.3.140),

$$\left\| \int_s^{\omega_d} \frac{h(r)}{r}\, \mathrm{d}r \right\|_{L^B(0,\omega_d)} = \left\| \int_{|x|}^1 \frac{h(\omega_d t^d)}{t}\, \mathrm{d}t \right\|_{L^B(\mathcal{B}_1)} = \|\rho(|x|)\|_{L^B(\mathcal{B}_1)} \tag{2.3.141}$$
$$\leq \|\nabla \mathbf{u}\|_{L^B(\mathcal{B}_1)} + \|h(\omega_d |x|^d)\|_{L^B(\mathcal{B}_1)} \leq C\|\boldsymbol{\varepsilon}(\mathbf{u})\|_{L^A(\mathcal{B}_1)} + \|h(\omega_d |x|^d)\|_{L^A(\mathcal{B}_1)}$$
$$\leq C'\|h(\omega_d |x|^d)\|_{L^A(\mathcal{B}_1)} = C'\|h(s)\|_{L^A(0,\omega_d)}$$

for suitable constants C and C'. Thanks to the arbitrariness of h, inequality (2.3.141) implies, via Lemma 1.2.1, that (5.3.12) holds for some c and t_0.

Part 2: Inequality (2.3.107) holds.

Let us preliminarily note that, if $A(t) = \infty$ for large t, then (2.3.107) holds trivially. We may thus assume that A is finite-valued, and hence continuous. By (2.3.140), the function B is also finite-valued and continuous.

For ease of notations, we hereafter focus on case when $d = 2$. An analogous argument carries over to any dimension along the lines of [52, Lemma 3]. Given $a, b \in \mathbb{R}$, we define the matrix $\mathbf{G}_{a,b}$ as

$$\mathbf{G}_{a,b} = \begin{pmatrix} 0 & a \\ b & 0 \end{pmatrix},$$

and set $\delta_{a,b} = \delta_{\mathbf{G}_{a,b}}$. Next, we define the sequence $\{\mu^{(m)}\}$ of laminates of order $2m$ iteratively by

$$\begin{cases} \mu^{(0)} = \delta_{t,t}, \\ \mu^{(m)} = \frac{1}{3}\delta_{2^{-m}t,-2^{-m}t} + \frac{1}{6}\delta_{-2^{1-m}t,2^{1-m}t} + \frac{1}{2}\mu^{(m-1)} \end{cases} \tag{2.3.142}$$

for $m \in \mathbb{N}$. We claim that $\mu^{(m)}$ is a laminate with average $\mathbf{G}_{2^{-m}t,2^{-m}t}$ for $m \in \mathbb{N}$. Indeed, the following holds

$$\mu^{(m)} = \tfrac{1}{4}\delta_{-2^{1-m}t,2^{1-m}t} + \tfrac{3}{4}\mu^{(m-1)}. \tag{2.3.143}$$

Since $\operatorname{rank}(\mathbf{G}_{-t,t} - \mathbf{G}_{t,t}) = 1$, the right-hand side of (2.3.143) is a laminate with average $\mathbf{G}_{2^{-1}t,t}$ for $m = 1$. Hence, $\mu^{(1)}$ is a laminate with average $\mathbf{G}_{2^{-1}t,2^{-1}t}$. An induction argument then proves our claim. Now, note the representation formula

$$\mu^{(m)} = 2^{-m}\delta_{t,t} + \sum_{k=1}^{m} \left(\tfrac{1}{3}2^{k-m}\delta_{2^{-k}t,-2^{-k}t} + \tfrac{1}{6}2^{k-m}\delta_{-2^{1-k}t,2^{1-k}t}\right) \tag{2.3.144}$$

for $m \in \mathbb{N}$. We remark that $\delta_{t,t}$ is concentrated at a symmetric matrix, whereas the sum in (2.3.144) is concentrated at skew-symmetric matrices. We define the functions $\Phi_j : \mathbb{R}^{2\times 2} \to [0, \infty)$, for $j = 1, 2$, by

$$\begin{aligned} \Phi_1(\mathbf{X}) &= A(|\mathbf{X}^{\text{sym}} - \mathbf{G}_{2^{-m}t,2^{-m}t}|), \\ \Phi_2(\mathbf{X}) &= B(C^{-1}|\mathbf{X} - \mathbf{G}_{2^{-m}t,2^{-m}t}|), \end{aligned}$$

for $\mathbf{X} \in \mathbb{R}^{2\times 2}$. Here, $\mathbf{X}^{\text{sym}} = \frac{1}{2}(\mathbf{X} + \mathbf{X}^T)$ is the symmetric part of $\mathbf{X}$, and C is the constant appearing in (2.3.138). Fix $m \in \mathbb{N}$. Without loss of generality, we may assume that $0 \in G$. We choose $r > 0$ so small that $(0, r)^2 \subset G$. Let $m \in \mathbb{N}$ be arbitrary. Due to Lemma 2.3.1 applied with $\nu = \mu^{(m)}$, there exists a sequence $\{\mathbf{u}_i\}$ of Lipschitz continuous functions $\mathbf{u}_i : (0, r)^2 \to \mathbb{R}^2$, such that $\mathbf{u}_i(x) = \mathbf{G}_{2^{-m}t,2^{-m}t}x$ on $\partial(0, r)^2$, and

$$\lim_{i\to\infty} \int_{(0,r)^2} \Phi_j(\nabla \mathbf{u}_i)\,\mathrm{d}x = r^2 \int_{\mathbb{R}^{2\times 2}} \Phi_j(\mathbf{X})\,\mathrm{d}\mu^{(m)}(\mathbf{X}) \quad \text{for} \quad j = 1, 2. \tag{2.3.145}$$

We define the sequence $\{\mathbf{v}_i\}$ of functions $\mathbf{v}_i : G \to \mathbb{R}$ by $\mathbf{v}_i(x) = \mathbf{u}_i(x) - \mathbf{G}_{2^{-m}t,2^{-m}t}x$ if $x \in (0, r)^2$, and $\mathbf{v}_i(x) = 0$ if $G \setminus (0, r)^2$. Then $\mathbf{v}_i \in W_0^{1,\infty}(G)$, and, by (2.3.145),

$$\lim_{i\to\infty}\int_G A(|\boldsymbol{\varepsilon}(\mathbf{v}_i)|)\,\mathrm{d}x = \lim_{i\to\infty}\int_{(0,r)^2} A(|\boldsymbol{\varepsilon}(\mathbf{v})_i|)\,\mathrm{d}x = r^2\int_{\mathbb{R}^{2\times 2}} A(|(\mathbf{X}^{\mathrm{sym}} - \mathbf{G}_{2^{-m}t,2^{-m}t})|)\,\mathrm{d}\mu^{(m)}(\mathbf{X}), \tag{2.3.146}$$

$$\lim_{i\to\infty}\int_\Omega B(C^{-1}|\nabla\mathbf{v}_i|)\,\mathrm{d}x = \lim_{i\to\infty}\int_{(0,r)^2} B(C^{-1}|\nabla\mathbf{v}_i|)\,\mathrm{d}x = r^2\int_{\mathbb{R}^{2\times 2}} B(C^{-1}|\mathbf{X} - \mathbf{G}_{2^{-m}t,2^{-m}t}|)\,\mathrm{d}\mu^{(m)}(\mathbf{X}). \tag{2.3.147}$$

We have the following chain

$$\begin{aligned}
&\int_{\mathbb{R}^{2\times 2}} A\big(|\mathbf{X}^{\mathrm{sym}} - \mathbf{G}_{2^{-m}t,2^{-m}t}|\big)\, d\mu^{(m)}(\mathbf{X}) \\
&\le \frac{1}{2}\int_{\mathbb{R}^{2\times 2}} A\big(2|\mathbf{X}^{\mathrm{sym}}|)\, d\mu^{(m)}(\mathbf{X}) + \frac{1}{2}\int_{\mathbb{R}^{2\times 2}} A\big(2|\mathbf{G}_{2^{-m}t,2^{-m}t}|)\,\mathrm{d}\mu^{(m)}(\mathbf{X}) \\
&= \frac{1}{2}2^{-m}A(2|\mathbf{G}_{t,t}|) + \frac{1}{2}A(2|\mathbf{G}_{2^{-m}t,2^{-m}t}|) \\
&= \frac{1}{2}2^{-m}A(2|\mathbf{G}_{t,t}|) + \frac{1}{2}A(2\,2^{-m}|\mathbf{G}_{t,t}|) \\
&\le 2^{-m}A(2|\mathbf{G}_{t,t}|).
\end{aligned} \tag{2.3.148}$$

Here the first inequality holds since A is convex, the first equality holds due to (2.3.144) and to the fact that $\mu^{(m)}$ is a probability measure, and the last inequality follows from (1.2.6). Combining (2.3.146) with (2.3.148) implies

$$\lim_{i\to\infty}\int_G A(|\boldsymbol{\varepsilon}(\mathbf{v}_i)|)\,\mathrm{d}x \le r^2 2^{-m}A(2|\mathbf{G}_{t,t}|). \tag{2.3.149}$$

Since A is a continuous function, there exists $t_m \in (0,\infty)$ such that

$$r^2 2^{-m}A(2|\mathbf{G}_{t_m,t_m}|) = \tfrac{1}{2}. \tag{2.3.150}$$

Thanks to (1.2.6), there exists $t_0 > 0$, independent of m, such that

$$t_m \le t_0 2^m. \tag{2.3.151}$$

Therefore, by neglecting, if necessary, a finite number of terms of the sequence $\{\mathbf{v}_i\}$, we can assume that

$$\int_G A(|\boldsymbol{\varepsilon}(\mathbf{v}_i)|)\,\mathrm{d}x \le 1$$

for $i\in\mathbb{N}$. Hence, $\|\boldsymbol{\varepsilon}(\mathbf{v}_i)\|_A \le 1$ for $i\in\mathbb{N}$, and, by (2.3.138), $\|\nabla\mathbf{v}_i\|_B \le C$ for $i\in\mathbb{N}$. Thus,

$$\int_G B(C^{-1}|\nabla\mathbf{v}_i|)\,\mathrm{d}x \le 1$$

for $i \in \mathbb{N}$. Combining the latter inequality with equation (2.3.147) implies us that

$$r^2 \int_{\mathbb{R}^{2\times 2}} B(C^{-1}|\mathbf{X} - \mathbf{G}_{2^{-m}t_m, 2^{-m}t_m}|)\, \mathrm{d}\mu^{(m)}(\mathbf{X}) \le 1. \tag{2.3.152}$$

Next, one can make use of (2.3.144) and obtain the following chain

$$\begin{aligned} r^{-2} &\ge \int_{\mathbb{R}^{2\times 2}} B(C^{-1}|\mathbf{X} - \mathbf{G}_{2^{-m}t_m, 2^{-m}t_m}|)\, \mathrm{d}\mu^{(m)}(\mathbf{X}) \\ &\ge 2^{-m} B\big(C^{-1}(1 - 2^{-m})|\mathbf{G}_{t_m,t_m}|\big) + \sum_{k=1}^{m} \tfrac{1}{3} 2^{k-m} B\big(C^{-1}(2^{-k} - 2^{-m})|\mathbf{G}_{t_m,t_m}|\big) \\ &\quad + \sum_{k=1}^{m} \tfrac{1}{6} 2^{k-m} B\big(C^{-1}(2^{1-k} - 2^{-m})|\mathbf{G}_{t_m,t_m}|\big) \\ &\ge \sum_{k=1}^{m-1} \tfrac{1}{3} 2^{k-m} B\big(C^{-1}(2^{-k} - 2^{-m})|\mathbf{G}_{t_m,t_m}|\big) \\ &\ge \sum_{k=1}^{m-1} \tfrac{1}{3} 2^{k-m} B\big(C^{-1} 2^{-k-1}|\mathbf{G}_{t_m,t_m}|\big) \\ &\ge \sum_{k=1}^{m-1} \tfrac{1}{3} \frac{1}{2C} 2^{-m} t_m \frac{B\big(\frac{1}{2C} 2^{-k} t_m\big)}{\frac{1}{2C} 2^{-k} t_m}. \end{aligned} \tag{2.3.153}$$

It follows from (2.3.150), (2.3.152) and (2.3.153) that

$$2 \cdot 2^{-m} A(2|\mathbf{G}_{t_m,t_m}|) \ge \sum_{k=1}^{m-1} \tfrac{1}{3} \tfrac{1}{2C} 2^{-m} t_m \frac{B\big(\frac{1}{2C} 2^{-k} t_m\big)}{\frac{1}{2C} 2^{-k} t_m}.$$

Hence, by (2.3.151),

$$\begin{aligned} A(c'' t_m) &\ge c\, t_m \sum_{k=1}^{m-1} \frac{B\big(\frac{1}{2C} 2^{-k} t_m\big)}{\frac{1}{2C} 2^{-k} t_m} \ge c'\, t_m \int_{2^{-m}\frac{t_m}{4C}}^{\frac{t_m}{4C}} \frac{B(s)}{s^2}\, \mathrm{d}s \\ &\ge c'\, t_m \int_{\frac{t_0}{2C}}^{\frac{t_m}{4C}} \frac{B(s)}{s^2}\, \mathrm{d}s \end{aligned} \tag{2.3.154}$$

for suitable positive constants c, c' c''. Since $\lim_{m\to\infty} t_m = \infty$, one can find $\hat{t} \ge \frac{t_0}{2C}$ such that, if $t > \hat{t}$, then there exists $m \in \mathbb{N}$ such that $t_m \le t < t_{m+1}$. Moreover, $\hat{t}$ can be chosen so large that A is invertible on $[\hat{t}, \infty)$ and

$$t_m = c_1 A^{-1}(c_2 2^m)$$

for some positive constants c_1, c_2. By (1.2.7), the latter equation ensures that $t_{m+1} \leq 2t_m$ for $m \in \mathbb{N}$. Thus, due to inequality (2.3.154),

$$\begin{aligned} A(2c''t) &\geq A(2c''t_m) \geq A(c''t_{m+1}) \\ &\geq c'\, t_{m+1} \int_{\frac{t_0}{2C}}^{\frac{t_{m+1}}{4C}} \frac{B(s)}{s^2}\, ds \geq c'\, t \int_{\frac{t_0}{2C}}^{\frac{t}{4C}} \frac{B(s)}{s^2}\, ds \end{aligned}$$

for $t \geq \hat{t}$. Hence, inequality (2.3.107) follows for suitable constants c and t_0. □

Corollary 2.3.2. *Let G be a Lipschitz domain in $\mathbb{R}^d$, $d \geq 2$. Let A and B be Young functions. Assume that there exists a constant C such that*

$$\|u - u_G\|_{L^B(G)} \leq C\|\nabla u\|_{W^{-1,A}(G)} \tag{2.3.155}$$

for every $u \in L^1(G)$. Then conditions (2.3.107) *and* (2.3.108) *hold.*

Proof. Let $\mathbf{u} \in W_0^{1,A}(G)$. As in the proof of Theorem (2.3.119) we can show the following chain

$$\begin{aligned} \|\nabla \mathbf{u} - (\nabla \mathbf{u})_G\|_{L^B(G)} &= \|\nabla \mathbf{u} - (\nabla \mathbf{u})_G\|_{L^B(G)} \leq C \sum_{i,j=1}^{d} \left\| \frac{\partial u_i}{\partial x_j} - \left(\frac{\partial u_i}{\partial x_j}\right)_G \right\|_{L^B(G)} \\ &\leq C \sum_{i,j=1}^{n} \left\| \nabla \frac{\partial u_i}{\partial x_j} \right\|_{W^{-1,A}(G)} \leq C \sum_{i,j,k=1}^{d} \left\| \frac{\partial^2 u_i}{\partial x_k \partial x_j} \right\|_{W^{-1,A}(G)} \\ &\leq C \sum_{i,j,k=1}^{d} \left(\left\| \frac{\partial (\boldsymbol{\varepsilon}(\mathbf{u}))_{ij}}{\partial x_k} \right\|_{W^{-1,A}(G)} + \left\| \frac{\partial (\boldsymbol{\varepsilon}(\mathbf{u}))_{ik}}{\partial x_j} \right\|_{W^{-1,A}(G)} + \left\| \frac{\partial (\boldsymbol{\varepsilon}(\mathbf{u}))_{jk}}{\partial x_i} \right\|_{W^{-1,A}(G)} \right) \\ &\leq C \sum_{i,j=1}^{d} \|\nabla (\boldsymbol{\varepsilon}(\mathbf{u}))_{ij}\|_{W^{-1,A}(G)} \leq C \sum_{i,j=1}^{n} \|(\boldsymbol{\varepsilon}(\mathbf{u}))_{ij} - ((\boldsymbol{\varepsilon}(\mathbf{u}))_{ij})_G\|_{L^A(G)} \\ &\leq C\, \|\boldsymbol{\varepsilon}(\mathbf{u}) - (\boldsymbol{\varepsilon}(\mathbf{u}))_G\|_{L^A(G)} = C\, \|\boldsymbol{\varepsilon}(\mathbf{u})\|_{L^A(G)}. \end{aligned}$$

Here the second inequality in line two holds by (2.3.155). The first inequality in line four is true for every Young function by the very definition of the negative norm. The conclusion follows via Theorem 2.3.14, due to the arbitrariness of $\mathbf{u}$. □

Corollary 2.3.3. *Let G be a bounded domain in $\mathbb{R}^d$, $d \geq 2$, which is star-shaped with respect to a ball, and let Bog_G be the Bogovskiĭ operator on G (see Section 2.1). Let A and B be Young functions such that*

$$\|\nabla \mathrm{Bog}_G f\|_{L^B(G,\mathbb{R}^d)} \leq C\|f\|_{L^A(G)} \tag{2.3.156}$$

for some constant C*, and for every* $f \in C^{\infty}_{0,\perp}(G)$*. Then conditions* (2.0.11) *and* (2.0.12) *hold.*

Proof. A close inspection of the proof of Theorem 2.0.5 reveals that inequality (2.3.156) implies inequality (2.3.155). The conclusion thus follows from Corollary 2.3.2. □

CHAPTER 3

Solenoidal Lipschitz truncation

Contents

Abstract

In this chapter we present the solenoidal Lipschitz truncation for stationary problems: we show how to construct a Lipschitz truncation which preserves the divergence-free character of a given Sobolev function. In the two-dimensional case we improve the method so that function spaces can be included which are not uniformly convex. This is of crucial importance for the Prandtl–Eyring fluid model. Finally, we present the $\mathcal{A}$-Stokes approximation for stationary problems. It aims at approximating almost solutions to the linear $\mathcal{A}$-Stokes system by exact solutions. Thanks to the solenoidal Lipschitz truncation we obtain an approximation result on the level of gradients.

In this chapter we develop a Lipschitz truncation which, in addition to the common properties explained in Section 1.3, preserves the solenoidal character of a given Sobolev function. This is motivated by PDEs in incompressible fluid mechanics for which it is often convenient to work with the so-called pressure-free formulation. This formulation is obtained by the use of solenoidal (i.e. divergence-free) test functions, since they are orthogonal to the pressure gradient. The difficulty with the standard Lipschitz truncation is, that it does not preserve the solenoidal property. The easiest strategy to overcome this defect is to correct the functions $\mathbf{u}_\lambda$ by means of the Bogovskiĭ operator (see Section 2.1). This operator works well in uniform convex settings, e.g. on L^p with $1 < p < \infty$. However, it cannot be used in the non-uniform convex setting, e.g. L^1, L^∞ or L^h with $h(t) = t\ln(1+t)$, since the (gradient of the) Bogovskiĭ correction is a singular integral operator. So in the limiting cases the Bogovskiĭ-corrected Lipschitz truncation loses some of its important fine properties. This is particular the case for Prandtl–Eyring fluids (see Chapter 4), for which the constitutive relation reads as

$$\mathbf{S} = \frac{\ln(1+|\boldsymbol{\varepsilon}(\mathbf{v})|)}{|\boldsymbol{\varepsilon}(\mathbf{v})|}\boldsymbol{\varepsilon}(\mathbf{v}). \tag{3.0.1}$$

Existence Theory for Generalized Newtonian Fluids.
DOI: http://dx.doi.org/10.1016/B978-0-12-811044-7.00004-5

To overcome these problems one needs a solenoidal Lipschitz truncation. In [33] a truncation method was developed which allows us to approximate the function $\mathbf{u}$ by a solenoidal Lipschitz function $\mathbf{u}_\lambda$ without losing the fine properties of the truncation. This method is based on a local projection into finite dimensional function spaces on which the Bogovskiĭ operator is continuous in L^∞ and whose dimensions are globally bounded. A much simpler idea works as follows: For $\mathbf{u} \in W^{1,p}_{0,\mathrm{div}}(G)$ let $\mathcal{O}_\lambda := \{M(\nabla^2 \operatorname{curl}^{-1} \mathbf{u}) > \lambda\}$ where M is the Hardy–Littlewood maximal operator. We define

$$\mathbf{u}_\lambda := \begin{cases} \mathbf{u} & \text{in} \quad G \setminus \mathcal{O}_\lambda, \\ \operatorname{curl}\Big(\sum_i \varphi_i \Pi_i\big(\operatorname{curl}^{-1} \mathbf{u}\big)\Big) & \text{in} \quad G \cap \mathcal{O}_\lambda, \end{cases} \tag{3.0.2}$$

where Π_i is a local linear approximation and $\{\varphi_i\} \subset C_0^\infty(\mathbb{R}^d)$ is a decomposition of unity with respect to the Whitney covering of $\mathcal{O}_\lambda$, cf. Lemma 1.3.6. The function $\mathbf{u}_\lambda$ is now obviously solenoidal. The idea is that curl and curl^{-1} should cancel in such a way that the estimates are in line with the standard truncation given by

$$\mathbf{u}_\lambda := \begin{cases} \mathbf{u} & \text{in} \quad G \setminus \mathcal{O}_\lambda, \\ \sum_i \varphi_i \Pi_i \mathbf{u} & \text{in} \quad G \cap \mathcal{O}_\lambda. \end{cases}$$

In Section 3.1 we will show that this is indeed true. The operator $\nabla \operatorname{curl}^{-1}$ is a singular integral operator and hence not continuous in borderline function spaces. As a consequence the truncation from (3.0.2) still does not work in general. However, in the two-dimensional case the curl-operator is just a rotation of ∇ and hence estimates for $\nabla \operatorname{curl}^{-1}$ even hold pointwise. So, a modification of the definition (3.0.2) can be used to study the Prandtl–Eyring model in two dimensions. We will present this truncation in Section 3.2 and apply it in Section 4.2.

Besides the existence theory for generalized Newtonian fluids a further application of the solenoidal Lipschitz truncation is the $\mathcal{A}$-Stokes approximation. Let $\mathcal{A} : \mathbb{R}^{d\times d}_{\mathrm{sym}} \to \mathbb{R}^{d\times d}_{\mathrm{sym}}$ be an elliptic tensor. We call a function $\mathbf{v} \in W^{1,q}_{0,\mathrm{div}}(G)$ an almost $\mathcal{A}$-Stokes solution iff

$$\left| \fint_G \mathcal{A}(\boldsymbol{\varepsilon}(\mathbf{v}), \boldsymbol{\varepsilon}(\boldsymbol{\xi}))\, dx \right| \le \delta \fint_G |\boldsymbol{\varepsilon}(\mathbf{v})|\, dx\, \|\nabla \boldsymbol{\xi}\|_\infty$$

for all $\boldsymbol{\xi} \in C^\infty_{0,\mathrm{div}}(G)$ and some small $\delta > 0$. In Theorem 3.3.19 we prove the existence of a function $\mathbf{h} \in \mathbf{v} + W^{1,q}_{0,\mathrm{div}}(G)$ such that

$$\int_G \mathcal{A}(\boldsymbol{\varepsilon}(\mathbf{h}), \boldsymbol{\varepsilon}(\boldsymbol{\xi}))\, dx = 0 \quad \forall \boldsymbol{\xi} \in C^\infty_{0,\mathrm{div}}(G)$$

and satisfies for all $s > 1$

$$\fint_G \left|\frac{\mathbf{v}-\mathbf{h}}{r}\right|^q \mathrm{d}x + \fint_G |\nabla(\mathbf{v}-\mathbf{h})|^q \,\mathrm{d}x \le \kappa \left(\fint_G |\nabla \mathbf{v}|^{qs}\,\mathrm{d}x\right)^{\frac{1}{s}}.$$

We have $\kappa = \kappa(q, s, \delta)$ and $\lim_{\delta\to 0} \kappa(q, s, \delta) = 0$. The function $\mathbf{h}$ is called the $\mathcal{A}$-Stokes approximation of $\mathbf{v}$. The $\mathcal{A}$-Stokes approximation is an adaption of the $\mathcal{A}$-harmonic approximation. This argument first appeared in [58]. For newer results we refer to the review article [67]. Usually the approximation is only stated on the level of functions. We obtain an approximation on the level of gradients which is much stronger. This argument goes back to [61] where it is used in the context of $\mathcal{A}$-harmonic approximation.

3.1 SOLENOIDAL TRUNCATION – STATIONARY CASE

In this section we show how the solenoidal Lipschitz truncation based on the curl-representation works in the stationary case.

Let us start with a ball $\mathcal{B} \subset \mathbb{R}^3$ and $\mathbf{u} \in W^{1,s}_{0,\operatorname{div}}(\mathcal{B})$ with $s \in (1, \infty)$. We can extend $\mathbf{u}$ by zero outside G so that $\mathbf{u} \in W^{1,s}_{\operatorname{div}}(\mathbb{R}^3)$. (We restrict our observations to the case $d = 3$; see Remark 6.1.10 for the general case.)

On the space $W^{1,s}_{\operatorname{div}}(\mathbb{R}^3)$ we define the inverse curl operator curl^{-1} by

$$\operatorname{curl}^{-1}\mathbf{g} := \operatorname{curl}(\Delta^{-1}\mathbf{g}) := \operatorname{curl}\left(\int_{\mathbb{R}^3} \frac{-1}{4\pi|x-y|}\mathbf{g}(y)\,\mathrm{d}y\right).$$

The definition is consistent, as in the sense of distributions

$$\begin{aligned}\operatorname{curl}(\operatorname{curl}^{-1}\mathbf{g}) &= \operatorname{curl}\operatorname{curl}(\Delta^{-1}\mathbf{g}) = (-\Delta + \nabla\operatorname{div})\Delta^{-1}\mathbf{g}\\ &= \mathbf{g} + \nabla\operatorname{div}\left(\frac{-1}{4\pi|\cdot|} * \mathbf{g}\right)\\ &= \mathbf{g} + \nabla\left(\frac{-1}{4\pi|\cdot|} * \operatorname{div}\mathbf{g}\right)\\ &= \mathbf{g},\end{aligned}$$

where we have used $\operatorname{div}\mathbf{g} = 0$ in the last step. Moreover,

$$\operatorname{div}(\operatorname{curl}^{-1}\mathbf{g}) = \operatorname{div}\operatorname{curl}(\Delta^{-1}\mathbf{g}) = 0. \tag{3.1.3}$$

Since $\mathbf{g} \mapsto \nabla^2(\Delta^{-1}\mathbf{g})$ is a singular integral operator, we have

$$\|\nabla\operatorname{curl}^{-1}\mathbf{g}\|_s \le \|\nabla^2(\Delta^{-1}\mathbf{g})\|_s \le c_s\,\|\mathbf{g}\|_s \tag{3.1.4}$$

for $s \in (1, \infty)$. Analogously, we have

$$\|\nabla^2\operatorname{curl}^{-1}\mathbf{g}\|_s \le c_s\,\|\nabla\mathbf{g}\|_s \tag{3.1.5}$$

for $s \in (1, \infty)$.

Since $\mathbf{u}$ is solenoidal, we can define $\mathbf{w} := \operatorname{curl}^{-1} \mathbf{u}$. According to (3.1.3), (3.1.4) and (3.1.5) we have $\mathbf{w} \in W^{2,s}(\mathbb{R}^3)$ with $\|\nabla \mathbf{w}\|_s \le \|\mathbf{u}\|_s$ and $\|\nabla^2 \mathbf{w}\|_s \le \|\nabla^2 \mathbf{u}\|_s$ and $\operatorname{div} \mathbf{w} = 0$. Since $\operatorname{curl}\operatorname{curl} \mathbf{w} = -\Delta \mathbf{w} + \nabla \operatorname{div} \mathbf{w} = -\Delta \mathbf{w}$ and $\operatorname{curl} \mathbf{w} = \mathbf{u} = 0$ on $\mathbb{R}^3 \setminus \mathcal{B}$, it follows that $\mathbf{w}$ is harmonic on $\mathbb{R}^3 \setminus \mathcal{B}$.

For $\lambda > 0$ define $\mathcal{O}_\lambda := \{M(\nabla^2 \mathbf{w}) > \lambda\}$, where M is the standard non-centered maximal operator, cf. Lemma 1.3.1. We do not have to truncate our function $\mathbf{w}$ if $\mathcal{O}_\lambda$ is empty. So we can assume in the following that $\mathcal{O}_\lambda \neq \emptyset$. In accordance to Lemma 1.3.5 we decompose the open set $\mathcal{O}_\lambda$ into a family of dyadic closed cubes $\{Q_j\}_j$ with side length $\ell(Q_j)$. We set $Q_j^* = \frac{9}{8}$ and $r_j := \ell(Q_j^*)$ (recall Corollary 1.3.1). Let $\{\varphi_j\}_j$ be the partition of unity subordinated to $\{Q_j\}_j$ from Lemma 1.3.6. Then the solenoidal Lipschitz truncation of $\mathbf{u}$ is pointwise defined as

$$\mathbf{u}_\lambda := \begin{cases} \sum_{j\in\mathbb{N}} \operatorname{curl}(\varphi_j \mathbf{w}_j) & \text{in} \quad \mathcal{O}_\lambda, \\ \mathbf{u} & \text{elsewhere,} \end{cases} \tag{3.1.6}$$

with $\mathbf{w}_j := \Pi^1_{Q_j^*} \mathbf{w}$, where $\Pi^1_{Q_j^*}$ denotes the first order averaged Taylor polynomial on Q_j^*, see [37,63]. We start with some estimates for $\mathbf{w}$.

Lemma 3.1.1. *For all $j \in \mathbb{N}$ and all $k \in \mathbb{N}$ with $Q_j^* \cap Q_k^* \neq \emptyset$ we have*

(a) $\fint_{Q_j^*} |\frac{\mathbf{w}-\mathbf{w}_j}{r_j^2}| \,dx + \fint_{Q_j^*} |\frac{\nabla(\mathbf{w}-\mathbf{w}_j)}{r_j}| \,dx \le c \fint_{Q_j} |\nabla^2 \mathbf{w}| \,dx$.

(b) $\fint_{Q_j^*} |\nabla^2 \mathbf{w}| \,dx \le c\lambda$.

(c) $\|\mathbf{w}_j - \mathbf{w}_k\|_{L^\infty(Q_j^*)} \le c \fint_{Q_j^*} |\mathbf{w} - \mathbf{w}_j| \,dx + c \fint_{Q_k^*} |\mathbf{w} - \mathbf{w}_k| \,dx$.

(d) $\|\mathbf{w}_j - \mathbf{w}_k\|_{L^\infty(Q_j^*)} \le c r_j^2 \lambda$.

Proof. The first part (a) is just a consequence of the classical Poincaré estimate and the properties of $\Pi^1_{Q_j}$, see [63, Lemma 3.1]. The second part (b) follows from $Q_j^* \subset c_3 Q_j$ and $c_3 Q_j \cap \mathcal{O}_\lambda^{\complement} \neq \emptyset$, so $\fint_{c_3 Q_j} |\nabla^2 \mathbf{w}| \,dx \le \lambda$.

Part (c) follows from the geometric property of the Q_j^*. If $Q_j^* \cap Q_k^* \neq \emptyset$, then $|Q_j^* \cap Q_k^*| \ge c \max\{|Q_j^*|, |Q_k^*|\}$. This and the norm equivalence for linear polynomials imply

$$\begin{aligned} \|\mathbf{w}_j - \mathbf{w}_k\|_{L^\infty(Q_j^*)} &\le c \fint_{Q_j^* \cap Q_k^*} |\mathbf{w}_j - \mathbf{w}_k| \,dx \\ &\le c \fint_{Q_j^*} \big|\mathbf{w}_j - \mathbf{w}\big| \,dx + c \fint_{Q_k^*} |\mathbf{w} - \mathbf{w}_k| \,dx. \end{aligned}$$

Finally, (d) is a consequence of (c), (a) and (b). □

Lemma 3.1.2. *There exists $c_0 > 0$ such that $\lambda \ge \lambda_0 := c_0 (\fint_{\mathcal{B}} |\nabla \mathbf{u}|^s \,dx)^{\frac{1}{s}}$ implies $\mathcal{O}_\lambda \subset 2\mathcal{B}$.*

Proof. Let $x \in \mathbb{R}^3 \setminus 2\mathcal{B}$. We have to show that $x \notin \mathcal{O}_\lambda$. We will show that $⨍_{\mathcal{B}'} |\nabla^2 \mathbf{w}| \,\mathrm{d}x \le c\,(⨍_{\mathcal{B}} |\nabla \mathbf{u}|^s \,\mathrm{d}x)^{\frac{1}{s}}$ for any ball $\mathcal{B}'$ containing x. We consider first the case where $\mathcal{B}' \cap \mathcal{B} \neq \emptyset$. Then $|\mathcal{B}'| \ge c\,|\mathcal{B}|$ and

$$⨍_{\mathcal{B}'} |\nabla^2 \mathbf{w}| \,\mathrm{d}x \le c \left(\frac{|\mathcal{B}|}{|\mathcal{B}'|}\right)^{\frac{1}{s}} \left(⨍_{\mathcal{B}} |\nabla \mathbf{u}|^s \,\mathrm{d}x\right)^{\frac{1}{s}} \le c \left(⨍_{\mathcal{B}} |\nabla \mathbf{u}|^s \,\mathrm{d}x\right)^{\frac{1}{s}},$$

where we have used Hölder's inequality and the fact that $\|\nabla^2 \mathbf{w}\|_s \le c\,\|\nabla \mathbf{u}\|_s$ by (3.1.5). In the alternate case, where $\mathcal{B}' \cap \mathcal{B} = \emptyset$, let $\mathcal{B}''$ denote the largest ball with the same center as $\mathcal{B}'$ such that $\mathcal{B}'' \cap \mathcal{B} = \emptyset$. Then $|\mathcal{B}''| \ge c\,|\mathcal{B}|$. Since $\mathbf{w}$ is harmonic on $\mathbb{R}^3 \setminus \mathcal{B}$, it follows by the interior estimates for harmonic functions, $\|\nabla^2 \mathbf{w}\|_s \le c\,\|\nabla \mathbf{u}\|_s$ and $|\mathcal{B}''| \ge c\,|\mathcal{B}|$ that

$$⨍_{\mathcal{B}'} |\nabla^2 \mathbf{w}| \,\mathrm{d}x \le c \left(⨍_{\mathcal{B}''} |\nabla^2 \mathbf{w}|^s \,\mathrm{d}x\right)^{\frac{1}{s}} \le c \left(⨍_{\mathcal{B}} |\nabla \mathbf{u}|^s \,\mathrm{d}x\right)^{\frac{1}{s}}.$$

This proves the claim. □

The following lemma enables us to conclude that $\mathbf{u}_\lambda$ is a global Sobolev function.

Lemma 3.1.3. *We have for* $\lambda \ge \lambda_0$

$$\mathbf{u}_\lambda - \mathbf{u} = \sum_{j \in \mathbb{N}} \operatorname{curl}(\varphi_j(\mathbf{w}_j - \mathbf{w})) \in W_0^{1,1}(2\mathcal{B}),$$

where the sum converges in $W_0^{1,1}(2\mathcal{B})$. *In particular,* $\mathbf{u}_\lambda \in W_0^{1,1}(2\mathcal{B})$.

Proof. The proof is similar to ones in [33,60]. Note that the convergence will be unconditional, i.e. irrespective of the order of summation. Obviously, the convergence holds pointwise. Since $\lambda \ge \lambda_0$, it follows by Lemma 3.1.2 that $Q_j \subset \mathcal{O}_\lambda \subset 2\mathcal{B}$. In particular, each summand is in $W_0^{1,1}(2\mathcal{B})$. It remains to prove convergence of the sum in $W_0^{1,1}(2\mathcal{B})$ in the gradient norm. We will show absolute convergence of the gradients in L^1. The estimates for φ_i and Lemma 3.1.1 (a) imply

$$\sum_j \int |\nabla \operatorname{curl}(\varphi_j(\mathbf{w}_j - \mathbf{w}))| \,\mathrm{d}x \le c \sum_j \int_{Q_j^*} |\nabla^2 \mathbf{w}| \,\mathrm{d}x \le c\,\|\nabla^2 \mathbf{w}\|_{L^1(2\mathcal{B})}.$$

Now the fact that $\nabla^2 \mathbf{w} \in L^s(\mathbb{R}^3)$ proves the claim. □

The following theorem describes the basic properties of the Lipschitz truncation. It is a combination of the techniques of [62,33,60].

Theorem 3.1.15. *Let* $1 < s < \infty$ *and let* $\mathcal{B} \subset \mathbb{R}^3$ *be a ball. If* $\mathbf{u} \in W_{0,\mathrm{div}}^{1,s}(\mathcal{B})$ *and* $\lambda \ge \lambda_0$, *then* $\mathbf{u}_\lambda \in W_{0,\mathrm{div}}^{1,\infty}(2\mathcal{B})$ *and*

(a) $\mathbf{u}_\lambda = \mathbf{u}$ *on* $\mathbb{R}^3 \setminus \mathcal{O}_\lambda = \mathbb{R}^3 \setminus \{M(\nabla^2\mathbf{w}) > \lambda\}$.
(b) $\|\mathbf{u}_\lambda\|_q \le c\,\|\mathbf{u}\|_q$ *for* $1 < q < \infty$ *provided that* $\mathbf{u} \in L^q(\mathcal{B})$.
(c) $\|\nabla\mathbf{u}_\lambda\|_q \le c\,\|\nabla\mathbf{u}\|_s$ *for* $1 < q < \infty$ *provided that* $\mathbf{u} \in W_0^{1,q}(\mathcal{B})$.
(d) $|\nabla\mathbf{u}_\lambda| \le c\lambda\chi_{\mathcal{O}_\lambda} + |\nabla\mathbf{u}|\chi_{\mathbb{R}^3\setminus\mathcal{O}_\lambda} \le c\lambda$ *a.e. for all* $\lambda > 0$.

Proof. We will use the representation of Lemma 3.1.3. The claim (a) follows from the fact that $\operatorname{supp}(\varphi_j) \subset \mathcal{O}_\lambda$. As in the proof of Lemma 3.1.2 we estimate

$$\Big\|\sum_{j\in\mathbb{N}} \operatorname{curl}(\varphi_j(\mathbf{w}_j - \mathbf{w}))\Big\|_q^q \le c\sum_j \int_{Q_j^*} \Big(\Big|\frac{\mathbf{w}_j - \mathbf{w}}{r_j}\Big|^q + |\nabla(\mathbf{w}_j - \mathbf{w})|^q\Big)\,dx$$
$$\le c\sum_j \int_{Q_j^*} |\nabla\mathbf{w}|^q\,dx \le c\,\|\mathbf{u}\|_q^q$$

using the properties of the averaged Taylor polynomial, see [63, Lemma 3.1], and (3.1.4). Similarly, using (3.1.5) we find that

$$\Big\|\nabla\sum_{j\in\mathbb{N}} \operatorname{curl}(\varphi_j(\mathbf{w}_j - \mathbf{w}))\Big\|_q^q$$
$$\le c\sum_j \int_{Q_j} \Big(\Big|\frac{\mathbf{w}_j - \mathbf{w}}{r_j^2}\Big|^q + \Big|\frac{\nabla(\mathbf{w}_j - \mathbf{w})}{r_j}\Big|^q + |\nabla^2\mathbf{w}|^q\Big)\,dx$$
$$\le c\sum_j \int_{Q_j^*} |\nabla^2\mathbf{w}|^q\,dx \le c\,\|\nabla\mathbf{u}\|_q^q.$$

This and the representation of Lemma 3.1.3 prove (b) and (c).

To prove (d) it suffices to verify $|\nabla\mathbf{u}_\lambda| \le c\lambda$ on $\mathcal{O}_\lambda$, since on $\mathbb{R}^3 \setminus \mathcal{O}_\lambda$ we have $|\nabla\mathbf{u}| \le M(\nabla\mathbf{u}) \le M(\nabla^2\mathbf{w}) \le \lambda$. For $k \in \mathbb{N}$ we estimate

$$\|\nabla\mathbf{u}_\lambda\|_{L^\infty(Q_k^*)} = \Big\|\sum_{j\in A_k} \nabla\operatorname{curl}(\varphi_j(\mathbf{w}_j - \mathbf{w}_k))\Big\|_{L^\infty(Q_k^*)}$$
$$\le c\sum_{j\in A_k}\Big(r_k^{-2}\|\mathbf{w}_j - \mathbf{w}_k\|_{L^\infty(Q_j)} + r_k^{-1}\|\nabla(\mathbf{w}_j - \mathbf{w}_k)\|_{L^\infty(Q_j^*)}\Big)$$
$$\le c\sum_{j\in A_k} r_k^{-2}\|\mathbf{w}_j - \mathbf{w}_k\|_{L^\infty(Q_j^*)} \le c\lambda,$$

where we used $\sum_{j\in A_k}\varphi_j = 1$ on Q_k^*, inverse estimates for linear polynomials and Lemma 3.1.1 (d). Now $\mathcal{O}_\lambda = \bigcup_k \overline{Q_k^*}$ proves (d). □

The following theorem is an application of the Lipschitz truncation to weak null sequences. It is similar to the results in [62,33,60], which were used to prove the existence of weak solutions.

Theorem 3.1.16. *Let $1 < s < \infty$ and let $\mathcal{B} \subset \mathbb{R}^3$ be a ball. Let $(\mathbf{u}_m) \subset W^{1,s}_{0,\mathrm{div}}(\mathcal{B})$ be a weak $W^{1,s}_{0,\mathrm{div}}(\mathcal{B})$ null sequence. Then there exist $j_0 \in \mathbb{N}$ and a double sequence $\lambda_{m,j} \in \mathbb{R}$ with $2^{2^j} \le \lambda_{m,j} \le 2^{2^{j+1}-1}$ such that the Lipschitz truncations $\mathbf{u}_{m,j} := \mathbf{u}_{\lambda_{m,j}}$ have the following properties for $j \ge j_0$.*

- **(a)** $\mathbf{u}_{m,j} \in W^{1,\infty}_{0,\mathrm{div}}(2\mathcal{B})$ *and* $\mathbf{u}_{m,j} = \mathbf{u}_m$ *on* $\mathbb{R}^3 \setminus \mathcal{O}_{m,j}$ *for all* $m \in \mathbb{N}$, *where* $\mathcal{O}_{m,j} := \{M(\nabla^2(\mathrm{curl}^{-1}\,\mathbf{u}_m)) > \lambda_{m,j}\}$.
- **(b)** $\|\nabla \mathbf{u}_{m,j}\|_\infty \le c\lambda_{m,j}$ *for all* $m \in \mathbb{N}$,
- **(c)** $\mathbf{u}_{m,j} \to 0$ *for* $m \to \infty$ *in* $L^\infty(G)$,
- **(d)** $\nabla \mathbf{u}_{m,j} \overset{*}{\rightharpoonup} 0$ *for* $m \to \infty$ *in* $L^\infty(G)$,
- **(e)** *For all* $m, j \in \mathbb{N}$ *holds* $\|\lambda_{m,j}\chi_{\mathcal{O}_{m,j}}\|_s \le c(q)\, 2^{-\frac{j}{s}} \|\nabla \mathbf{u}_m\|_s$.

Proof. The proof follows by applying Theorem 3.1.15 to sequences and the continuity properties of curl^{-1}, see (3.1.4) and (3.1.5). We will construct below a double sequence $\lambda_{m,j}$ with $2^{2^j} \le \lambda_{m,j} \le 2^{2^{j+1}}$ and define $\mathbf{u}_{m,j} := (\mathbf{u}_m)^{\lambda_{m,j}}$. Choose j_0 such that $\sup_m c_0 \big(⨍_{\mathcal{B}} |\nabla(\mathbf{u}_m)|^s \,\mathrm{d}x\big)^{\frac{1}{s}} \le 2^{2^{j_0}}$. Properties (a) and (b) follow immediately from Theorem 3.1.15, for $j \ge j_0$. Moreover, $\mathbf{u}_{m,j}$ is bounded in $W^{1,\infty}_0(2\mathcal{B})$. Therefore, there exists a subsequence such that $\nabla \mathbf{u}_{m,j}$ converges $*$-weakly. As in the argument used above, this implies that the whole sequence $\nabla \mathbf{u}_{m,j}$ converges weakly* to zero, which proves (d). Applying Kondrachov's Theorem gives (d).

Finally, (e) is a consequence of Lemma 1.3.2 applied of $\nabla \mathbf{u}_m$. □

3.2 SOLENOIDAL LIPSCHITZ TRUNCATION IN 2D

Let $\mathbf{u} \in E^h_{0,\mathrm{div}}(\mathcal{B})$, $h(t) = t\ln(1+t)$, for some ball $\mathcal{B} \subset \mathbb{R}^2$. We define the "bad set" as $\mathcal{O}_\lambda := \{M(\boldsymbol{\varepsilon}(\mathbf{u})) > \lambda\}$. We do not have to truncate our function $\mathbf{w}$ if $\mathcal{O}_\lambda$ is empty. So we can assume in the following that $\mathcal{O}_\lambda \ne \emptyset$. We decompose the open set $\mathcal{O}_\lambda$ into a family of dyadic closed cubes $\{Q_j^*\}$ as in Section 3.1, such that (W5)–(W8) from Corollary 1.3.1 hold. For brevity we define $r_j := \ell(Q_j^*)$. Moreover we consider a partition of unity $(\varphi_j) \subset C_0^\infty(\mathbb{R}^2)$ such that we have (U1)–(U3) from Lemma 1.3.6.

We define $\Pi^{\mathcal{R}}_{Q_j^*}\mathbf{u}$ as the $L^2(Q_j^*)$-orthonormal projection of $\mathbf{u}$ onto the space of rigid motions $\mathcal{R}$, i.e.

$$\big(\Pi^{\mathcal{R}}_{Q_j^*}\mathbf{u}\big)(x) := \sum_l \bigg(\int_{Q_j^*} \mathbf{R}^j_l \cdot \mathbf{u}\,\mathrm{d}y\bigg)\mathbf{R}^j_l(x),$$

where $\{\mathbf{R}_l^j\}$ is an $L^2(Q_j^*)$-orthonormal base of $\mathcal{R}(Q_j^*)$. This means that every $\mathbf{R}_l^j$ has the structure

$$\mathbf{R}_l^j(x) = \mathbf{A}_l^j x + \mathbf{b}_l^j,\ \mathbf{A}_l^j = \begin{pmatrix} 0 & -\alpha_l^j \\ \alpha_l^j & 0 \end{pmatrix} \in \mathbb{R}^{2\times 2},\ \mathbf{b}_l^j = \begin{pmatrix} (b_1)_l^j \\ (b_2)_l^j \end{pmatrix} \in \mathbb{R}^2, \tag{3.2.7}$$

where all entries are real-valued. The operator $\Pi_{Q_j^*}^{\mathcal{R}}$ is also well defined for $\mathbf{u} \in L^1(Q_j^*)$. Moreover, it is continuous from L^1 to $W^{1,\infty}$ and

$$\|\Pi_{Q_j^*}^{\mathcal{R}} \mathbf{u}\|_{L^\infty(Q_j^*)} + r_j \|\nabla \Pi_{Q_j^*}^{\mathcal{R}} \mathbf{u}\|_{L^\infty(Q_j^*)} \le c \fint_{Q_j^*} |\mathbf{u}|\,dx \tag{3.2.8}$$

for all $\mathbf{u} \in L^1(Q_j^*)$. Since $\Pi_{Q_j^*}^{\mathcal{R}}$ acts on constant vectors as the identity it follows easily from (3.2.8) using Poincaré's inequality that $\Pi_{Q_j^*}^{\mathcal{R}}$ is also $W^{1,1}$-stable in the sense that

$$\fint_{Q_j^*} |\nabla \Pi_{Q_j^*}^{\mathcal{R}} \mathbf{u}|\,dx \le c \fint_{Q_j^*} |\nabla \mathbf{u}|\,dx. \tag{3.2.9}$$

Moreover, it follows from (3.2.8) and the fact that $\Pi_{Q_j^*}^{\mathcal{R}}$ is the identity on $\mathcal{R}$ that

$$\fint_{Q_j^*} |\mathbf{u} - \Pi_{Q_j^*}^{\mathcal{R}} \mathbf{u}|\,dx \le c \inf_{\mathbf{R}\in\mathcal{R}} \fint_{Q_j^*} |\mathbf{u} - \mathbf{R}|\,dx. \tag{3.2.10}$$

We consider $\mathbf{u} \in E_{0,\mathrm{div}}^h(\mathcal{B})$ and extend it by zero outside $\mathcal{B}$ to $\mathbf{u} \in E_{0,\mathrm{div}}^h(\mathbb{R}^2)$. We set $w := \operatorname{curl}^{-1}\mathbf{u}$ and obtain $w \in W^{2,1}(\mathbb{R}^2)$ using $|\nabla^2 \operatorname{curl}^{-1}\mathbf{u}| = |\nabla \mathbf{u}|$ as well as Korn's inequality from $L^h(\mathcal{B})$ to $L^1(\mathcal{B})$, see Theorem 2.3.11. Here we have

$$\operatorname{curl}^{-1}\mathbf{u}(x) := \operatorname{rot}\left(\int_{\mathbb{R}^2} \frac{\ln(|x-y|)}{2\pi}\mathbf{u}(y)\,dy\right),$$

where $\operatorname{rot}\mathbf{w} = \partial_1 w_2 - \partial_2 w_1$ for $\mathbf{w} : \mathbb{R}^2 \to \mathbb{R}^2$ and $\operatorname{curl} w = (\partial_2 w, -\partial_1 w)^T$ for $w : \mathbb{R}^2 \to \mathbb{R}$. In order to take into account the rotation from ∇ to curl which holds in two dimensions, we define on Q_j^*

$$\widetilde{R}_l^j(x) = \frac{1}{2}\widetilde{\mathbf{A}}_l^j(x,x) + \widetilde{\mathbf{b}}_l^j x,\quad \widetilde{\mathbf{A}}_l^j = \begin{pmatrix} -\alpha_l^j & 0 \\ 0 & -\alpha_l^j \end{pmatrix},\quad \widetilde{\mathbf{b}}_l^j = \begin{pmatrix} -(b_2)_l^j \\ (b_1)_l^j \end{pmatrix},$$

for all $\mathbf{R}_l^j \in \mathcal{R}(Q_j^*)$ as in (3.2.7). This means we have

$$\operatorname{curl} \widetilde{R}_l^j = \mathbf{R}_l^j.$$

Now we set

$$(\widetilde{\Pi}_{Q_j^*}^{\mathcal{R}} w)(x) := \sum_l \left(\int_{Q_j^*} \mathbf{R}_l^j \cdot \operatorname{curl} w\,dy\right) \widetilde{R}_l^j(x),$$

so that

$$\operatorname{curl}\left(\widetilde{\Pi}^{\mathcal{R}}_{Q_j^*} w\right) = \Pi^{\mathcal{R}}_{Q_j^*}\left(\operatorname{curl} w\right). \tag{3.2.11}$$

We define the Lipschitz truncation w_λ of w by

$$w_\lambda = \begin{cases} w & \text{in} \quad \mathbb{R}^2 \setminus \mathcal{O}_\lambda \\ \sum_j \varphi_j w_j & \text{in} \quad \mathcal{O}_\lambda. \end{cases}$$

Here w_j is defined by

$$w_j = \widetilde{\Pi}^{\mathcal{R}}_{Q_j^*} w - \left(\widetilde{\Pi}^{\mathcal{R}}_{Q_j^*} w\right)_{Q_j^*} + (w)_{Q_j^*}.$$

We will see later that $w_\lambda \in W^{2,1}(\mathbb{R}^2)$. The following Lemma provides some important estimates for w_λ.

Lemma 3.2.1. **(a)** *For all $j \in \mathbb{N}$ and all $y \in Q_j^*$ we have*

$$\fint_{Q_j^*} \left| \frac{w - w_j}{r_j^2} \right| \mathrm{d}x \le c \fint_{Q_j^*} \left| \frac{\nabla(w - w_j)}{r_j} \right| \mathrm{d}x \le c \fint_{Q_j^*} |\boldsymbol{\varepsilon}(\operatorname{curl} w)| \, \mathrm{d}x.$$

(b) *For all j the following holds*

$$\fint_{Q_j^*} |\nabla^2 (w - w_j)| \, \mathrm{d}x \le c \fint_{Q_j^*} |\nabla(\operatorname{curl} w)| \, \mathrm{d}x.$$

(c) *For all j the following holds*

$$\fint_{Q_j^*} |\boldsymbol{\varepsilon}(\operatorname{curl} w)| \, \mathrm{d}x \le c \fint_{2Q_j^*} |\boldsymbol{\varepsilon}(\operatorname{curl} w)| \, \mathrm{d}x \le c\,\lambda.$$

(d) *For all j and k with $Q_j^* \cap Q_k^* \neq \emptyset$ we have*

$$\begin{aligned} r_j^{-1} \| w_j - w_k \|_{L^\infty(Q_j^*)} &\sim \fint_{Q_j^*} \left| \frac{w_j - w_k}{r_j} \right| \mathrm{d}x \\ &\le c \fint_{Q_j^*} \left| \frac{w - w_j}{r_j} \right| \mathrm{d}x + c \fint_{Q_k^*} \left| \frac{w - w_j}{r_j} \right| \mathrm{d}x. \end{aligned}$$

Proof. (a): The claimed inequality follows by applying Poincaré's inequality and Korn's inequality (2.3.114) from Theorem 2.3.13. We also take into account the fact that $|\nabla g| = |\operatorname{curl} g|$, the definition of w_j, (3.2.11) as well (3.2.10) to obtain the following

$$\begin{aligned} \fint_{Q_j^*} \left| \frac{w - w_j}{r_j^2} \right| \mathrm{d}x &\le c \fint_{Q_j^*} \left| \frac{\nabla(w - w_j)}{r_j} \right| \mathrm{d}x = c \fint_{Q_j^*} \left| \frac{\operatorname{curl}(w - w_j)}{r_j} \right| \mathrm{d}x \\ &= c \fint_{Q_j^*} \left| \frac{\operatorname{curl} w - \Pi^{\mathcal{R}}_{Q_j^*}(\operatorname{curl} w)}{r_j} \right| \mathrm{d}x \le c \fint_{Q_j^*} |\boldsymbol{\varepsilon}(\operatorname{curl} w)| \, \mathrm{d}x. \end{aligned}$$

(b): Using the continuity of $\Pi^{\mathcal{R}}_{Q_j^*}$ from (3.2.9), $|\nabla^2 g| = |\nabla \operatorname{curl} g|$ and (3.2.11) we find

$$\begin{aligned}\fint_{Q_j^*} |\nabla^2(w - w_j)|\,\mathrm{d}x &= \fint_{Q_j^*} |\nabla \operatorname{curl}(w - w_j)|\,\mathrm{d}x\\ &= \fint_{Q_j^*} \big|\nabla(\operatorname{curl} w - \Pi^{\mathcal{R}}_{Q_j^*}(\operatorname{curl} w))\big|\,\mathrm{d}x\\ &\leq \fint_{Q_j^*} |\nabla \operatorname{curl} w|\,\mathrm{d}x + \fint_{Q_j^*} \big|\nabla \Pi^{\mathcal{R}}_{Q_j^*}(\operatorname{curl} w)\big|\,\mathrm{d}x \leq c \fint_{Q_j^*} |\nabla(\operatorname{curl} w)|\,\mathrm{d}x.\end{aligned}$$

(c) and (d) follow as in Lemma 3.1.1 (note that the w_j's are still elements of a finite dimensional function space). □

The next Lemma shows that although w_λ is defined differently on two different sets it is a global Sobolev function.

Lemma 3.2.2. *Let $w \in W^{2,1}(\mathbb{R}^2)$, then $w_\lambda - w \in W_0^{2,1}(\mathcal{O}_\lambda)$ and $w_\lambda \in W^{2,1}(\mathbb{R}^2)$.*

Proof. Because of the definition of w_λ it suffices to show that $w_\lambda - w \in W_0^{2,1}(\mathcal{O}_\lambda)$. We have pointwise

$$\begin{aligned}\nabla^2(w_\lambda - w) &= \nabla^2\Big(\sum_{j\in\mathbb{N}} \varphi_j(w_j - w)\Big)\\ &= \sum_{j\in\mathbb{N}} \big(\varphi_j \nabla^2(w_j - w) + 2(\nabla\varphi_j)\otimes\nabla(w_j - w) + \nabla^2\varphi_j(w_j - w)\big).\end{aligned}$$

Since every summand in the last sum belongs to $W_0^{2,1}(\mathcal{O}_\lambda)$, it is enough to show that the last sum converges absolutely in $L^1(\mathbb{R}^2)$. We obtain for finite $J \subset \mathbb{N}$

$$\begin{aligned}(I) &:= \int \sum_{j\in\mathbb{N}\setminus J} \big|\varphi_j\nabla^2(w_j - w) + \nabla\varphi_j\otimes\nabla(w_j - w) + (\nabla^2\varphi_j)(w_j - w)\big|\,\mathrm{d}x\\ &\leq \sum_{j\in\mathbb{N}\setminus J}\int_{Q_j^*}\left|\frac{w_j - w}{r_j^2}\right|\mathrm{d}x + \sum_{j\in\mathbb{N}\setminus J}\int_{Q_j^*}\left|\frac{\nabla w_j - \nabla w}{r_j}\right|\mathrm{d}x\\ &\quad + \sum_{j\in\mathbb{N}\setminus J}\int_{Q_j^*}|\nabla^2 w_j - \nabla^2 w|\,\mathrm{d}x\end{aligned}$$

Now, by Lemma 3.2.1 (a) and (b) it follows that

$$(I) \leq c\sum_{j\in\mathbb{N}\setminus J}\int_{Q_j^*}|\nabla^2 w|\,\mathrm{d}x \leq c\int \chi_{\cup_{j\in\mathbb{N}\setminus J}}|\nabla^2 w|\,\mathrm{d}x.$$

Since $\chi_{\bigcup_{j\in\mathbb{N}\setminus J}} \to 0$ for $J \to \mathbb{N}$ and $\nabla^2 w \in L^1(\mathbb{R}^2)$, dominated convergence yields $(I) \to 0$ for $J \to \mathbb{N}$. In particular, we have shown that $\sum_{j\in\mathbb{N}} \varphi_j(w_j - w)$ converges unconditionally in the norm $\|\nabla^2 \cdot\|_1$ and therefore in $W_0^{2,1}(\mathcal{O}_\lambda)$. □

As a consequence of the previous lemma we have

$$\nabla^2 w_\lambda = \chi_{\mathbb{R}^2\setminus\mathcal{O}_\lambda} \nabla^2 w + \chi_{\mathcal{O}_\lambda} \sum_j \nabla^2(\varphi_j w_j). \tag{3.2.12}$$

Lemma 3.2.3. *If $w \in W^{2,1}(\mathbb{R}^2)$ with $\boldsymbol{\varepsilon}(\operatorname{curl} w) \in L^h(\mathbb{R}^2)$, then the following holds.*

- **(a)** $\|w_\lambda\|_1 \le c\,\|w\|_1$, $\|\nabla w_\lambda\|_1 \le c\,\|\nabla w\|_1$ *and* $\|\nabla^2 w_\lambda\|_1 \le c\,\|\nabla^2 w\|_1$.
- **(b)** $|\boldsymbol{\varepsilon}(\operatorname{curl} w_\lambda)| \le c\lambda\chi_{\mathcal{O}_\lambda} + |\boldsymbol{\varepsilon}(\operatorname{curl} w)|\chi_{\mathbb{R}^2\setminus\mathcal{O}_\lambda}$ *and* $|\boldsymbol{\varepsilon}(\operatorname{curl} w_\lambda)| \le c\lambda$ *a.e.*
- **(c)** $w_\lambda = w$ *a.e. in* $\mathbb{R}^d \setminus \mathcal{O}_\lambda$.
- **(d)** $\|\boldsymbol{\varepsilon}(\operatorname{curl} w_\lambda)\|_h \le c\,\|\boldsymbol{\varepsilon}(\operatorname{curl} w)\|_h$ *and* $\int h(|\boldsymbol{\varepsilon}(\operatorname{curl} w_\lambda)|)\,dx \le c \int h(|\boldsymbol{\varepsilon}(\operatorname{curl} w)|)\,dx$.

Proof. (a): By the definition of w_λ we have

$$\int |w_\lambda|\,dx \le \int_{\mathbb{R}^2\setminus\mathcal{O}_\lambda} |w|\,dx + \sum_j \int_{Q_j^*} |\varphi_j w_j|\,dx.$$

Now it follows with the help of (3.2.8) and the local finiteness of the Q_j^*'s from (W7) that

$$\int |w_\lambda|\,dx \le \int_{\mathbb{R}^d\setminus\mathcal{O}_\lambda} |w|\,dx + \sum_j \int_{Q_j^*} |w|\,dx \le c \int_{\mathbb{R}^2} |w|\,dx.$$

For the gradients we have

$$\nabla(w_\lambda - w) = \sum_j \nabla\varphi_j(w_j - w) + \sum_j \varphi_j \nabla(w_j - w).$$

Thus, using Lemma 3.2.1 (a) we obtain

$$\begin{aligned}
\int |\nabla w_\lambda|\,dx &\le \int |\nabla w|\,dx + c\sum_j \int_{Q_j^*} \left|\frac{w_j - w}{r_j}\right| dx + \sum_j \int_{Q_j^*} |\nabla(w_j - w)|\,dx \\
&\le \int |\nabla w|\,dx + c\sum_j \int_{Q_j^*} |\nabla(w_j - w)|\,dx \\
&\le \int |\nabla w|\,dx + c\sum_j \int_{Q_j^*} |\nabla w_j|\,dx + c\sum_j \int_{Q_j^*} |\nabla w|\,dx.
\end{aligned}$$

The desired estimate is a consequence of (3.2.8) and the finite intersection property of the Q_j^*'s, cf. (W7). Note that for every $j \in \mathbb{N}$ we have

$$\begin{aligned}\int_{Q_j^*} |\nabla w_j|\,dx &= \int_{Q_j^*} |\operatorname{curl} w_j|\,dx = \int_{Q_j^*} |\Pi^{\mathcal{R}}_{Q_j^*}(\operatorname{curl} w)|\,dx \\ &\le c\int_{Q_j^*} |\operatorname{curl} w|\,dx = c\int_{Q_j^*} |\nabla w|\,dx.\end{aligned}$$

Finally, we compute

$$\begin{aligned}\nabla^2(w_\lambda - w) &= \sum_k \varphi_k \nabla^2(w_k - w) + 2\sum_k \nabla\varphi_k \otimes \nabla(w_k - w) \\ &\quad + \sum_k \nabla^2\varphi_k(w_k - w).\end{aligned}$$

If we combine $|\nabla^2 g| = |\nabla \operatorname{curl} g|$ with Lemma 3.2.1 (a) we obtain the last inequality in the same fashion.

(b): Fix $j \in \mathbb{N}$. Since $\boldsymbol{\varepsilon}(\operatorname{curl} w_k) = 0$ for all k we have on Q_j^*

$$\begin{aligned}\boldsymbol{\varepsilon}(\operatorname{curl} w_\lambda) &= \sum_{k\in A_j} \boldsymbol{\varepsilon}(\operatorname{curl}\varphi_k)(w_k - w_j) + \sum_{k\in A_j} \operatorname{curl}\varphi_k \odot \nabla(w_k - w_j) \\ &\quad + \sum_{k\in A_j} \nabla\varphi_k \odot \operatorname{curl}(w_k - w_j).\end{aligned}$$

Using the local finiteness of the Q_k^*, Lemma 3.2.1 (a) and (d) it follows that

$$\begin{aligned}&\|\boldsymbol{\varepsilon}(\operatorname{curl} w_\lambda)\|_{L^\infty(Q_j^*)} \\ &\le c\sum_{k\in A_j} \frac{\|w_j - w_k\|_{L^\infty(Q_j^*\cap Q_k^*)}}{r_j^2} + c\sum_{k\in A_j} \frac{\|\nabla w_j - \nabla w_k\|_{L^\infty(Q_j^*\cap Q_k^*)}}{r_j} \\ &\le c\sum_{k\in A_j} \frac{\|w_j - w_k\|_{L^\infty(Q_j^*\cap Q_k^*)}}{r_j^2} \le c\sum_{k\in A_j} \fint_{Q_k^*} \frac{|w - w_k|}{r_k^2}\,dx \\ &\le c\sum_{k\in A_j} \fint_{Q_k^*} |\boldsymbol{\varepsilon}(\operatorname{curl} w)|\,dx \le c\fint_{c_2 Q_j^*} |\boldsymbol{\varepsilon}(\operatorname{curl} w)|\,dx.\end{aligned}$$

In the third line we also took into account the equivalence of norms on finite dimensional spaces in the second line. It follows by Lemma 3.2.1 (c) that $|\boldsymbol{\varepsilon}(\operatorname{curl} w_\lambda)| \le c\lambda$ in Q_j^*. Since $\bigcup_k Q_k^* = \mathcal{O}_\lambda$ we get $|\boldsymbol{\varepsilon}(\operatorname{curl} w_\lambda)| \le c\lambda$ in $\mathcal{O}_\lambda$. As a consequence $|\boldsymbol{\varepsilon}(\operatorname{curl} w_\lambda)| \le c\lambda\chi_{\mathcal{O}_\lambda} + |\nabla^2 w|\chi_{\mathbb{R}^2\setminus\mathcal{O}_\lambda}$. In $\mathbb{R}^2 \setminus \mathcal{O}_\lambda$ we have $|\boldsymbol{\varepsilon}(\operatorname{curl} w_\lambda)| = |\boldsymbol{\varepsilon}(\operatorname{curl} w)| \le M(\boldsymbol{\varepsilon}(\operatorname{curl} w)) \le \lambda$. So we conclude that $|\boldsymbol{\varepsilon}(\operatorname{curl} w_\lambda)| \le c\lambda$ in $\mathbb{R}^2$.

(d): We estimate using (b)

$$\begin{aligned}\|\boldsymbol{\varepsilon}(\operatorname{curl} w_\lambda)\|_h &\le \|\chi_{\mathbb{R}^2\setminus\mathcal{O}_\lambda}\boldsymbol{\varepsilon}(\operatorname{curl} w)\|_h + \|\chi_{\mathcal{O}_\lambda}\boldsymbol{\varepsilon}(\operatorname{curl} w_\lambda)\|_h\\ &\le \|\boldsymbol{\varepsilon}(\operatorname{curl} w)\|_h + c\,\|\chi_{\mathcal{O}_\lambda}\lambda\|_h.\\ &\le \|\boldsymbol{\varepsilon}(\operatorname{curl} w)\|_h + c\,\|\chi_{\{M(\boldsymbol{\varepsilon}(\operatorname{curl} w))>\lambda\}}\lambda\|_h.\end{aligned}$$

Now the weak-type estimate for the norm of the maximal function on L^h (see [134]) proves

$$\|\boldsymbol{\varepsilon}(\operatorname{curl} w_\lambda)\|_h \le c\,\|\boldsymbol{\varepsilon}(\operatorname{curl} w)\|_h.$$

The estimate for the modular follows analogously using the weak-type estimate for the modular of the maximal function (see [134]). □

The final Theorem of this section follows now by combining Lemma 3.2.3 with the definition $\mathbf{u}_\lambda = \operatorname{curl} w_\lambda$ and continuity properties of the inverse curl-operator (for $d=2$).

Theorem 3.2.17. *Let $\mathcal{B}\subset\mathbb{R}^2$ be a ball and $\mathbf{u}\in E^h_{0,\mathrm{div}}(\mathcal{B})$. Then there is a function $\mathbf{u}_\lambda \in E^\infty_{0,\mathrm{div}}(\mathbb{R}^2)$, called the Lipschitz truncation of $\mathbf{u}$, with the following properties:*

(a) $\|\mathbf{u}_\lambda\|_1 \le c\,\|\mathbf{u}\|_1$ *and* $\|\nabla\mathbf{u}_\lambda\|_1 \le c\,\|\nabla\mathbf{u}\|_1$.

(b) $|\boldsymbol{\varepsilon}(\mathbf{u}_\lambda)| \le c\lambda\chi_{\mathcal{O}_\lambda} + |\boldsymbol{\varepsilon}(\mathbf{u})|\chi_{\mathcal{B}\setminus\mathcal{O}_\lambda}$ *and* $|\boldsymbol{\varepsilon}(\mathbf{u}_\lambda)| \le c\lambda$ *a.e.*

(c) $\mathbf{u} = \mathbf{u}_\lambda$ *a.e. in* $\mathcal{B}\setminus\mathcal{O}_\lambda$.

(d) $\|\boldsymbol{\varepsilon}(\mathbf{u}_\lambda)\|_h \le c\,\|\boldsymbol{\varepsilon}(\mathbf{u})\|_h$ *and* $\int h(|\boldsymbol{\varepsilon}(\mathbf{u}_\lambda)|)\,dx \le c\int h(|\boldsymbol{\varepsilon}(\mathbf{u})|)\,dx$.

Theorem 3.2.18. *Let $\mathcal{B}\subset\mathbb{R}^2$ be a ball. Let $(\mathbf{u}_n)\subset E^h_{0,\mathrm{div}}(\mathcal{B})$ be a bounded sequence which converges strongly to zero in $L^1(\mathcal{B})$. Then there is a double sequence $(\lambda_{n,j})\subset\mathbb{R}$ and a number $j_0\in\mathbb{N}$, null sequences $(\kappa_j)(\widetilde{\kappa}_j)$ such that the sequence $\mathbf{u}_{n,j}\in E^\infty_{0,\mathrm{div}}(\mathbb{R}^2)$ has the following properties for $j\ge j_0$.*

(a) $\mathbf{u}_{n,j}\in E^\infty_{0,\mathrm{div}}(\mathbb{R}^2)$ *and* $\mathbf{u}_{n,j} = \mathbf{u}_n$ *in* $\mathbb{R}^2\setminus\{M(\boldsymbol{\varepsilon}(\mathbf{u}_n)) > \lambda_{n,j}\}$,

(b) $\|\boldsymbol{\varepsilon}(\mathbf{u}_{n,j})\|_\infty \le c\lambda_{n,j}$ *where* $2^{2^j} \le \lambda_{n,j} \le 2^{2^{j+1}}$,

(c) $\boldsymbol{\varepsilon}(\mathbf{u}_{n,j}) \overset{*}{\rightharpoonup} 0$ *for* $n\to\infty$ *in* $L^\infty(\mathbb{R}^2)$,

(d) *there exists a (non-relabelled) subsequence of* $\mathbf{u}_n$ *which satisfies* $\limsup\limits_{n\to\infty}\int h(|\lambda_{n,j}\chi_{\{\mathbf{u}_{n,j}\neq\mathbf{u}_n\}}|)\,dx \le \kappa_j$ *and* $\limsup\limits_{n\to\infty}\|\lambda_{n,j}\chi_{\{\mathbf{u}_{n,j}\neq\mathbf{u}_n\}}\|_h \le \widetilde{\kappa}_j$.

Proof. We will construct below a double sequence $\lambda_{n,j}$ with $2^{2^j}\le\lambda_{n,j}\le 2^{2^{j+1}}$ and define $\mathbf{u}_{n,j} := (\mathbf{u}_n)_{\lambda_{n,j}}$. Choose j_0 such that $\sup_n \fint_{\mathcal{B}} |\boldsymbol{\varepsilon}(\mathbf{u}_n)|\,dx \le 2^{2^{j_0}}$. Properties (a)–(c) now follow as in Section 3.1 as a consequence of Theorem 3.2.17. Since M is bounded from $L^h(2\mathcal{B})$ to $L^1(2\mathcal{B})$ (see [134] I 8.14(a)), we have

$$K := \sup_n \|M(\boldsymbol{\varepsilon}(\mathbf{u}_n))\|_{L^1(2\mathcal{B})} < \infty.$$

Next, we observe as in (1.3.26) that

$$\sum_{j\in\mathbb{N}}\sum_{k=2^j}^{2^{j+1}-1}\int_{2\mathcal{B}} 2^k\,\chi_{\{|M(\boldsymbol{\varepsilon}(\mathbf{u}_n))|>2\cdot 2^k\}}\,\mathrm{d}x \le K.$$

We can rewrite the last inequality as

$$\sum_{j\in\mathbb{N}} b_j^n \le K$$

with an obvious definition for b_j^n.

Since the sum in the definition of b_j contains 2^j summands, there is at least one index $k_{n,j}$ such that

$$\int_{2\mathcal{B}} 2^{k_{n,j}}\,\chi_{\{|M(\boldsymbol{\varepsilon}(\mathbf{u}_n))|>2\cdot 2^{k_{n,j}}\}}\,\mathrm{d}x \le 2^{-j}\,b_j^n, \tag{3.2.13}$$

which is equivalent to

$$\int_{2\mathcal{B}} h\left(2^{k_{n,j}}\right)\chi_{\{|M(\boldsymbol{\varepsilon}(\mathbf{u}_n))|>2\cdot 2^{k_{n,j}}\}}\,\mathrm{d}x \le \ln(1+2^{k_{n,j}})\,2^{-j}\,b_j^n. \tag{3.2.14}$$

Note that $\ln(1+2^{k_{n,j}})\,2^{-j}\le 3$ on account of $k_{n,j}\le 2^{j+1}$; thus we get

$$\int_{2\mathcal{B}} h\left(2^{k_{n,j}}\right)\chi_{\{|M(\boldsymbol{\varepsilon}(\mathbf{u}_n))|>2\cdot 2^{k_{n,j}}\}}\,\mathrm{d}x \le 3\,b_j^n. \tag{3.2.15}$$

Define $\delta_1 := \liminf_n b_1^n$. Then there exists a subsequence (not relabelled) with

$$\limsup_n b_1^n = \liminf_n b_1^n = \delta_1.$$

This proves

$$\limsup_n \int_{2\mathcal{B}} h\left(2^{k_{n,1}}\right)\chi_{\{|M(\boldsymbol{\varepsilon}(\mathbf{u}_n))|>2\cdot 2^{k_{n,1}}\}}\,\mathrm{d}x \le 3\limsup_n b_1^n = 3\,\delta_1.$$

Next, define $\delta_2 := \liminf_n b_2^n$ and by passing to a further subsequence we get

$$\limsup_n \int_{2\mathcal{B}} h\left(2^{k_{n,2}}\right)\chi_{\{|M(\boldsymbol{\varepsilon}(\mathbf{u}_n))|>2\cdot 2^{k_{n,2}}\}}\,\mathrm{d}x \le 3\limsup_n b_2^n = 3\,\delta_2.$$

Using this iterative argument we can construct a diagonal sequence (not relabelled) such that for every j

$$\limsup_n \int_{2\mathcal{B}} h\left(2^{k_{n,j}}\right)\chi_{\{|M(\boldsymbol{\varepsilon}(\mathbf{u}_n))|>2\cdot 2^{k_{n,j}}\}}\,\mathrm{d}x \le 3\limsup_n b_j^n = 3\,\delta_j. \tag{3.2.16}$$

From now on we will use the diagonal sequence. Fatou's Lemma gives

$$K \ge \liminf_n \sum_j b_j^n \ge \sum_j \liminf_n b_j^n = \sum_j \delta_j,$$

hence, δ_j is a null sequence. Define $\kappa_j := 3\,\delta_j$ and $\lambda_{n,j} := 2^{k_{n,j}}$. Then (3.2.16) proves the integral estimate of (d). The norm estimate is a direct consequence. □

Remark 3.2.8. Note that it is not possible to show (d) of Theorem 3.2.18 by the technique of [62], since there the boundedness of the maximal function is used, which does not hold in L^h. Therefore, we must apply a more subtle weak type argument.

3.3 $\mathcal{A}$-STOKES APPROXIMATION – STATIONARY CASE

A major argument in the regularity theory for nonlinear PDE's is the comparison with solutions of linear equations: If a solution is close to a harmonic function, then it inherits some of the harmonic function's regularity properties. A refinement of this argument is the method of almost $\mathcal{A}$-harmonicity, which requires the closeness to the $\mathcal{A}$-harmonic function only in a very weak sense (namely in $W^{-1,1}$-sense). This argument first appeared in [58]. For newer results we refer the reader to the review article [67]. In the following, we extend this principle to the Stokes problem involving solenoidal functions.

By $\mathcal{A}$ we denote a symmetric, elliptic tensor, i.e.

$$c_0|\boldsymbol{\tau}|^2 \leq \mathcal{A}(\boldsymbol{\tau},\boldsymbol{\tau}) \leq c_1|\boldsymbol{\tau}|^2 \quad \text{for all} \quad \boldsymbol{\tau} \in \mathbb{R}^{d\times d}.$$

We define $|\mathcal{A}| := c_1/c_0$.

We begin with a variational inequality for the $\mathcal{A}$-Stokes system.

Lemma 3.3.1. *For all balls $\mathcal{B}$ and $\mathbf{u} \in W_0^{1,q}(\mathcal{B})$ we have*

$$\fint_{\mathcal{B}} |\boldsymbol{\varepsilon}(\mathbf{u})|^q \, dx \leq c \sup_{\boldsymbol{\xi}\in C^\infty_{0,\mathrm{div}}(\mathcal{B})} \left[\fint_{\mathcal{B}} \mathcal{A}(\boldsymbol{\varepsilon}(\mathbf{u}),\boldsymbol{\varepsilon}(\boldsymbol{\xi}))\,dx - \fint_{\mathcal{B}} |\nabla\boldsymbol{\xi}|^{q'}\,dx\right], \tag{3.3.17}$$

where c depends only on $\mathcal{A}$.

Proof. Duality arguments show that

$$\fint_{\mathcal{B}} |\boldsymbol{\varepsilon}(\mathbf{u})|^q \, dx = \sup_{\mathbf{H}\in L^{q'}(\mathcal{B})} \left[\fint_{\mathcal{B}} \boldsymbol{\varepsilon}(\mathbf{u}) : \mathbf{H}\,dx - \fint_{\mathcal{B}} |\mathbf{H}|^{q'}\,dx\right].$$

For a given $\mathbf{H}$ let $\mathbf{z}_\mathbf{H}$ be the unique $W^{1,q'}_{0,\mathrm{div}}(\mathcal{B})$-solution of

$$\int_{\mathcal{B}} \mathcal{A}(\boldsymbol{\varepsilon}(\mathbf{z}_\mathbf{H}),\boldsymbol{\varepsilon}(\boldsymbol{\xi}))\,dx = \int_{\mathcal{B}} \mathbf{H} : \nabla\boldsymbol{\xi}\,dx$$

for all $\boldsymbol{\xi} \in C^\infty_{0,\mathrm{div}}(\mathcal{B})$. Due to Lemma B.1.1 this solution satisfies

$$\fint_{\mathcal{B}} |\nabla \mathbf{z}_{\mathbf{H}}|^{q'} \,\mathrm{d}x \le c \fint_{\mathcal{B}} |\mathbf{H}|^{q'} \,\mathrm{d}x.$$

In other words, the mapping $L^{q'}(\mathcal{B}) \ni \mathbf{H} \mapsto \mathbf{z}_{\mathbf{H}} \in W^{1,q'}_{0,\mathrm{div}}(\mathcal{B})$ is continuous. This and the density of $C^\infty_{0,\mathrm{div}}(\mathcal{B})$ in $W^{1,q'}_{0,\mathrm{div}}(\mathcal{B})$ imply

$$\begin{aligned}\fint_{\mathcal{B}} |\boldsymbol{\varepsilon}(\mathbf{u})|^q \,\mathrm{d}x &\le c \sup_{\mathbf{H} \in L^{q'}(\mathcal{B})} \left[\fint_{\mathcal{B}} \mathcal{A}(\boldsymbol{\varepsilon}(\mathbf{u}), \boldsymbol{\varepsilon}(\mathbf{z}_{\mathbf{H}})) \,\mathrm{d}x - \fint_{\mathcal{B}} |\nabla \mathbf{v}_{\mathbf{H}}|^{q'} \,\mathrm{d}x \right] \\ &\le c \sup_{\boldsymbol{\xi} \in C^\infty_{0,\mathrm{div}}(\mathcal{B})} \left[\fint_{\mathcal{B}} \mathcal{A}(\boldsymbol{\varepsilon}(\mathbf{u}), \boldsymbol{\varepsilon}(\boldsymbol{\xi})) \,\mathrm{d}x - \fint_{\mathcal{B}} |\nabla \boldsymbol{\xi}|^{q'} \,\mathrm{d}x \right]. \qquad \square\end{aligned}$$

Let us now state the $\mathcal{A}$-Stokes approximation.

Theorem 3.3.19. *Let $\mathcal{B}$ be a ball with radius r and let $\tilde{\mathcal{B}}$ denote either $\mathcal{B}$ or $2\mathcal{B}$. Let $\mathbf{v} \in W^{1,qs}_{\mathrm{div}}(2\tilde{\mathcal{B}})$, $q, s > 1$ be an almost $\mathcal{A}$-Stokes solution in the sense that*

$$\left| \fint_{2\mathcal{B}} \mathcal{A}(\boldsymbol{\varepsilon}(\mathbf{v}), \boldsymbol{\varepsilon}(\boldsymbol{\xi})) \,\mathrm{d}x \right| \le \delta \fint_{2\tilde{\mathcal{B}}} |\boldsymbol{\varepsilon}(\mathbf{v})| \,\mathrm{d}x \, \|\nabla \boldsymbol{\xi}\|_\infty$$

for all $\boldsymbol{\xi} \in C^\infty_{0,\mathrm{div}}(2\mathcal{B})$ and some small $\delta > 0$. Then the unique solution $\mathbf{w} \in W^{1,q}_{0,\mathrm{div}}(\mathcal{B})$ of

$$\int_{\mathcal{B}} \mathcal{A}(\boldsymbol{\varepsilon}(\mathbf{w}), \boldsymbol{\varepsilon}(\boldsymbol{\xi})) \,\mathrm{d}x = \int_{\mathcal{B}} \mathcal{A}(\boldsymbol{\varepsilon}(\mathbf{v}), \boldsymbol{\varepsilon}(\boldsymbol{\xi})) \,\mathrm{d}x \tag{3.3.18}$$

for all $\boldsymbol{\xi} \in C^\infty_{0,\mathrm{div}}(\mathcal{B})$ satisfies

$$\fint_{\mathcal{B}} \left| \frac{\mathbf{w}}{r} \right|^q \,\mathrm{d}x + \fint_{\mathcal{B}} |\nabla \mathbf{w}|^q \,\mathrm{d}x \le \kappa \left(\fint_{2\tilde{\mathcal{B}}} |\nabla \mathbf{u}|^{qs} \,\mathrm{d}x \right)^{\frac{1}{s}}.$$

It holds $\kappa = \kappa(q, s, \delta)$ and $\lim_{\delta \to 0} \kappa(q, s, \delta) = 0$. The function $\mathbf{h} := \mathbf{v} - \mathbf{w}$ is called the $\mathcal{A}$-Stokes approximation of $\mathbf{v}$.

The use of $\tilde{\mathcal{B}} = 2\mathcal{B}$ enables a better combination of the $\mathcal{A}$-Stokes approximation with Caccioppoli type estimates, which usually increase the domains of integration.

Proof. Combining Korn's inequality and Poincaré's inequality with (3.3.17) shows

$$\begin{aligned}&\fint_{\mathcal{B}} \left| \frac{\mathbf{w}}{r} \right|^q \,\mathrm{d}x + \fint_{\mathcal{B}} |\nabla \mathbf{w}|^q \,\mathrm{d}x \\ &\quad \le c \sup_{\boldsymbol{\xi} \in C^\infty_{0,\mathrm{div}}(\mathcal{B})} \left[\fint_{\mathcal{B}} \mathcal{A}(\boldsymbol{\varepsilon}(\mathbf{v}), \boldsymbol{\varepsilon}(\boldsymbol{\xi})) \,\mathrm{d}x - \fint_{\mathcal{B}} |\nabla \boldsymbol{\xi}|^{q'} \,\mathrm{d}x \right].\end{aligned} \tag{3.3.19}$$

In the following let us fix $\boldsymbol{\xi} \in C^{\infty}_{0,\operatorname{div}}(\mathcal{B})$ and consider $\fint_{\mathcal{B}} \mathcal{A}(\boldsymbol{\varepsilon}(\mathbf{v}), \boldsymbol{\varepsilon}(\boldsymbol{\xi}))\,\mathrm{d}x$. Let

$$\gamma := \left(\fint_{\mathcal{B}} |\nabla \boldsymbol{\xi}|^{q'}\,\mathrm{d}x\right)^{\frac{1}{q'}} \tag{3.3.20}$$

and $m_0 \in \mathbb{N}$, $m_0 \gg 1$. Owing to Theorem 3.1.15 applied with $p = q'$ we find $\lambda \in [\gamma, 2^{m_0}\gamma]$ and $\boldsymbol{\xi}_\lambda \in W^{1,\infty}_{0,\operatorname{div}}(2\mathcal{B})$ such that

$$\|\nabla \boldsymbol{\xi}_\lambda\|_\infty \le c\lambda, \tag{3.3.21}$$

$$\lambda^{q'} \frac{\mathcal{L}^d\left(\{\boldsymbol{\xi}_\lambda \neq \boldsymbol{\xi}\}\right)}{|\mathcal{B}|} \le \frac{c}{m_0} \fint_{\mathcal{B}} |\nabla \boldsymbol{\xi}|^{q'}\,\mathrm{d}x \tag{3.3.22}$$

$$\fint_{2\mathcal{B}} |\nabla \boldsymbol{\xi}_\lambda|^{q'}\,\mathrm{d}x \le c \fint_{\mathcal{B}} |\nabla \boldsymbol{\xi}|^{q'}\,\mathrm{d}x. \tag{3.3.23}$$

We calculate

$$\begin{aligned}\int_{\mathcal{B}} \mathcal{A}(\boldsymbol{\varepsilon}(\mathbf{v}), \boldsymbol{\varepsilon}(\boldsymbol{\xi}))\,\mathrm{d}x &= 2^n \fint_{2\mathcal{B}} \mathcal{A}(\boldsymbol{\varepsilon}(\mathbf{v}), \boldsymbol{\varepsilon}(\boldsymbol{\xi}_\lambda))\,\mathrm{d}x + 2^n \fint_{2\mathcal{B}} \mathcal{A}(\boldsymbol{\varepsilon}(\mathbf{v}), \boldsymbol{\varepsilon}(\boldsymbol{\xi} - \boldsymbol{\xi}_\lambda))\,\mathrm{d}x \\ &=: I + II.\end{aligned}$$

Using Young's inequality and (3.3.23) we estimate

$$\begin{aligned} II &= 2^n \fint_{2\mathcal{B}} \mathcal{A}(\boldsymbol{\varepsilon}(\mathbf{v}), \boldsymbol{\varepsilon}(\boldsymbol{\xi} - \boldsymbol{\xi}_\lambda))\chi_{\{\boldsymbol{\xi} \neq \boldsymbol{\xi}_\lambda\}}\,\mathrm{d}x \\ &\le c \fint_{2\mathcal{B}} |\boldsymbol{\varepsilon}(\mathbf{v})|^q \chi_{\{\boldsymbol{\xi} \neq \boldsymbol{\xi}_\lambda\}}\,\mathrm{d}x + \frac{1}{2}\fint_{2\mathcal{B}} |\nabla \boldsymbol{\xi}|^{q'}\,\mathrm{d}x =: II_1 + II_2,\end{aligned}$$

where c depends on $|\mathcal{A}|$, p and p'. Then, using Hölder's inequality we obtain

$$II_1 \le c\left(\fint_{2\mathcal{B}} |\nabla \mathbf{v}|^{qs}\,\mathrm{d}x\right)^{\frac{1}{s}} \left(\frac{\mathcal{L}^d\left(\{\boldsymbol{\xi}_\lambda \neq \boldsymbol{\xi}\}\right)}{|\mathcal{B}|}\right)^{1-\frac{1}{s}}.$$

If follows from (3.3.22), by the choice of γ in (3.3.20) and $\lambda \ge \gamma$ that

$$\frac{\mathcal{L}^d\left(\{\boldsymbol{\xi}_\lambda \neq \boldsymbol{\xi}\}\right)}{|\mathcal{B}|} \le \frac{c\gamma^{q'}}{m_0\lambda^{q'}} \le \frac{c}{m_0}.$$

Thus

$$II_1 \le c\left(\fint_{2\mathcal{B}} |\nabla \mathbf{v}|^{qs}\,\mathrm{d}x\right)^{\frac{1}{s}} \left(\frac{c}{m_0}\right)^{1-\frac{1}{s}}.$$

We choose m_0 so large that

$$II_1 \le \frac{\kappa}{2}\left(\fint_{2\mathcal{B}} |\nabla \mathbf{v}|^{qs}\,\mathrm{d}x\right)^{\frac{1}{s}}.$$

Since $\mathbf{v}$ is almost $\mathcal{A}$-harmonic and $\|\nabla\boldsymbol{\xi}_\lambda\|_\infty \le c\lambda \le c2^{m_0}\gamma$ we have

$$|I| \le \delta \fint_{2\tilde{\mathcal{B}}} |\nabla\mathbf{v}|\,\mathrm{d}x\,\|\nabla\boldsymbol{\xi}_\lambda\|_\infty \le \delta \fint_{2\tilde{\mathcal{B}}} |\nabla\mathbf{v}|\,\mathrm{d}x\, c\,2^{m_0}\gamma.$$

We apply Young's inequality and Jensen's inequality to get

$$\begin{aligned} |I| &\le \delta 2^{m_0} c\left(\fint_{2\tilde{\mathcal{B}}} |\nabla\mathbf{v}|^q\,dx + \gamma^{q'}\right) \\ &\le \delta 2^{m_0} c\left(\fint_{2\tilde{\mathcal{B}}} |\nabla\mathbf{v}|^{qs}\,\mathrm{d}x\right)^{\frac{1}{s}} + \delta 2^{m_0} c \fint_{\mathcal{B}} |\nabla\boldsymbol{\xi}|^{q'}\,\mathrm{d}x. \end{aligned}$$

Now, we choose $\delta > 0$ so small that $\delta 2^{m_0} c \le \varepsilon/2$. Thus

$$|I| \le \frac{\kappa}{2}\left(\fint_{2\tilde{\mathcal{B}}} |\nabla\mathbf{v}|^{qs}\,\mathrm{d}x\right)^{\frac{1}{s}} + \frac{1}{2}\fint_{\mathcal{B}} |\nabla\boldsymbol{\xi}|^{q'}\,\mathrm{d}x.$$

Combining the estimates for I, II and II_1 we get

$$\fint_{2\mathcal{B}} \mathcal{A}(\boldsymbol{\varepsilon}(\mathbf{v}), \boldsymbol{\varepsilon}(\boldsymbol{\xi}))\,\mathrm{d}x \le \kappa\left(\fint_{2\tilde{\mathcal{B}}} |\nabla\mathbf{v}|^{qs}\,\mathrm{d}x\right)^{\frac{1}{s}} + \fint_{\mathcal{B}} |\nabla\boldsymbol{\xi}|^{q'}\,\mathrm{d}x.$$

The claim follows by inserting this in (3.3.19). □

CHAPTER 4

Prandtl–Eyring fluids

Contents

Abstract

We study the stationary flow of Prandtl–Eyring fluids in two dimensions. Based on the solenoidal Lipschitz truncation from chapter 3 we show the existence of weak solutions to the equations of motion. The proof benefits from the improved smallness estimate for the level-sets of the Lipschitz truncation and Korn's inequality from section 2.3. Our approach is completely pressure-free. However, we can recover the pressure based on the results from section 2.2.

The stationary flow of a homogeneous incompressible fluid in a bounded body $G \subset \mathbb{R}^d$ $(d = 2, 3)$ is described by the equations

$$\begin{cases} \operatorname{div} \mathbf{S}(\boldsymbol{\varepsilon}(\mathbf{v})) = \rho(\nabla \mathbf{v})\mathbf{v} + \nabla \pi - \rho \mathbf{f} & \text{in } G, \\ \operatorname{div} \mathbf{v} = 0 & \text{in } G, \\ \mathbf{v} = 0 & \text{on } \partial G. \end{cases} \tag{4.0.1}$$

The constitutive law of a Prandtl–Eyring fluid is

$$\mathbf{S} = \eta_0 \frac{\operatorname{arsinh}(\lambda|\boldsymbol{\varepsilon}(\mathbf{v})|)}{\lambda|\boldsymbol{\varepsilon}(\mathbf{v})|}\boldsymbol{\varepsilon}(\mathbf{v}) \tag{4.0.2}$$

with physical constants $\eta_0, \lambda > 0$. Eyring [69] obtained this relation from a molecular theory and similar relations were given by Prandtl (see [41] for an overview). It follows from equation (4.0.2) that the viscosity $\nu : G \to \mathbb{R}$ of the fluid can be described by the function

$$\nu = \nu_0 \frac{\operatorname{arsinh}(\lambda|\boldsymbol{\varepsilon}(\mathbf{v})|)}{\lambda|\boldsymbol{\varepsilon}(\mathbf{v})|}. \tag{4.0.3}$$

Equation (4.0.3) shows that the fluid is very shear thinning. Such behaviour can be observed, for example, in the motion of lubricants. Furthermore one can use the model as an approximation for perfectly plastic fluids, for which the constitutive law reads as

$$\boldsymbol{\varepsilon}(\mathbf{v}) = 0 \quad \Rightarrow \quad |\mathbf{S}| \le g, \qquad \boldsymbol{\varepsilon}(\mathbf{v}) \neq 0 \quad \Rightarrow \quad \mathbf{S} = \frac{g}{|\boldsymbol{\varepsilon}(\mathbf{v})|}\boldsymbol{\varepsilon}(\mathbf{v}), \tag{4.0.4}$$

Existence Theory for Generalized Newtonian Fluids.
DOI: http://dx.doi.org/10.1016/B978-0-12-811044-7.00005-7

see [114]. Incompressible flows whose constitutive law is given by (4.0.4) are called perfectly plastic fluids (or von Mises solids) with yield value $g > 0$. These media have been further studied by Prager [121–123]. Similar approximations are used in the study of plastic material behaviour (see [81] and [76] for a mathematical approach).

Letting

$$W(\boldsymbol{\varepsilon}) := \eta_0 \int_0^{|\boldsymbol{\varepsilon}|} \frac{1}{\lambda} \operatorname{arsinh}(\lambda t)\, \mathrm{d}t \tag{4.0.5}$$

for $\boldsymbol{\varepsilon} \in \mathbb{R}^{d\times d}_{\mathrm{sym}}$ we can replace (4.0.2) by the equation

$$\mathbf{S} = DW(\boldsymbol{\varepsilon}(\mathbf{v})). \tag{4.0.6}$$

If the flow is slow we can neglect the convective term $(\nabla\mathbf{v})\mathbf{v}$. Under this additional assumption it is shown in [76] inspired by ideas of Frehse and Seregin [81], how to reduce (4.0.1), (4.0.6) to a variational problem. A weak solution $\mathbf{v}$ can be obtained in the natural function space

$$E^h_{0,\mathrm{div}}(G) := \left\{ \mathbf{w} \in L^1(G) : \int_G h(|\boldsymbol{\varepsilon}(\mathbf{w})|)\, \mathrm{d}x < \infty,\ \operatorname{div}\mathbf{w} = 0,\ \mathbf{w}|_{\partial G} = 0 \right\},$$
$$h(t) := t\ln(1+t),\ t \geq 0,$$

see Appendix A.2 for a precise definition. The weak solution is a smooth function if $d = 2$ and partially of class C^1 in the three-dimensional case. Note that we can replace the energy W from (4.0.5) by the more convenient expression

$$W(\boldsymbol{\varepsilon}) = h(|\boldsymbol{\varepsilon}|). \tag{4.0.7}$$

All mathematical arguments actually work for potentials of the form $g(|\boldsymbol{\varepsilon}|)$, where g is C^2-close to the function h.

For the natural case $(\nabla\mathbf{v})\mathbf{v} \neq 0$ it is not immediate how to find a solution to (4.0.1), (4.0.6) with W defined in (4.0.5) or (4.0.7). In order to obtain an idea of how to proceed, let us have a look at the weak formulation (for simplicity we set $\rho = 1$)

$$\int_G DW(\boldsymbol{\varepsilon}(\mathbf{v})) : \boldsymbol{\varepsilon}(\boldsymbol{\varphi})\, \mathrm{d}x = \int_G \mathbf{f}\cdot\boldsymbol{\varphi}\, \mathrm{d}x + \int_G \mathbf{v}\otimes\mathbf{v} : \boldsymbol{\varepsilon}(\boldsymbol{\varphi})\, \mathrm{d}x \tag{4.0.8}$$

for $\boldsymbol{\varphi} \in C^\infty_{0,\mathrm{div}}(G)$. In accordance with the results of Appendix A.2 we see that in the case $d = 2$ all terms in (4.0.8) are well-defined, provided we choose $\mathbf{v}$ from the space $E^h_{0,\mathrm{div}}(G)$ and require

$$\mathbf{f} \in L^2(G). \tag{4.0.9}$$

The main result of this chapter states that such a weak solution actually exists.

Theorem 4.0.20. *Suppose that $G \subset \mathbb{R}^2$ is a bounded Lipschitz domain and consider volume forces* $\mathbf{f}$ *satisfying* (4.0.9). *Moreover, let W be defined according to* (4.0.5) *or* (4.0.7). *Then there exists a velocity field* $\mathbf{v} \in E^h_{0,\mathrm{div}}(G)$ *satisfying* (4.0.8) *for all fields* $\boldsymbol{\varphi} \in C^\infty_{0,\mathrm{div}}(G)$.

Corollary 4.0.1. *Under the assumptions of Theorem 4.0.20 there exists a pressure* $\pi \in L^h_\perp(G)$ *such that*

$$\begin{aligned}\int_G DW(\boldsymbol{\varepsilon}(\mathbf{v})) : \boldsymbol{\varepsilon}(\boldsymbol{\varphi})\,\mathrm{d}x \\ = \int_G \pi \operatorname{div} \boldsymbol{\varphi}\,\mathrm{d}x + \int_G \mathbf{f} \cdot \boldsymbol{\varphi}\,\mathrm{d}x + \int_G \mathbf{v} \otimes \mathbf{v} : \boldsymbol{\varepsilon}(\boldsymbol{\varphi})\,\mathrm{d}x \qquad (4.0.10)\end{aligned}$$

holds for all fields $\boldsymbol{\varphi} \in C^\infty_0(G)$.

We will prove Theorem 4.0.20 by approximation, i.e., by replacing (4.0.8) with a sequence of more regular problems with corresponding solutions $\mathbf{v}_n$. It turns out that the sequence $(\mathbf{v}_n)$ is bounded in the space $E^h_{0,\mathrm{div}}(G)$, and in Theorem A.2.47 we will investigate spaces like $E^h_{0,\mathrm{div}}(G)$ with the result that $E^h_{0,\mathrm{div}}(G)$ is compactly embedded in the space $L^2(G)$. Hence it holds (for a subsequence)

$$\int_G \mathbf{v}_n \otimes \mathbf{v}_n : \boldsymbol{\varepsilon}(\boldsymbol{\varphi})\,\mathrm{d}x \longrightarrow \int_G \mathbf{v} \otimes \mathbf{v} : \boldsymbol{\varepsilon}(\boldsymbol{\varphi})\,\mathrm{d}x, \quad n \to \infty,$$

with a suitable limit function $\mathbf{v}$, which turns out to be in the class $E^h_{0,\mathrm{div}}(G)$. The main task is the proof of

$$\int_G DW(\boldsymbol{\varepsilon}(\mathbf{v}_n)) : \boldsymbol{\varepsilon}(\boldsymbol{\varphi})\,\mathrm{d}x \longrightarrow \int_G DW(\boldsymbol{\varepsilon}(\mathbf{v})) : \boldsymbol{\varepsilon}(\boldsymbol{\varphi})\,\mathrm{d}x, \quad n \to \infty.$$

A main tool in our approach is the Lipschitz truncation method by Acerbi and Fusco [3]. It was firstly used in the context of fluid mechanics in [79] and advanced in [62] (see also Section 1.3 for further background and references). The two latter papers deal with the situation of a power law fluid, i.e.

$$W(\boldsymbol{\varepsilon}) \approx |\boldsymbol{\varepsilon}|^p$$

for a power $p > 1$ (but in 2D arbitrarily close to 1). This situation is much better than our case. The spaces L^p (for $p > 1$) feature a nicer behaviour than the space L^h, which is the natural space (for the symmetric gradient) in our setting. Due to the lack of a Korn-type inequality in L^h (see Theorem 2.3.14), we are not able to bound $M(\nabla \mathbf{v}_n)$ in L^1. This means that an ordinary Lipschitz-truncation is not possible. The main idea to overcome this difficulty is instead of approximating $\mathbf{v}_n$ by a sequence of Lipschitz continuous functions, to use functions only having a bounded symmetric gradient (instead of a bounded gradient), cf. Section 3.2.

In equation (4.0.8) only solenoidal test functions are admissible. Since the Lipschitz truncation is based on a nonlinear extension operator it does not preserve the incompressibility condition of the solution. In the power law fluid situation there are two ways to overcome this difficulty:

- Introducing the pressure function π which belongs to L^s for some $s > 1$;
- Correcting the divergence by means of the Bogovskiĭ operator.

In case of p-fluids, both methods are applicable but they fail for Prandtl–Eyring fluids. Neither the pressure belongs to the correct space (see Theorem 2.2.10) nor is the Bogovskiĭ operator continuous (see Corollary 2.3.3). This strongly motivates the construction of a solenoidal Lipschitz-truncation which was done in Sections 3.1 for the uniformly convex case in dimensions three. In Section 3.2 the approach is extended to our setting by using the continuity of $\nabla\,\mathrm{curl}^{-1}$ in L^1 which holds in the case $d = 2$. The situation in higher dimensions is completely different. For a much more complicated construction which also works for $d \geq 3$ we refer to [33]. The approach there is based on local projections by the Bogovskiĭ operator.

In connection with Theorem 4.0.20 we mention three problems.

i) What are the smoothness properties of the specific weak solution $\mathbf{v}$ constructed in the proof of Theorem 4.0.20?

ii) The logarithmic potential $|\boldsymbol{\varepsilon}(\mathbf{v})|\ln(1 + |\boldsymbol{\varepsilon}(\mathbf{v})|)$ serves as an approximation for perfectly plastic fluids with potential $|\boldsymbol{\varepsilon}(\mathbf{v})|$. Is it possible to handle the linear-growth case with similar arguments?

iii) Can we obtain similar results for non-stationary Prandtl–Eyring fluids?

In the paper [65] a parabolic version of the Lipschitz-truncation was developed in order to consider unsteady flows of power-law fluids. This method is improved in section 6.1 and a solenoidal Lipschitz truncation for parabolic PDEs is constructed. However it is still based on singular integrals and does not seem to extend to the setting of Prandtl–Eyring fluids.

4.1 THE APPROXIMATED SYSTEM

Throughout this section we assume that $\mathbf{S} \in C^0(\mathbb{R}^{d\times d}_{sym}) \cap C^1(\mathbb{R}^{d\times d}_{sym} \setminus \{0\})$ and for some $\kappa \geq 0$

$$\lambda\big(\kappa + |\boldsymbol{\varepsilon}|\big)^{q-2}|\boldsymbol{\sigma}|^2 \leq D\mathbf{S}(\boldsymbol{\varepsilon})(\boldsymbol{\sigma},\boldsymbol{\sigma}) \leq \Lambda\big(\kappa + |\boldsymbol{\varepsilon}|\big)^{q-2}|\boldsymbol{\sigma}|^2 \tag{4.1.11}$$

for all $\boldsymbol{\varepsilon}, \boldsymbol{\sigma} \in \mathbb{R}^{d\times d}_{sym} \setminus \{0\}$ with some positive constants λ, Λ.

Theorem 4.1.21. *Let (4.1.11) hold with $q > \frac{3d}{d+2}$ and assume $\mathbf{f} \in L^{q'}(G)$. Then there is a solution $\mathbf{v} \in W^{1,q}_{0,\mathrm{div}}(G)$ to*

$$\int_G \mathbf{S}(\boldsymbol{\varepsilon}(\mathbf{v})) : \boldsymbol{\varepsilon}(\boldsymbol{\varphi})\,\mathrm{d}x = \int_G \mathbf{f}\cdot\boldsymbol{\varphi}\,\mathrm{d}x + \int_G \mathbf{v}\otimes\mathbf{v} : \boldsymbol{\varepsilon}(\boldsymbol{\varphi})\,\mathrm{d}x \tag{4.1.12}$$

for all $\boldsymbol{\varphi} \in C^\infty_{0,\mathrm{div}}(G)$.

Proof. We want to apply Brezis' Theorem (see [143]), which generalizes the Theorem of Browder and Minty, to the operator equation

$$\big(\mathcal{A}_1 + \mathcal{A}_2\big)\mathbf{u} = b \quad \text{in } W^{-1,q'}_{\mathrm{div}}(G),$$

where, for $\boldsymbol{\varphi} \in W^{1,q}_{0,\mathrm{div}}(G)$, we abbreviated

$$\mathcal{A}_1\mathbf{u}(\boldsymbol{\varphi}) := \int_G \mathbf{S}(\boldsymbol{\varepsilon}(\mathbf{u})) : \boldsymbol{\varepsilon}(\boldsymbol{\varphi})\,\mathrm{d}x,$$
$$\mathcal{A}_2\mathbf{u}(\boldsymbol{\varphi}) := \int_G \mathbf{u}\otimes\mathbf{u} : \boldsymbol{\varepsilon}(\boldsymbol{\varphi})\,\mathrm{d}x,$$
$$b(\boldsymbol{\varphi}) := \int_G \mathbf{f}\cdot\boldsymbol{\varphi}\,\mathrm{d}x.$$

It holds:

- On account of the assumption on $\mathbf{S}$ in (4.1.11) we have that $\mathcal{A}_1 : W^{1,q}_{0,\mathrm{div}}(G) \to W^{-1,q'}_{\mathrm{div}}(G)$ is bounded, strictly monotone, coercive and continuous (we use Korn's inequality).
- For $q > \frac{3d}{d+2}$ we have that $\mathcal{A}_2 : W^{1,q}_{0,\mathrm{div}}(G) \to W^{-1,q'}_{\mathrm{div}}(G)$ is bounded and strongly continuous.
- Since $\mathbf{f} \in L^{q'}(G)$ the operator b belongs to $W^{-1,q'}_{\mathrm{div}}(G)$.

Finally we use $\mathcal{A}_2\mathbf{u}(\mathbf{u}) = 0$ for all $\mathbf{u} \in W^{1,q}_{0,\mathrm{div}}(G)$ such that $\mathcal{A}_1 + \mathcal{A}_2$ is coercive. Brezis' Theorem gives a solution $\mathbf{v} \in W^{1,q}_{0,\mathrm{div}}(G)$ to $\big(\mathcal{A}_1 + \mathcal{A}_2\big)\mathbf{v} = b$ which is equivalent to (4.1.12). □

If we apply De Rahm's Theorem to the functional $\big(\mathcal{A}_1 + \mathcal{A}_2\big)\mathbf{v} - b$ we can recover the pressure (see Theorem 2.2.10 for a general statement in this setting).

Corollary 4.1.1. *Under the assumptions of Theorem 4.1.21 there is a function* $\pi \in L_0^{q'}(G)$ *such that*

$$\int_G \mathbf{S}(\boldsymbol{\varepsilon}(\mathbf{v})) : \boldsymbol{\varepsilon}(\boldsymbol{\varphi})\,\mathrm{d}x = \int_G \mathbf{f}\cdot\boldsymbol{\varphi}\,\mathrm{d}x + \int_G \mathbf{v}\otimes\mathbf{v} : \boldsymbol{\varepsilon}(\boldsymbol{\varphi})\,\mathrm{d}x + \int_G \pi\,\operatorname{div}\boldsymbol{\varphi}\,\mathrm{d}x$$

for all $\boldsymbol{\varphi} \in C_0^\infty(G)$.

4.2 STATIONARY FLOWS

In this section we prove Theorem 4.0.20. In particular, we show the existence of a weak solution $\mathbf{v} \in E^h_{0,\mathrm{div}}(G)$ to the equation

$$\int_G DW(\boldsymbol{\varepsilon}(\mathbf{v})) : \boldsymbol{\varepsilon}(\boldsymbol{\varphi})\,\mathrm{d}x = \int_G \mathbf{f}\cdot\varphi\,\mathrm{d}x + \int_G \mathbf{v}\otimes\mathbf{v} : \boldsymbol{\varepsilon}(\boldsymbol{\varphi})\,\mathrm{d}x \tag{4.2.13}$$

for all $\boldsymbol{\varphi} \in C^\infty_{0,\mathrm{div}}(G)$, where $G \subset \mathbb{R}^2$ is a bounded Lipschitz domain. We start by approximating this equation. We consider solutions $\mathbf{v}_n \in W^{1,2}_{0,\mathrm{div}}(G)$ of the system

$$\int_G \big(DW(\boldsymbol{\varepsilon}(\mathbf{v})) + n^{-1}\boldsymbol{\varepsilon}(\mathbf{v})\big) : \boldsymbol{\varepsilon}(\boldsymbol{\varphi})\,\mathrm{d}x = \int_G \mathbf{f}\cdot\varphi\,\mathrm{d}x + \int_G \mathbf{v}\otimes\mathbf{v} : \boldsymbol{\varepsilon}(\boldsymbol{\varphi})\,\mathrm{d}x. \tag{4.2.14}$$

The existence of solutions to this system can easily be verified due to the quadratic growth of the main part by means of monotone operators (see Theorem 4.1.21). An important advantage of this approximation consists in the fact that the space of test functions coincides with the space where the solution is constructed.

All $\mathbf{v}_n$ satisfy the uniform estimate

$$\int_G h(|\boldsymbol{\varepsilon}(\mathbf{v}_n)|)\,\mathrm{d}x + n^{-1}\int_G |\boldsymbol{\varepsilon}(\mathbf{v}_n)|^2\,\mathrm{d}x \le c,$$

which follows from testing (4.2.14) by $\mathbf{v}_n$. Consequently, we get

$$\|\boldsymbol{\varepsilon}(\mathbf{v}_n)\|_h \le c,$$
$$\|n^{-1/2}\boldsymbol{\varepsilon}(\mathbf{v}_n)\|_2 \le c.$$

This estimate and Theorem A.2.47 imply the existence of $\mathbf{v} \in E^h_{0,\mathrm{div}}(G)$, and a (not relabelled) subsequence of $(\mathbf{v}_n)$ such that

$$\begin{aligned}
\mathbf{v}_n &\to \mathbf{v} && \text{in } L^2(G),\\
\boldsymbol{\varepsilon}(\mathbf{v}_n) &\rightharpoonup \boldsymbol{\varepsilon}(\mathbf{v}) && \text{in } L^1(G),\\
n^{-1}\boldsymbol{\varepsilon}(\mathbf{v}_n) &\to 0 && \text{in } L^2(G).
\end{aligned}$$

It follows from these convergences that

$$\frac{1}{n}(\boldsymbol{\varepsilon}(\mathbf{v}_n), \boldsymbol{\varepsilon}(\boldsymbol{\varphi})) \to 0 \qquad \text{and}$$
$$(\mathbf{v}_n \otimes \mathbf{v}_n, \boldsymbol{\varepsilon}(\boldsymbol{\varphi})) \to (\mathbf{v} \otimes \mathbf{v}, \boldsymbol{\varepsilon}(\boldsymbol{\varphi})) \qquad \text{for all} \quad \boldsymbol{\varphi} \in C^\infty_{0,\mathrm{div}}(G).$$

Clearly these statements extend to $\boldsymbol{\varphi} \in E^\infty_{0,\mathrm{div}}(G)$.

Next, to prove that also

$$(DW(\boldsymbol{\varepsilon}(\mathbf{v}_n)), \boldsymbol{\varepsilon}(\boldsymbol{\varphi})) \to (DW(\boldsymbol{\varepsilon}(\mathbf{v})), \boldsymbol{\varepsilon}(\boldsymbol{\varphi})) \quad \text{for all} \quad \boldsymbol{\varphi} \in C^\infty_{0,\mathrm{div}}(G) \tag{4.2.15}$$

it suffices, by virtue of $\|\boldsymbol{\varepsilon}(\mathbf{v}_n)\|_h \le c$ and Vitali's theorem, to show at least for a subsequence that $\boldsymbol{\varepsilon}(\mathbf{v}^n) \to \boldsymbol{\varepsilon}(\mathbf{v})$ almost everywhere. This follows, see for example [54] for details, from the strict monotonicity of the operator DW provided that for a certain $\theta \in (0,1]$ and every ball $\mathcal{B} \subset G$ with $4\mathcal{B} \subset G$

$$\limsup_n \int_{\mathcal{B}} \Big((DW(\boldsymbol{\varepsilon}(\mathbf{v}_n)) - DW(\boldsymbol{\varepsilon}(\mathbf{v}))) : (\boldsymbol{\varepsilon}(\mathbf{v}_n) - \boldsymbol{\varepsilon}(\mathbf{v})) \Big)^\theta \,\mathrm{d}x = 0. \tag{4.2.16}$$

To verify equation (4.2.16), let $\eta \in C^\infty_0(2\mathcal{B})$ with $\chi_{\mathcal{B}} \le \eta \le \chi_{2\mathcal{B}}$ and $|\nabla \eta| \le cR^{-1}$, where R is the radius of $\mathcal{B}$. We define

$$\mathbf{u}_n := \eta(\mathbf{v}_n - \mathbf{v}) - \mathrm{Bog}_{2\mathcal{B}}(\nabla\eta \cdot (\mathbf{v}_n - \mathbf{v})),$$

where $\mathrm{Bog}_{2\mathcal{B}}$ is the Bogovskiĭ operator on $2\mathcal{B}$ from $L^2_\perp(2\mathcal{B})$ to $W^{1,2}_0(2\mathcal{B})$. Since $\nabla\eta \cdot (\mathbf{v}_n - \mathbf{v})$ is bounded in $L^2_\perp(2\mathcal{B})$, we have that $\mathbf{u}_n$ is bounded in $E^h_{0,\mathrm{div}}(2\mathcal{B})$. Moreover, $\mathbf{v}_n \to \mathbf{v}$ in L^2 and the continuity of Bog implies $\mathbf{u}_n \to 0$ in L^1. In particular, we can apply our solenoidal Lipschitz truncation of Theorem 3.2.18 to gain a suitable double sequence $\mathbf{u}_{n,j} \in E^\infty_{0,\mathrm{div}}(4\mathcal{B})$.

The weak formulation of the approximate problem (4.2.14) with $\mathbf{u}_{n,j}$ as a test function can be rewritten as

$$\begin{aligned}(DW(\boldsymbol{\varepsilon}(\mathbf{v}_n)) - DW(\boldsymbol{\varepsilon}(\mathbf{v})), \boldsymbol{\varepsilon}(\mathbf{u}_{n,j})) = &-(DW(\boldsymbol{\varepsilon}(\mathbf{v})), \boldsymbol{\varepsilon}(\mathbf{u}_{n,j})) \\ &- \frac{1}{n}(\boldsymbol{\varepsilon}(\mathbf{v}_n), \boldsymbol{\varepsilon}(\mathbf{u}_{n,j})) + (\mathbf{f}, \mathbf{u}_{n,j}) \\ &+ (\mathbf{v}^n \otimes \mathbf{v}_n, \boldsymbol{\varepsilon}(\mathbf{u}_{n,j})).\end{aligned}$$

It follows from the properties of $\mathbf{u}_{n,j}$ and $\mathbf{v}_n$ that the right-hand side converges for fixed j to zero as $n \to \infty$. So we obtain

$$\lim_{n\to\infty} (DW(\boldsymbol{\varepsilon}(\mathbf{v}_n)) - DW(\boldsymbol{\varepsilon}(\mathbf{v})), \boldsymbol{\varepsilon}(\mathbf{u}_{n,j})) = 0. \tag{4.2.17}$$

We decompose the set $4\mathcal{B}$ into $\{\mathbf{u} \ne \mathbf{u}_{n,j}\}$ and $4\mathcal{B} \cap \{\mathbf{u} = \mathbf{u}_{n,j}\}$ to get

$$(I) := \limsup_n \left| \int_{4\mathcal{B}\cap\{\mathbf{w}=\mathbf{w}_{n,j}\}} \eta \left(DW(\boldsymbol{\varepsilon}(\mathbf{v}_n))\right) - DW(\boldsymbol{\varepsilon}(\mathbf{v})\right) : (\boldsymbol{\varepsilon}(\mathbf{v}_n) - \boldsymbol{\varepsilon}(\mathbf{v}))\,\mathrm{d}x \right|$$

$$= \limsup_n \left| \int_{\{\mathbf{w} \neq \mathbf{w}_{n,j}\}} \big(DW(\boldsymbol{\varepsilon}(\mathbf{v}_n))\big) - DW(\boldsymbol{\varepsilon}(\mathbf{v})\big) : \boldsymbol{\varepsilon}(\mathbf{u}_{n,j})\, \mathrm{d}x \right|$$

$$+ \limsup_n \left| \int_{4\mathcal{B} \cap \{\mathbf{w} = \mathbf{w}_{n,j}\}} \big(DW(\boldsymbol{\varepsilon}(\mathbf{v}_n))\big) - DW(\boldsymbol{\varepsilon}(\mathbf{v})\big) : \big(\nabla \eta \odot (\mathbf{v}_n - \mathbf{v})\big)\, \mathrm{d}x \right|$$

$$+ \limsup_n \left| \int_{4\mathcal{B} \cap \{\mathbf{u} = \mathbf{u}_{n,j}\}} \big(DW(\boldsymbol{\varepsilon}(\mathbf{v}_n))\big) - DW(\boldsymbol{\varepsilon}(\mathbf{v})\big) : \boldsymbol{\varepsilon}\big(\mathrm{Bog}_{2\mathcal{B}}(\nabla \eta \cdot (\mathbf{v}_n - \mathbf{v}))\big)\, \mathrm{d}x \right|$$

$$=: (II) + (III) + (IV).$$

Since $\nabla \eta \otimes (\mathbf{v}_n - \mathbf{v}) \overset{n}{\to} 0$ in L^2, we have $(III) + (IV) \overset{n}{\to} 0$, where we also used the continuity of $\mathrm{Bog}_{2\mathcal{B}}$ from $L^2_\perp(2\mathcal{B})$ to $W^{1,2}_0(2\mathcal{B})$.

By Young's inequality

$$(II) \leq \limsup_n \big(\| DW(\boldsymbol{\varepsilon}(\mathbf{v}_n)) \|_{h^*} + \| DW(\boldsymbol{\varepsilon}(\mathbf{v})) \|_{h^*} \big) \, \| \chi_{\{\mathbf{u}_n \neq \mathbf{u}_{n,j}\}} \boldsymbol{\varepsilon}(\mathbf{u}_{n,j}) \|_h,$$

where h^* is the conjugate N-function of h. Since

$$h^*(|DW(\boldsymbol{\varepsilon})|) \leq h^*(h'(|\boldsymbol{\varepsilon}|)) \leq h(2|\boldsymbol{\varepsilon}|) \leq c\, h(|\boldsymbol{\varepsilon}|),$$

we deduce from the uniform boundedness of $\mathbf{u}^n$ and $\mathbf{u}$ in $E^h_0(G)$ that $DW(\boldsymbol{\varepsilon}(\mathbf{u}_n))$ and $DW(\boldsymbol{\varepsilon}(\mathbf{u}))$ are uniformly bounded in $L^{h^*}(G)$. On the other hand by Theorem 3.2.18

$$\| \chi_{\{\mathbf{u}_n \neq \mathbf{u}_{n,j}\}} \boldsymbol{\varepsilon}(\mathbf{u}_{n,j}) \|_h \leq c \, \| \chi_{\{\mathbf{u}_n \neq \mathbf{u}_{n,j}\}} \lambda \|_h \leq c \widetilde{\kappa}_j$$

for a null sequence $\widetilde{\kappa}_j$. This proves $(II) \leq c\widetilde{\kappa}_j$. Overall we get

$$\limsup_n \left| \int_{4\mathcal{B} \cap \{\mathbf{u} = \mathbf{u}_{n,j}\}} \eta \big(DW(\boldsymbol{\varepsilon}(\mathbf{v}_n))\big) - DW(\boldsymbol{\varepsilon}(\mathbf{v})\big) : (\boldsymbol{\varepsilon}(\mathbf{v}_n) - \boldsymbol{\varepsilon}(\mathbf{v}))\, \mathrm{d}x \right| \leq c \widetilde{\kappa}_j. \tag{4.2.18}$$

Let $\theta \in (0,1)$. We claim that the previous estimate implies

$$\limsup_n \int_{4\mathcal{B}} \Big(\eta \big(DW(\boldsymbol{\varepsilon}(\mathbf{v}_n))\big) - DW(\boldsymbol{\varepsilon}(\mathbf{v})\big) : (\boldsymbol{\varepsilon}(\mathbf{v}_n) - \boldsymbol{\varepsilon}(\mathbf{v}))\Big)^\theta \, \mathrm{d}x = 0. \tag{4.2.19}$$

Let z^n denote the integrand of the integral in (4.2.18). Then

$$\limsup_n \left| \int_{4\mathcal{B} \cap \{\mathbf{u} = \mathbf{u}_{n,j}\}} z_n \, dx \right| \leq c \widetilde{\kappa}_j. \tag{4.2.20}$$

Hölder's inequality implies

$$\int_{4\mathcal{B}} (z_n)^\theta \, \mathrm{d}x \leq \bigg(\int_{4\mathcal{B} \cap \{\mathbf{u} = \mathbf{u}_{n,j}\}} z^n \, \mathrm{d}x \bigg)^\theta |4\mathcal{B}|^{1-\theta} + \bigg(\int_{\{\mathbf{u} \neq \mathbf{u}_{n,j}\}} z_n \, \mathrm{d}x \bigg)^\theta |\{\mathbf{u} \neq \mathbf{u}_{n,j}\}|^{1-\theta}.$$

From $\limsup_{n\to\infty} \rho_h(\lambda_{n,j}\chi_{\{\mathbf{u}_{n,j}\neq\mathbf{u}_n\}}) \le \kappa_j$ we deduce $|\{\mathbf{u}\neq\mathbf{u}_{n,j}\}| \le \kappa_j 2^{-2^j} \le \kappa_j$. Overall, using (4.2.20) and passing to the limit $j\to\infty$ we obtain

$$\limsup_n \int_{4\mathcal{B}} (z_n)^\theta \, dx = 0.$$

This proves (4.2.19). Now, (4.2.16) is a consequence of $\eta \ge \chi_B$, which in turn implies the almost every convergence of $\boldsymbol{\varepsilon}(\mathbf{v}_n)\to\boldsymbol{\varepsilon}(\mathbf{v})$. So we can pass to the limit in (4.2.15) as desired, which shows that $\mathbf{v}$ is a weak solution of (4.2.13). The proof for the existence of $\mathbf{v}$ is complete.

Proof of Corollary 4.0.1. It remains to reconstruct the pressure. Equation (4.2.13) can be written as

$$\int_G \mathbf{H} : \boldsymbol{\varepsilon}(\boldsymbol{\varphi})\,\mathrm{d}x = 0, \qquad \mathbf{H} := DW(\boldsymbol{\varepsilon}(\mathbf{v})) - \mathbf{F} - \mathbf{v}\otimes\mathbf{v}, \tag{4.2.21}$$

for all $\boldsymbol{\varphi}\in C^\infty_{0,\mathrm{div}}(G)$ (where $\mathbf{F} = \nabla(\Delta^{-1}\mathbf{f}) \in L^{p_0}(G)$). Since $\mathbf{v}\in E^{1,h}(G)$ we see that $\mathbf{v}\otimes\mathbf{v}$ is bounded in $L^{t\ln^2(t)}(G)$ as a consequence of Lemma A.2.4 and Lemma A.2.1. This means $\mathbf{H}\in L^{t\ln^2(t)}(G)$ and Theorem 2.2.10 implies the existence of a function $\pi\in L^h(G)$ such that

$$\langle DW(\boldsymbol{\varepsilon}(\mathbf{v})) - (\mathbf{v}\otimes\mathbf{v}), \boldsymbol{\varepsilon}(\boldsymbol{\varphi})\rangle - \langle \mathbf{f}, \boldsymbol{\varphi}\rangle = \langle \pi, \operatorname{div}\boldsymbol{\varphi}\rangle \qquad \text{for all } \boldsymbol{\varphi}\in C_0^\infty(G),$$

which proves (4.0.10). □

PART 2

Non-stationary problems

CHAPTER 5

Preliminaries

Contents

Abstract

In this chapter we present some preliminary material which will be needed in order to study non-stationary models for generalized Newtonian fluids. We begin with the functional analytic framework and define Bochner-spaces. After this we present the Lipschitz truncation method for non-stationary problems. Finally, we give a historical overview about the mathematical theory of weak solutions for non-stationary flows of power law fluids.

5.1 BOCHNER SPACES

In the study of parabolic PDEs it is very useful to work with Banach space-valued mappings (see [143]). Let $(\mathscr{V}, \|\cdot\|_{\mathscr{V}})$ be a Banach space. A mapping $u:[0,T]\rightarrow\mathscr{V}$ is called a step function iff for some $N\in\mathbb{N}$

$$u(t)=\sum_{k=1}^{N}\chi_{A_k}(t)x_k, \qquad t\in[0,T],$$

where $\cup_k A_k=[0,T]$, $A_k\cap A_j=\emptyset$ for $k\neq j$ and $x_k\in\mathscr{V}$ for $k=1,\ldots,N$. A function $u:(0,T)\rightarrow\mathscr{V}$ is called Bochner measurable iff there is a sequence (u_n) of step functions such that

$$u_n(t)\rightarrow u(t) \quad \text{in} \quad \mathscr{V} \tag{5.1.1}$$

for a.e. t. A function $u:(0,T)\rightarrow\mathscr{V}$ is called Bochner integrable iff there is a sequence (u_n) of step functions such that (5.1.1) holds and

$$\int_0^T \|u_n(t)-u(t)\|_{\mathscr{V}}\,\mathrm{d}t\rightarrow 0, \qquad n\rightarrow\infty. \tag{5.1.2}$$

The Bochner integral (as an element of $\mathscr{V}$) is defined as

$$\int_0^T u(t)\,\mathrm{d}t:=\lim_n\int_0^T u_n(t)\,\mathrm{d}t=\lim_n\sum_{k=1}^{n}\mathcal{L}^1(A_k^n)x_k^n.$$

Existence Theory for Generalized Newtonian Fluids.
DOI: http://dx.doi.org/10.1016/B978-0-12-811044-7.00007-0

We define for $T > 0$ and $p \in [1, \infty)$, the space $L^p(0, T; \mathscr{V})$ to be the set of all Bochner measurable functions $u : (0, T) \to \mathscr{V}$ such that

$$\|u\|_{L^p(0,T;\mathscr{V})} := \left(\int_0^T \|u(t)\|_{\mathscr{V}}^p \, \mathrm{d}t \right)^{\frac{1}{p}} < \infty.$$

The space $L^\infty(0, T; \mathscr{V})$ is the set of all Bochner measurable functions such that

$$\|u\|_{L^\infty(0,T;\mathscr{V})} := \inf_{\mathcal{L}^1(A)=0} \sup_{(0,T)\setminus A} \|u(t)\|_{\mathscr{V}} < \infty.$$

The space $L^p(0, T; \mathscr{V})$ for $p \in [1, \infty]$ is a Banach space together with the norm given above.

Lemma 5.1.1. *Let $\mathscr{V}$ be a separable and reflexive Banach space.*

a) *If $p \in (1, \infty)$ then $L^p(0, T; \mathscr{V})$ is reflexive and we have*

$$L^p(0, T; \mathscr{V})' \cong L^{p'}(0, T; \mathscr{V}').$$

b) *For $p = 1$ we still have*

$$L^1(0, T; \mathscr{V})' \cong L^\infty(0, T; \mathscr{V}').$$

Lemma 5.1.2. *Let $1 \le p < \infty$ and $\mathscr{V}$ be a Banach space. Let $\mathscr{C}$ be dense in $L^p(0, T; \mathbb{R})$ and $\mathscr{V}_0$ dense in $\mathscr{V}$. Then the set*

$$\operatorname{span}\left\{ g x_0;\ g \in \mathscr{C},\ x_0 \in \mathscr{V}_0 \right\}$$

is dense in $L^p(0, T; \mathscr{V})$.

For $u \in L^1(0, T; \mathscr{V})$ we consider the distribution

$$C_0^\infty(0, T) \ni \varphi \mapsto \int_0^T u(t)\varphi'(t) \, \mathrm{d}t \in \mathscr{V}.$$

Let $\mathscr{Y}$ be a Banach space with $\mathscr{V} \hookrightarrow \mathscr{Y}$ continuously. If there is $v \in L^1(0, T; \mathscr{Y})$ such that

$$\int_0^T u(t)\varphi'(t) \, \mathrm{d}t = -\int_0^T v(t)\varphi(t) \, \mathrm{d}t \quad \text{for all} \quad \varphi \in C_0^\infty(0, T)$$

we say that v is the weak derivative of u in $\mathscr{Y}$ and write $v = \partial_t u$. The space $W^{1,p}(0, T; \mathscr{V})$ consists of those functions from $L^p(0, T; \mathscr{V})$ having weak derivatives in $L^p(0, T; \mathscr{V})$. It is a Banach function space together with the norm

$$\|u\|_{W^{1,p}(0,T;\mathscr{V})}^p := \|u\|_{L^p(0,T;\mathscr{V})}^p + \|\partial_t u\|_{L^p(0,T;\mathscr{V})}^p.$$

Obviously this can be iterated to define the space $W^{k,p}(0, T; \mathscr{V})$.

In order to study the time regularity of functions from Bochner spaces we need to define different notations of continuity.

Definition 5.1.1. Let $(\mathcal{V}, \|\cdot\|_{\mathcal{V}})$ be a Banach space, $T > 0$ and $\alpha \in (0, 1]$.

a) $C([0, T]; \mathcal{V})$ denotes the set of functions $u : [0, T] \to \mathcal{V}$ being continuous with respect to the norm topology, i.e.

$$u(t_k) \to u(t_0) \quad \text{in} \quad \mathcal{V}$$

for any sequence $(t_k) \subset [0, T]$ with $t_k \to t_0$.

b) $C_w([0, T]; \mathcal{V})$ denotes the set of functions $u : [0, T] \to \mathcal{V}$ being continuous with respect to the weak topology, i.e.

$$u(t_k) \rightharpoonup u(t_0) \quad \text{in} \quad \mathcal{V}$$

for any sequence $(t_k) \subset [0, T]$ with $t_k \to t_0$.

c) $C^\alpha([0, T]; \mathcal{V})$ denotes the set of functions $u : [0, T] \to \mathcal{V}$ being α-Hölder-continuous with respect to the norm topology, i.e.

$$\sup_{t,s \in [0,T]; t \neq s} \frac{\|u(t) - u(s)\|_{\mathcal{V}}}{|t - s|^\alpha} < \infty.$$

Obviously, we have the inclusions

$$C^\alpha([0, T]; \mathcal{V}) \subset C([0, T]; \mathcal{V}) \subset C_w([0, T]; \mathcal{V})$$

for any $\alpha \in (0, 1]$. The following variant of Sobolev's Theorem holds.

Lemma 5.1.3. *Let X be a Banach space and $1 \leq p < \infty$. The embedding*

$$W^{1,p}(0, T; \mathcal{V}) \hookrightarrow C^{1-\frac{1}{p}}([0, T]; \mathcal{V})$$

is continuous.

The following theorem shows how to obtain compactness in Bochner spaces. The original version is due to Aubin and Lions (see [14] and [109]) but does not include the case $p = 1$. The following more general version can be found in [128].

Theorem 5.1.22. *Let $(\mathcal{V}, \mathcal{X}, \mathcal{Y})$ be a triple of separable and reflexive Banach spaces such that the embedding $\mathcal{V} \hookrightarrow \mathcal{X}$ is compact and the embedding $\mathcal{X} \hookrightarrow \mathcal{Y}$ is continuous. Then the embedding*

$$\{u \in L^p(0, T; \mathcal{V}) : \partial_t u \in L^p(0, T; \mathcal{Y})\} \hookrightarrow L^p(0, T; \mathcal{X})$$

is compact for $1 \leq p < \infty$.

In the context of stochastic PDEs we will be confronted with functions having only fractional derivatives in time. We define for $p \in (1, \infty)$ and $\alpha \in (0, 1)$ the norm

$$\|u\|^p_{W^{\alpha,p}(0,T;\mathcal{V})} := \|u\|^p_{L^p(0,T;\mathcal{V})} + \int_0^T \int_0^T \frac{\|u(\sigma_1) - u(\sigma_2))\|^p_{\mathcal{V}}}{|\sigma_1 - \sigma_2|^{1+\alpha p}} \, d\sigma_1 \, d\sigma_2.$$

The space $W^{\alpha,p}(0,T;\mathcal{V})$ is now defined as the subspace of $L^p(0,T;\mathcal{V})$ consisting of the functions having finite $W^{\alpha,p}(0,T;\mathcal{V})$-norm. It can be shown that this is a complete space and we have $W^{1,p}(0,T;\mathcal{V}) \subset W^{\alpha,p}(0,T;\mathcal{V}) \subset L^p(0,T;\mathcal{V})$. The following version of Theorem 5.1.22 holds (see [73]).

Theorem 5.1.23. *Let $(\mathcal{V},\mathcal{X},\mathcal{Y})$ be a triple of separable and reflexive Banach spaces such that the embedding $\mathcal{V} \hookrightarrow \mathcal{X}$ is compact and the embedding $\mathcal{X} \hookrightarrow \mathcal{Y}$ is continuous. Then the embedding*

$$L^p(0,T;\mathcal{V}) \cap W^{\alpha,p}(0,T;\mathcal{Y}) \hookrightarrow L^p(0,T;\mathcal{X})$$

is compact for $1<p<\infty$ and $0<\alpha<1$.

The following interpolation result is a special case of [8, Thm. 3.1]

Lemma 5.1.4. *Let $p \in [1,\infty)$, and $s_0, s_1, r_0, r_1 \in \mathbb{R}$ such that $s_0 < s_1$ and $r_0 > r_1$. Let $\theta \in (0,1)$ and define $s_\theta \in (s_0,s_1)$ and $r_\theta \in (r_1,r_0)$ by*

$$\frac{1}{s_\theta} = \frac{\theta}{s_0} + \frac{1-\theta}{s_1}, \qquad \frac{1}{r_\theta} = \frac{\theta}{r_0} + \frac{1-\theta}{r_1}.$$

Then the following embedding is continuous

$$W^{s_0,p}(0,T;W^{r_0,p}(G)) \cap W^{s_1,p}(0,T;W^{r_1,p}(G)) \hookrightarrow W^{s_\theta,p}(0,T;W^{r_\theta,p}(G)).$$

5.2 BASICS ON PARABOLIC LIPSCHITZ TRUNCATION

In this section we show how a Lipschitz truncation for non-stationary problems can be constructed. We follow the ideas of [65] (see [101] for a similar approach). Let $Q_0 = I_0 \times \mathcal{B}_0 \subset \mathbb{R} \times \mathbb{R}^d$ be a space time cylinder. Let $\mathbf{u} \in L^\sigma(I_0, W^{1,\sigma}(\mathcal{B}_0))$ and $\mathbf{H} \in L^\sigma(Q_0)$ satisfy

$$\int_{Q_0} \mathbf{u} \cdot \partial_t \boldsymbol{\xi} \, dx\, dt = \int_{Q_0} \mathbf{H} : \nabla \boldsymbol{\xi} \, dx\, dt \qquad \text{for all} \quad \boldsymbol{\xi} \in C_0^\infty(Q_0). \tag{5.2.3}$$

Let $\alpha > 0$. We say that $Q' = I' \times \mathcal{B}' \subset \mathbb{R} \times \mathbb{R}^d$ is an α-parabolic cylinder if $r_{I'} = \alpha\, r_{\mathcal{B}'}^2$. For $\kappa > 0$ we define the scaled cylinder $\kappa Q' := (\kappa I') \times (\kappa \mathcal{B}')$. By $\mathcal{Q}^\alpha$ we denote the set of all α-parabolic cylinders. We define the α-parabolic maximal operators $\mathcal{M}^\alpha$ and $\mathcal{M}_s^\alpha$ for $s \in [1,\infty)$ by

$$(\mathcal{M}^\alpha f)(t,x) := \sup_{Q' \in \mathcal{Q}^\alpha \,:\, (t,x) \in Q'} \fint_{Q'} |f(\tau,y)| \, d\tau\, dy,$$

$$\mathcal{M}_s^\alpha f(t,x) := \big(\mathcal{M}^\alpha(|f|^s(t,x))\big)^{\frac{1}{s}}.$$

It is standard [134] that for all $f \in L^p(\mathbb{R}^{d+1})$

$$\|\mathcal{M}_s^\alpha f\|_{L^q(\mathbb{R}^{d+1})} \le c\,\|f\|_{L^q(\mathbb{R}^{d+1})} \quad \forall q \in (s,\infty], \tag{5.2.4}$$

$$\mathcal{L}^{d+1}\Big(\{x \in \mathbb{R}^d : |\mathcal{M}_s^\alpha f(x)| > \lambda\}\Big) \le c\frac{\|f\|_q}{\lambda^q} \quad \forall q \in [s,\infty),\ \forall \lambda > 0, \tag{5.2.5}$$

where the constants are independent of α. Another important tool is a parabolic Poincaré estimate for $\mathbf{u}$ in terms of $\nabla\mathbf{u}$ and $\mathbf{H}$, see Theorem B.1 of [65]: Let $Q_r^\alpha = I_r \times \mathcal{B}_r \in \mathcal{Q}^\alpha$ and any $\mathbf{u} \in L^1(I_r; W^{1,1}(\mathcal{B}_r))$ with $\partial_t\mathbf{u} = \operatorname{div}\mathbf{H}$ in $\mathcal{D}'(Q_r^\alpha)$ for some $\mathbf{H} \in L^1(Q_r^\alpha)$. Then the following holds

$$\fint_{Q_r^\alpha} |\mathbf{u} - \mathbf{u}_{Q_r^\alpha}|\,\mathrm{d}x\,\mathrm{d}t \le cr\fint_{Q_r^\alpha} \Big(|\nabla\mathbf{u}| + \alpha|\mathbf{H}|\Big)\,\mathrm{d}x\,\mathrm{d}t, \tag{5.2.6}$$

where c only depends on the dimension.

In order to define the Lipschitz truncation we have to cut large values of $\nabla\mathbf{u}$ and $\mathbf{H}$. We define the "bad set" as

$$\mathcal{O}_\lambda^\alpha := \{\mathcal{M}_\sigma^\alpha(\chi_{Q_0}|\nabla\mathbf{u}|) > \lambda\} \cup \{\alpha\mathcal{M}_\sigma^\alpha(\chi_{Q_0}|\mathbf{H}|) > \lambda\}. \tag{5.2.7}$$

This is the set where we have to change $\mathbf{u}$. In contrast to the stationary case discussed in Section 1.3, a straightforward argument is not available. So we follow the more flexible strategy based on a Whitney covering as done at the end of Section 3.1. According to Lemma 3.1 of [65] there exists an α-parabolic Whitney covering $\{Q_i\}$ of $\mathcal{O}_\lambda^\alpha$.

Lemma 5.2.1. *There is an α-parabolic Whitney covering $\{Q_i\}$ of $\mathcal{O}_\lambda^\alpha$ with the following properties.*

(PW1) $\bigcup_i \frac{1}{2}Q_i = \mathcal{O}_\lambda^\alpha$,
(PW2) *for all $i \in \mathbb{N}$ we have $8Q_i \subset \mathcal{O}_\lambda^\alpha$ and $16Q_i \cap (\mathbb{R}^{d+1} \setminus \mathcal{O}_\lambda^\alpha) \neq \emptyset$,*
(PW3) *if $Q_i \cap Q_j \neq \emptyset$ then $\frac{1}{2}r_j \le r_i < 2r_j$,*
(PW4) *at every point at most 120^{d+2} of the sets $4Q_i$ intersect,*
where $r_i := r_{\mathcal{B}_i}$, the radius of $\mathcal{B}_i$ and $Q_i = I_i \times \mathcal{B}_i$.

For each Q_i we define $A_i := \{j : Q_j \cap Q_i \neq \emptyset\}$. Note that $\#A_i \le 120^{d+2}$ and $r_j \sim r_i$ for all $j \in A_i$.

Lemma 5.2.2. *There exists a partition of unity $\{\varphi_i\} \subset C_0^\infty(\mathbb{R}^{d+1})$ with respect to the covering $\{Q_i\}$ from Lemma 5.2.1 such that*

(PP1) $\chi_{\frac{1}{2}Q_i} \le \varphi_i \le \chi_{\frac{2}{3}Q_i}$,
(PP2) $\sum_j \varphi_j = \sum_{j \in A_i} \varphi_j = 1$ *in Q_i,*
(PP3) $|\varphi_i| + r_i|\nabla\varphi_i| + r_i^2|\nabla^2\varphi_i| + \alpha\, r_i^2|\partial_t\varphi_i| \le c.$

Now we define

$$\mathbf{u}_\lambda^\alpha := \mathbf{u} - \sum_{i\in\mathcal{I}} \varphi_i(\mathbf{u} - \mathbf{u}_i), \tag{5.2.8}$$

where $\mathbf{u}_i := \mathbf{u}_{Q_i} := \fint_{Q_i} \mathbf{u}\, \mathrm{d}x\, \mathrm{d}t$. (In order to obtain a truncation with suitable properties up to the boundary one has to involve cut-off function as can be seen in [65] and Chapter 6. We neglect this for brevity.) We show first that the sum in (5.2.8) converges absolutely in $L^1(Q_0)$:

$$\int_{Q_0} |\mathbf{u} - \mathbf{u}_\lambda^\alpha|\, \mathrm{d}x \le c \sum_i \int_{Q_i} |\mathbf{u} - \mathbf{u}_i|\, \mathrm{d}x\, \mathrm{d}t \le c \sum_i \int_{Q_i} |\mathbf{u}|\, \mathrm{d}x\, \mathrm{d}t \le c \int_{Q_0} |\mathbf{u}|\, \mathrm{d}x\, \mathrm{d}t,$$

where we used (PP1) and the finite intersection property of Q_i (PW4). We proceed by showing the estimate for the gradient

$$\begin{aligned}
\int_{\mathcal{O}_\lambda^\alpha} |\nabla(\mathbf{u} - \mathbf{u}_\lambda^\alpha)|\, \mathrm{d}x\, \mathrm{d}t &\le c \sum_i \int_{Q_i} \big|\nabla(\varphi_i(\mathbf{u} - \mathbf{u}_i))\big|\, \mathrm{d}x\, \mathrm{d}t \\
&\le c \sum_i \int_{Q_i} |\nabla \mathbf{u}| + \left|\frac{\mathbf{u} - \mathbf{u}_i}{r_i}\right| \mathrm{d}x\, \mathrm{d}t \le c \sum_i \int_{Q_i} |\nabla \mathbf{u}| + \alpha\big|\mathbf{H}\big|\, \mathrm{d}x\, \mathrm{d}t \\
&\le c \int_{Q_0} |\nabla \mathbf{u}| + \alpha\big|\mathbf{H}\big|\, \mathrm{d}x\, \mathrm{d}t,
\end{aligned}$$

where we used (PP3), (5.2.6) and (PW4). This shows that the definition in (5.2.8) makes sense. In particular we have

$$\mathbf{u}_\lambda^\alpha = \begin{cases} \mathbf{u} & \text{in } \; Q_0 \setminus \mathcal{O}_\lambda^\alpha, \\ \sum_i \varphi_i \mathbf{u}_i & \text{in } \; Q_0 \cap \mathcal{O}_\lambda^\alpha. \end{cases} \tag{5.2.9}$$

The truncation $\mathbf{u}_\lambda^\alpha$ has better regularity properties than $\mathbf{u}$; indeed, $\nabla \mathbf{u}_\lambda^\alpha$ is bounded by λ.

Lemma 5.2.3. *For $\lambda > \lambda_0$ we have*

$$\|\nabla \mathbf{u}_\lambda^\alpha\|_{L^\infty(Q_0)} \le c\lambda.$$

Proof. Let $(t, x) \in Q_i$, then

$$\begin{aligned}
|\nabla \mathbf{u}_\lambda^\alpha(t, x)| &= \left|\sum_{j\in A_i} \nabla(\varphi_j \mathbf{u}_j)(t, x)\right| \le \sum_{j\in A_i} |\nabla(\varphi_j(\mathbf{u}_j - \mathbf{u}_i))(t, x)| \\
&\le c \sum_{j\in A_i} \left|\frac{\mathbf{u}_j - \mathbf{u}_i}{r_i}\right| \le c \fint_{Q_i} \left|\frac{\mathbf{u} - \mathbf{u}_i}{r_i}\right| \mathrm{d}x\, \mathrm{d}t
\end{aligned}$$

because $\{\varphi_j\}$ is a partition of unity with (PP3), $r_i \sim r_j$ and $\mathbf{u}_i$ is constant. We also used that $|Q_j \cap Q_k| \ge c \max\{|Q_j|, |Q_k|\}$ if $Q_j \cap Q_k \ne \emptyset$ as well as $\# A_j \le c$.

By (5.2.6), (PW2) and the definition of $\mathcal{O}_\lambda^\alpha$ we have

$$\begin{aligned} |\nabla \mathbf{u}_\lambda^\alpha(t,x)| &\le c \fint_{Q_i} |\nabla \mathbf{u}| + \alpha |\mathbf{H}| \,\mathrm{d}x\,\mathrm{d}t \\ &\le c \fint_{16Q_i} |\nabla \mathbf{u}| + \alpha |\mathbf{H}| \,\mathrm{d}x\,\mathrm{d}t \le c\lambda. \end{aligned} \tag{5.2.10}$$

As the $\{Q_i\}$ cover $\mathcal{O}_\lambda^\alpha$ and $|\nabla \mathbf{u}_\lambda^\alpha| = |\nabla \mathbf{u}| \le \lambda$ outside $\mathcal{O}_\lambda^\alpha$ the claim follows. □

The next lemma will control the time error we obtain when we use the truncation as a test function. We will only consider this from a formal point of view ignoring the technical difficulties connected with the distributional character of the time-derivative of $\mathbf{u}$. We refer to [65, Thm. 3.9.(iii)] for a rigorous treatment.

Lemma 5.2.4. *For all $\lambda \ge \lambda_0$ we have*

$$\left| \int_{Q_0} \partial_t \mathbf{u}_\lambda^\alpha \cdot (\mathbf{u} - \mathbf{u}_\lambda^\alpha) \,\mathrm{d}x\,\mathrm{d}t \right| \le c\alpha^{-1}\lambda^2 \mathcal{L}^{d+1}(\mathcal{O}_\lambda^\alpha),$$

where the constant c is independent of α and λ.

Proof. We use Hölder's inequality, (PP3) and Lemma 6.1.10 to obtain

$$\begin{aligned} (I) &:= \left| \int_{Q_0} \partial_t \mathbf{u}_\lambda^\alpha \cdot (\mathbf{u} - \mathbf{u}_\lambda^\alpha) \,\mathrm{d}x\,\mathrm{d}t \right| = \left| \sum_i \sum_{j \in A_i} \int_{Q_i \cap Q_j} \partial_t (\varphi_i \mathbf{u}_i) \cdot \varphi_j (\mathbf{u} - \mathbf{u}_j) \,\mathrm{d}x\,\mathrm{d}t \right| \\ &= \left| \sum_i \sum_{j \in A_i} \int_{Q_i \cap Q_j} \partial_t \varphi_i (\mathbf{u}_i - \mathbf{u}_j)) \cdot \varphi_j (\mathbf{u} - \mathbf{u}_j) \,\mathrm{d}x\,\mathrm{d}t \right| \\ &\le \frac{c}{\alpha} \sum_i \sum_{j \in A_i} \left| \frac{\mathbf{u}_i - \mathbf{u}_j}{r_i} \right| \int_{Q_j} \left| \frac{\mathbf{u} - \mathbf{u}_j}{r_j} \right| \,\mathrm{d}x\,\mathrm{d}t. \end{aligned}$$

Note that we also took into account $r_i \sim r_j$ and that $\mathbf{u}_j$ is constant. Recalling the estimates in (5.2.10) and (5.2.6) we find that

$$\begin{aligned} (I) &\le \frac{c\lambda}{\alpha} \sum_i \sum_{j \in A_i} \int_{Q_j} \left| \frac{\mathbf{u} - \mathbf{u}_j}{r_j} \right| \,\mathrm{d}x\,\mathrm{d}t \le \frac{c\lambda}{\alpha} \sum_i \sum_{j \in A_i} r_j^{d+2} \fint_{Q_j} |\nabla \mathbf{u}| + \alpha |\mathbf{H}| \,\mathrm{d}x\,\mathrm{d}t \\ &\le \frac{c\lambda^2}{\alpha} \sum_i r_i^{d+2} \fint_{Q_i} |\nabla \mathbf{u}| + \alpha |\mathbf{H}| \,\mathrm{d}x\,\mathrm{d}t \le c\alpha^{-1}\lambda^2 \mathcal{L}^{d+1}(\mathcal{O}_\lambda^\alpha) \end{aligned}$$

using (PW2), the definition of $\mathcal{O}_\lambda^\alpha$, $r_i \sim r_j$ and the local finiteness of the $\{Q_i\}$. □

Remark 5.2.9. As in Lemma 1.3.2 it is possible to have smallness of the level-sets in the sense that

$$\lambda^p \mathcal{L}^{d+1}(\mathcal{O}_\lambda^\alpha) \le \kappa(\lambda)$$

with $\kappa(\lambda) \to 0$ if $\lambda \to \infty$. This and the choice $\alpha = \lambda^{2-p}$ implies the smallness of the time error from Lemma 5.2.4. See [65, Section 4] for details.

5.3 EXISTENCE RESULTS FOR POWER LAW FLUIDS

The flow of a homogeneous incompressible fluid in a bounded body $G \subset \mathbb{R}^d$ $(d = 2, 3)$ during the time interval $(0, T)$ is described by the following set of equations

$$\begin{cases} \rho\partial_t \mathbf{v} + \rho(\nabla\mathbf{v})\mathbf{v} = \operatorname{div}\mathbf{S} - \nabla\pi + \rho\mathbf{f} & \text{in } \; Q, \\ \operatorname{div}\mathbf{v} = 0 & \text{in } \; Q, \\ \mathbf{v} = 0 & \text{on } \; \partial G, \\ \mathbf{v}(0,\cdot) = \mathbf{v}_0 & \text{in } \; G. \end{cases} \tag{5.3.11}$$

See for instance [23]. Here the unknown quantities are the velocity field $\mathbf{v} : Q \to \mathbb{R}^d$ and the pressure $\pi : Q \to \mathbb{R}$. The function $\mathbf{f}: Q \to \mathbb{R}^d$ represents a system of volume forces and $\mathbf{v}_0 : G \to \mathbb{R}^d$ the initial datum, while $\mathbf{S} : Q \to \mathbb{R}^{d\times d}_{sym}$ is the stress deviator and $\rho > 0$ is the density of the fluid. Equation $(5.3.11)_1$ and $(5.3.11)_2$ describe the conservation of balance and the conservation of mass respectively. Both are valid for all homogeneous incompressible liquids and gases. In order to describe a specific fluid one needs a constitutive law relating the viscous stress tensor $\mathbf{S}$ to the symmetric gradient $\boldsymbol{\varepsilon}(\mathbf{v}) := \frac{1}{2}\left(\nabla\mathbf{v} + \nabla\mathbf{v}^T\right)$ of the velocity. In the simplest case this relation is linear, i.e.,

$$\mathbf{S} = \mathbf{S}(\boldsymbol{\varepsilon}(\mathbf{v})) = \nu\boldsymbol{\varepsilon}(\mathbf{v}), \tag{5.3.12}$$

where $\nu > 0$ is the viscosity of the fluid. In this case we have $\operatorname{div}\mathbf{S} = \nu\Delta\mathbf{v}$ and (5.3.11) is the famous Navier–Stokes equation. Its mathematical treatment started with the work of Leray and Ladyshenskaya (see [106]). The existence of a weak solution (where derivatives are to be understood in a distributional sense) can be established by nowadays standard arguments. However the regularity issue (i.e. the existence of a strong solution) is still open.

As already motivated at the beginning of Chapter 4, a much more flexible model is

$$\mathbf{S}(\boldsymbol{\varepsilon}(\mathbf{v})) = \nu(|\boldsymbol{\varepsilon}(\mathbf{v})|)\boldsymbol{\varepsilon}(\mathbf{v}), \tag{5.3.13}$$

where ν is the generalized viscosity function. Of particular interest is the power law model

$$\mathbf{S}(\boldsymbol{\varepsilon}(\mathbf{v})) = \nu_0\big(1 + |\boldsymbol{\varepsilon}(\mathbf{v})|\big)^{p-2}\boldsymbol{\varepsilon}(\mathbf{v}) \tag{5.3.14}$$

where $\nu_0 > 0$ and $p \in (1, \infty)$, cf. [13,23]. We recall that the case $p \in [\frac{3}{2}, 2]$ covers many interesting applications.

In the following we give a historical overview concerning the theory of weak solutions to (5.3.11) and sketch the proofs, cf. [29]. It can be understood as the non-stationary counterpart to Section 1.4.

Monotone operator theory (1969).

Due to the appearance of the convective term $(\nabla\mathbf{v})\mathbf{v} = \operatorname{div}\big(\mathbf{v}\otimes\mathbf{v}\big)$ the equations for power law fluids (the constitutive law is given by (5.3.11)) highly depend on the value of p. The first results were achieved by Ladyshenskaya and Lions for $p \geq \frac{3d+2}{d+2}$ (see [106] and [109]). They showed the existence of a weak solution in the space

$$L^p(0, T; W^{1,p}_{0,\mathrm{div}}(G)) \cap L^\infty(0, T; L^2(G)).$$

The weak formulation reads as

$$\begin{aligned}\int_Q \mathbf{S}(\boldsymbol{\varepsilon}(\mathbf{v})) : \boldsymbol{\varepsilon}(\boldsymbol{\varphi})\,\mathrm{d}x\,\mathrm{d}t &= \int_Q \mathbf{f}\cdot\boldsymbol{\varphi}\,\mathrm{d}x\,\mathrm{d}t - \int_Q (\nabla\mathbf{v})\mathbf{v}\cdot\boldsymbol{\varphi}\,\mathrm{d}x\,\mathrm{d}t\\ &\quad + \int_Q \mathbf{v}\cdot\partial_t\boldsymbol{\varphi}\,\mathrm{d}x\,\mathrm{d}t + \int_G \mathbf{v}_0\cdot\boldsymbol{\varphi}(0)\,\mathrm{d}x\end{aligned} \tag{5.3.15}$$

for all $\boldsymbol{\varphi} \in C^\infty_{0,\mathrm{div}}([0, T) \times G)$ with $\mathbf{S}$ given by (5.3.14). In the case $p \geq \frac{3d+2}{d+2}$ it follows from parabolic interpolation that $(\nabla\mathbf{v})\mathbf{v}\cdot\mathbf{v} \in L^1(Q)$. So the weak solution is also a test-function and the existence proof is based on monotone operator theory and compactness arguments.

Let us assume that

$$p > \frac{3d+2}{d+2} \tag{5.3.16}$$

and that we have a sequence of approximated solutions, i.e.,

$$(\mathbf{v}_n) \subset L^p(0, T; W^{1,p}_{0,\mathrm{div}}(G)) \cap L^\infty(0, T; L^2(G))$$

solving (5.3.15). A sequence of approximated solutions can be obtained, for instance via a Galerkin–Ansatz (see [111], Chapter 5). We want to pass to the limit. Assume further that

$$\partial_t\mathbf{v}_n \in L^{p'}(0, T; W^{-1,p'}_{\mathrm{div}}(G)) \cong L^p(0, T; W^{1,p}_{0,\mathrm{div}}(G))'.$$

Then $\mathbf{v}_n$ is also an admissible test-function (using (5.3.16)). We gain uniform a priori estimates and (after choosing an appropriate subsequence and applying Korn's inequality)

$$\begin{aligned} \mathbf{v}_n &\rightharpoonup: \mathbf{v} \quad \text{in} \quad L^p(0,T; W^{1,p}_{0,\operatorname{div}}(G)), \\ \mathbf{v}_n &\rightharpoonup^* \mathbf{v} \quad \text{in} \quad L^\infty(0,T; L^2(G)). \end{aligned} \tag{5.3.17}$$

A parabolic interpolation implies

$$\mathbf{v}_n \rightharpoonup \mathbf{v} \quad \text{in} \quad L^{p\frac{d+2}{d}}(Q). \tag{5.3.18}$$

As in the stationary case, (5.3.14) yields together with (5.3.17)

$$\mathbf{S}(\boldsymbol{\varepsilon}(\mathbf{v}_n)) \rightharpoonup: \tilde{\mathbf{S}} \quad \text{in} \quad L^{p'}(Q). \tag{5.3.19}$$

A main difference to the stationary problem is the compactness of the velocity. Due to (5.3.17), (5.3.18) and (5.3.15) we can control the time derivative and have

$$\partial_t \mathbf{v}_n \rightharpoonup \partial_t \mathbf{v} \quad \text{in} \quad L^{p'}(0,T; W^{-1,p'}_{\operatorname{div}}(\Omega)). \tag{5.3.20}$$

Combining (5.3.17) and (5.3.20) the Aubin–Lions Compactness Theorem (cf. Theorem 5.1.22) yields

$$\mathbf{v}_n \to \mathbf{v} \quad \text{in} \quad L^{\min\{p',p\}}(0,T; L^2_{\operatorname{div}}(\Omega))$$

and together with (5.3.18)

$$\mathbf{v}_n \to \mathbf{v} \quad \text{in} \quad L^q(Q) \quad \forall q < p\frac{d+2}{d}. \tag{5.3.21}$$

Plugging the convergences (5.3.17)–(5.3.21) together we can pass to the limit in the approximate equation in all terms except for $\mathbf{S}(\boldsymbol{\varepsilon}(\mathbf{v}_n))$. As done in section 1.4 we have to apply arguments from monotone operator theory and show

$$\int_Q \big(\mathbf{S}(\boldsymbol{\varepsilon}(\mathbf{v}_n)) - \mathbf{S}(\boldsymbol{\varepsilon}(\mathbf{v}))\big) : \big(\boldsymbol{\varepsilon}(\mathbf{v}_n) - \boldsymbol{\varepsilon}(\mathbf{v})\big)\,\mathrm{d}x\,\mathrm{d}t \longrightarrow 0, \quad n \to \infty. \tag{5.3.22}$$

This follows along the same line as in the stationary case; the only term which needs a comment is the integral involving the time derivative. Here we have in addition to the terms from the stationary case the integral

$$\begin{aligned} &-\int_0^T \big\langle \partial_t \mathbf{v}_n, \mathbf{v}_n - \mathbf{v}\big\rangle\,\mathrm{d}t \\ &= -\int_0^T \frac{d}{dt}\Big(\int_\Omega |\mathbf{v}_n - \mathbf{v}|^2\,\mathrm{d}x\Big)\,\mathrm{d}t - \int_0^T \big\langle \partial_t \mathbf{v}, \mathbf{v}_n - \mathbf{v}\big\rangle\,\mathrm{d}t \\ &\leq -\int_0^T \big\langle \partial_t \mathbf{v}, \mathbf{v}_n - \mathbf{v}\big\rangle\,\mathrm{d}t \longrightarrow 0, \quad n \to \infty, \end{aligned}$$

using $\mathbf{v}_n(0) = \mathbf{v}(0) = \mathbf{v}_0$ a.e., (5.3.17) and (5.3.20). As the integrand in (5.3.22) is non-negative the claim follows.

L^∞-truncation (2007).

The classical results have been improved by Wolf to the case $p > \frac{2d+2}{d+2}$ via L^∞-truncation. In this situation we have $(\nabla\mathbf{v})\mathbf{v} \in L^1(Q)$ and therefore we can test with functions from $L^\infty(Q)$. The basic idea (which was already used in the stationary case in [78] together with the bound $p \geq \frac{2d}{d+1}$) is to approximate $\mathbf{v}$ by a bounded function $\mathbf{v}_L$ which is equal to $\mathbf{v}$ on a large set and whose L^∞-norm can be controlled by L. Now we will present the approach developed in [140]. Note that the L^∞-truncation has been used in the parabolic context before in [80] and [39]. Different from [140], both these papers deal with periodic and Navier's slip boundary conditions, respectively. So, the problems connected with the harmonic pressure do not occur.

Let us assume that

$$p > \frac{2d+2}{d+2} \tag{5.3.23}$$

and the existence of approximate solutions $\mathbf{v}_n$ to (5.3.15) with uniform a priori estimates in

$$L^p(0,T; W^{1,p}_{0,\mathrm{div}}(G)) \cap L^\infty(0,T; L^2(G)).$$

Note that test-functions have to be bounded as we only have $(\nabla\mathbf{v})\mathbf{v} \in L^1(Q)$ due to (5.3.23) and a parabolic interpolation. We have again the convergences (5.3.17)–(5.3.21) so we only have to establish the limit in $\mathbf{S}(\boldsymbol{\varepsilon}(\mathbf{v}_n))$. As the solution is not a test-function anymore we have to use some truncation. The L^∞-truncation destroys the solenoidal character of a function and a correction via the Bogovskiĭ operator does not give the right sign when testing the time-derivative. So one has to introduce the pressure. In [140] this is done locally for the difference of approximate equation and limit equation. Due to the localization one has to use cut-off functions which we neglect in the following as they only produce additional terms of lower order. We have

$$-\int_Q \mathbf{u}_n \cdot \partial_t \boldsymbol{\varphi}\,\mathrm{d}x\,\mathrm{d}t = -\int_Q \mathbf{H}^1_n : \nabla\boldsymbol{\varphi}\,\mathrm{d}x\,\mathrm{d}t + \int_Q \operatorname{div}\mathbf{H}^2_n \cdot \boldsymbol{\varphi}\,\mathrm{d}x\,\mathrm{d}t \tag{5.3.24}$$

for all $\boldsymbol{\varphi} \in C^\infty_{0,\mathrm{div}}(Q)$ with

$$\begin{aligned} \mathbf{H}^1_n &:= \mathbf{S}(\boldsymbol{\varepsilon}(\mathbf{v}_n)) - \tilde{\mathbf{S}} \rightharpoonup 0 \quad \text{in} \quad L^{p'}(Q), \\ \mathbf{H}^2_n &:= \mathbf{v}_n \otimes \mathbf{v}^n - \mathbf{v} \otimes \mathbf{v} \rightharpoonup 0 \quad \text{in} \quad L^\sigma(Q), \\ \nabla\mathbf{H}^2_n &= \mathbf{v}_n \otimes \mathbf{v}_n - \mathbf{v} \otimes \mathbf{v} \rightharpoonup 0 \quad \text{in} \quad L^\sigma(Q), \end{aligned} \tag{5.3.25}$$

where $\sigma := \frac{p(d+2)}{p(d+2)-(2d+2)} \in (1,\infty)$, cf. (5.3.23). Now we can introduce a pressure π_n and decompose it into $\pi_n = \pi_n^h + \pi_n^1 + \pi_n^2$ such that

$$\begin{aligned} &-\int_Q \left(\mathbf{u}_n - \nabla\pi_n^h\right)\cdot\partial_t\boldsymbol{\varphi}\,\mathrm{d}x \\ &\quad= -\int_Q \left(\mathbf{H}_n^1 - \pi_n^1 I\right) : \nabla\boldsymbol{\varphi}\,\mathrm{d}x\,\mathrm{d}t + \int_Q \operatorname{div}\left(\mathbf{H}_n^2 - \pi_n^2 I\right)\cdot\boldsymbol{\varphi}\,\mathrm{d}x\,\mathrm{d}t \end{aligned} \tag{5.3.26}$$

for all $\boldsymbol{\varphi} \in C_0^\infty(Q)$. The pressure π_n^h is harmonic whereas π_n^1 and π_n^2 feature the same convergences properties as $\mathbf{H}_n^1$ and $\mathbf{H}_n^2$ respectively (see (5.3.25)). Now we test (5.3.15) with the L^∞-truncation of $\mathbf{u}_n - \nabla\pi_n^h$. The result is the same as in the stationary case (cf. Section 1.4) since the term involving the time-derivative has the right sign. Finally we have

$$\int_\Omega \left(\mathbf{S}(\boldsymbol{\varepsilon}(\mathbf{v}_n)) - \tilde{\mathbf{S}}\right) : \psi_1(|\mathbf{u}_n - \nabla\pi_n^h|)\boldsymbol{\varepsilon}(\mathbf{u}_n)\,\mathrm{d}x \longrightarrow 0, \quad n \to \infty,$$

and due to (5.3.17) and $\tilde{\mathbf{S}}, \mathbf{S}(\boldsymbol{\varepsilon}(\mathbf{v})) \in L^{p'}(Q)$

$$\int_\Omega \left(\mathbf{S}(\boldsymbol{\varepsilon}(\mathbf{v}_n)) - \mathbf{S}(\boldsymbol{\varepsilon}(\mathbf{v}))\right) : \psi_1(|\mathbf{u}_n - \nabla\pi_n^h|)\boldsymbol{\varepsilon}(\mathbf{u}_n)\,\mathrm{d}x \longrightarrow 0, \quad n \to \infty.$$

We can finish the proof as in the stationary case; the additional function $\nabla\pi_n^h$ is compact (harmonic in space and bounded in time).

Lipschitz truncation (2010).

Wolf's result was improved to

$$p > \frac{2d}{d+2} \tag{5.3.27}$$

in [65] by the Lipschitz truncation method. Under this restriction to p we have $\mathbf{v}\otimes\mathbf{v} \in L^1(Q)$ which means we can test by functions having bounded gradients. So one has to approximate $\mathbf{v}$ by a Lipschitz continuous function $\mathbf{v}_\lambda$. The best result so far has been shown in [65] by a parabolic Lipschitz truncation, see Section 5.2 for more details. Let us assume that (5.3.27) holds and that there is a sequence of approximate solutions $\mathbf{v}_n$ to (5.3.15) with uniform estimates in $L^p(0,T;W^{1,p}_{0,\mathrm{div}}(G)) \cap L^\infty(0,T;L^2(G))$. On account of (5.3.27) we have $\mathbf{v}_n \otimes \mathbf{v}_n \in L^1(Q)$ such that test-functions must have bounded gradients. We have again the convergences (5.3.17)–(5.3.21) so we only have to establish the limit in $\mathbf{S}(\boldsymbol{\varepsilon}(\mathbf{v}_n))$.

In contrast to the stationary Lipschitz truncation explained in Section 1.4, the parabolic version requires a suitable scaling of the Whitney cubes Q_i. To be precise, they shall be of the form

$$Q_i = Q_i(t_0^i, x_0^i) = (t_i^0 - \alpha r^2, t_i^0 + \alpha r^2) \times \mathcal{B}_r(x_i^0) \tag{5.3.28}$$

with $\alpha = \lambda^{2-p}$ (λ is the Lipschitz constant of the truncation). The reason for this is the control of the distributional time derivative. Despite the L^∞-truncation the Lipschitz truncation is not only nonlinear but also non-local. So the term involving the time derivative does not have a sign. But due to (5.3.28) it is possible to show that

$$\left| \int_0^T \langle \partial_t \mathbf{u}, \mathbf{u}_\lambda - \mathbf{u} \rangle \, \mathrm{d}t \right| \leq \kappa(\lambda) \to 0, \quad \lambda \to \infty,$$

recalling Lemma 5.2.4. On account of this the Lipschitz truncation can be roughly speaking applied as in the stationary case in Section 1.4. However, there are certain technical difficulties. First of all, the known parabolic versions of the Lipschitz truncation work only locally. So, one has to involve bubble functions in order to localize the arguments. The approach in [65] introduces the pressure function as explained in (5.3.26) for the parabolic L^∞-truncation. In fact, the authors use the Lipschitz truncation of the function $\mathbf{u}_n - \nabla \pi_n^h$.

CHAPTER 6

Solenoidal Lipschitz truncation

Contents

Abstract

In this chapter we present the solenoidal Lipschitz truncation for non-stationary problems: we show how to construct a Lipschitz truncation which preserves the divergence-free character of a given Sobolev function. As a matter of fact, it suffices to have distributional time-derivatives in the sense of divergence-free test-functions. After this, we present the $\mathcal{A}$-Stokes approximation for non-stationary problems. It aims at approximating almost solutions to the non-stationary $\mathcal{A}$-Stokes system by exact solutions. Thanks to the solenoidal Lipschitz truncation this can be done on the level of gradients.

In this chapter we develop a non-stationary counterpart of the solenoidal Lipschitz truncation from Chapter 3. Here, the main difficulty is to handle problems connected with the distributional time derivative of the function we aim to truncate. Let us be a little bit more precise. Let $Q_0 = I_0 \times \mathcal{B}_0 \subset \mathbb{R} \times \mathbb{R}^3$ be a space time cylinder and $\sigma \in (1, \infty)$. Let $\mathbf{u} \in L^\sigma(I_0, W^{1,\sigma}_{\operatorname{div}}(\mathcal{B}_0))$ and $\mathbf{G} \in L^\sigma(Q_0)$ satisfy

$$\int_{Q_0} \partial_t \mathbf{u} \cdot \boldsymbol{\xi} \,\mathrm{d}x\,\mathrm{d}t = \int_{Q_0} \mathbf{G} : \nabla \boldsymbol{\xi} \,\mathrm{d}x\,\mathrm{d}t \quad \text{for all} \quad \boldsymbol{\xi} \in C^\infty_{0,\operatorname{div}}(Q_0). \tag{6.0.1}$$

The main purpose of the solenoidal Lipschitz truncation is to avoid the appearance of the pressure function. Hence we start in (6.0.1) with an equation on the level of divergence-free test-functions. Unfortunately, this is not enough information on the time derivative for a Poincaré-type inequality as in (5.2.6). Hence the approach from [65] as explained in Section 5.2 will not give L^∞-estimates for the gradient of the truncation, cf. the proof of Lemma 5.2.3. Our aim is to construct a truncation which preserves the properties from [65] and is, in addition, divergence-free.

We will show that there is a truncation $\mathbf{u}_\lambda$ of $\mathbf{u}$ with roughly the following properties (see Theorem 6.1.25 for a precise formulation).

Existence Theory for Generalized Newtonian Fluids.
DOI: http://dx.doi.org/10.1016/B978-0-12-811044-7.00008-2

(a) $\nabla \mathbf{u}_\lambda \in L^\infty(Q_0)$ with $\|\nabla \mathbf{u}_\lambda\|_\infty \leq c\lambda$ and $\operatorname{div} \mathbf{u}_\lambda = 0$.
(b) $\mathbf{u}_\lambda = \mathbf{u}$ a.e. outside a suitable set $\mathcal{O}_\lambda$.
(c) There holds

$$\left|\langle \partial_t \mathbf{u}, \mathbf{u}_\lambda - \mathbf{u}\rangle\right| + \|\chi_{\mathcal{O}_\lambda^\alpha} \nabla \mathbf{u}_\lambda\|_p^p \leq c\lambda^p |\mathcal{O}_\lambda| \leq \delta(\lambda),$$

with $\delta(\lambda) \to 0$ if $\lambda \to \infty$.

In the following we sketch the construction on a heuristic level. In fact, the rigorous approach which we shall present in the next section requires a series of localization arguments, so it is quite technical. Let us start with a function

$$\mathbf{u} \in L^\infty(I_0; L^2(\mathcal{B}_0)) \cap L^p(I_0; W^{1,p}_{0,\operatorname{div}}(\mathcal{B}_0))$$

with $\partial_t \mathbf{u} = \operatorname{div} \mathbf{H}$ in $\mathcal{D}'_{\operatorname{div}}(\mathcal{B}_0)$, where $\mathbf{H} \in L^\sigma(\mathcal{B})$ for some $\sigma > 1$. We define

$$\mathbf{w} := \operatorname{curl}^{-1} \mathbf{u} \in L^\infty(I_0; W^{1,2}(\mathcal{B}_0)) \cap L^p(I_0; W^{2,p}_{\operatorname{div}}(\mathcal{B}_0)).$$

It follows that $\partial_t \Delta \mathbf{w} = \operatorname{curl} \operatorname{div} \mathbf{H}$ in $\mathcal{D}'(\mathcal{B}_0)$. Also we can obtain an information about the time derivative of $\mathbf{w}$ as a distribution acting on all test-functions. However, we do not have control about a possible harmonic part of $\mathbf{w}$. Hence we decompose $\mathbf{w}$ into a harmonic and anti-harmonic part. To do this we define, pointwise in time,

$$\mathbf{w}(t) = \mathbf{z}(t) + \mathbf{h}(t),$$

where $\mathbf{z}(t) \in \Delta W^{2,p}_0(\mathcal{B}_0)$ and $\Delta \mathbf{h}(t) = 0$. This decomposition is based on a singular integral operator which is continuous on L^p-spaces such that

$$\mathbf{z}, \mathbf{w} \in L^\infty(I_0; W^{1,2}(\mathcal{B}_0)) \cap L^p(I_0; W^{2,p}(\mathcal{B}_0)). \tag{6.0.2}$$

Moreover, we have

$$\partial_t \Delta \mathbf{z} = \partial_t \mathbf{w} = \operatorname{curl} \operatorname{div} \mathbf{H}$$

in $\mathcal{D}'(\mathcal{B}_0)$. As $\mathbf{z}$ is anti-harmonic by construction this yields

$$\|\partial_t \mathbf{z}\|_\sigma \leq c\|\mathbf{H}\|_\sigma.$$

In fact, $\partial_t \mathbf{z}$ is a measurable function. Now, we truncate $\mathbf{z}$ to $\mathbf{z}_\lambda$ with an approach similar to (5.2.8). This truncation satisfies with $\|\nabla^2 \mathbf{z}_\lambda\|_\infty \leq c\lambda$ as well as $\mathbf{z}_\lambda = \mathbf{z}$ in $\mathcal{O}_\lambda$, where $\mathcal{O}_\lambda = \mathcal{O}_\lambda(\mathcal{M}(\nabla^2 \mathbf{z}); \mathcal{M}(\partial_t \mathbf{z}))$. Finally, we set

$$\mathbf{u}_\lambda := \operatorname{curl} \mathbf{z}_\lambda + \operatorname{curl} \mathbf{h}.$$

Obviously, we have $\operatorname{div} \mathbf{u}_\lambda = 0$. Due to (6.0.2) and the properties of harmonic functions we have $\mathbf{h} \in L^\infty(I_0; W^{k,2}(\mathcal{B}_0))$ for any $k \in \mathbb{N}$ (at least locally in space). Hence $\mathbf{u}_\lambda$ has the same regularity as $\operatorname{curl}\mathbf{z}_\lambda$. In particular, $\nabla\mathbf{u}_\lambda$ is bounded.

In Section 6.2 we develop the $\mathcal{A}$-Stokes approximation for non-stationary problems, see [29]. This is, on the one hand, a non-stationary variant of the $\mathcal{A}$-Stokes approximation from Section 3.3. On the other hand it is a fluid-mechanical counterpart of the $\mathcal{A}$-caloric approximation from [68] which is concerned with the $\mathcal{A}$-heat equation.

6.1 SOLENOIDAL TRUNCATION – EVOLUTIONARY CASE

In this section we examine solenoidal functions, whose time derivative is only a distribution acting on solenoidal test-functions. Let $\mathbf{u} \in L^\sigma(I_0, W^{1,\sigma}_{\operatorname{div}}(\mathcal{B}_0))$ be such that (6.0.1) holds for some $\mathbf{G} \in L^\sigma(Q_0)$. So the time derivative is only well defined via the duality with solenoidal test functions. The goal of this section is to construct a solenoidal truncation $\mathbf{u}_\lambda$ of $\mathbf{u}$ which preserves the properties of the truncation in [65].

First we extend our function $\mathbf{u}$ in a suitable way to the whole space and then apply the inverse curl operator. Let $\gamma \in C^\infty_0(\mathcal{B}_0)$ with $\chi_{\frac{1}{2}\mathcal{B}_0} \leq \gamma \leq \chi_{\mathcal{B}_0}$, where $\mathcal{B}_0$ is a ball. Let $\mathcal{C}_0$ denote the annulus $\mathcal{B}_0 \setminus \frac{1}{2}\mathcal{B}_0$. Then according to Theorem 2.1.6 (with $A(t) = B(t) = t^q$) there exists a Bogovskiĭ operator $\operatorname{Bog}_{\mathcal{C}_0} : C^\infty_{0,\perp}(\mathcal{C}_0) \to C^\infty_0(\mathcal{C}_0)$ which is bounded from $L^q_\perp(\mathcal{C}_0) \to W^{1,q}_0(\mathcal{C}_0)$ for all $q \in (1,\infty)$, and such that $\operatorname{div}\operatorname{Bog}_{\mathcal{C}_0} = \operatorname{Id}$. Define

$$\tilde{\mathbf{u}} := \gamma\mathbf{u} - \operatorname{Bog}_{\mathcal{C}_0}(\operatorname{div}(\gamma\mathbf{u})) = \gamma\mathbf{u} - \operatorname{Bog}_{\mathcal{C}_0}(\nabla\gamma\cdot\mathbf{u}).$$

Then $\operatorname{div}\tilde{\mathbf{u}} = 0$ on $I_0 \times \mathcal{B}_0$ and $\tilde{\mathbf{u}}(t) \in W^{1,\sigma}_0(\mathcal{B}_0)$, so we can extend $\tilde{\mathbf{u}}$ by zero in space to $\tilde{\mathbf{u}} \in L^\sigma(I_0, W^{1,\sigma}_{\operatorname{div}}(\mathbb{R}^3))$. Since $\tilde{\mathbf{u}} = \mathbf{u}$ on $I_0 \times \frac{1}{2}\mathcal{B}_0$, we have

$$\int_{Q_0} \partial_t\tilde{\mathbf{u}}\cdot\boldsymbol{\xi}\,\mathrm{d}x\,\mathrm{d}t = \int_{Q_0} \mathbf{G} : \nabla\boldsymbol{\xi}\,\mathrm{d}x\,\mathrm{d}t \quad \text{for all} \quad \boldsymbol{\xi} \in C^\infty_{0,\operatorname{div}}(\tfrac{1}{2}Q_0). \tag{6.1.3}$$

Now, we define, pointwise in time,

$$\mathbf{w} := \operatorname{curl}^{-1}(\tilde{\mathbf{u}}) = \operatorname{curl}^{-1}\big(\gamma\mathbf{u} - \operatorname{Bog}_{\mathcal{C}_0}(\nabla\gamma\cdot\mathbf{u})\big).$$

Overall, we get the following lemma.

Lemma 6.1.1. *We have*

$$\begin{aligned} \operatorname{curl}\mathbf{w} = \tilde{\mathbf{u}} = \mathbf{u} \quad & \textit{in} \quad \tfrac{1}{2}Q_0 \\ \operatorname{div}\mathbf{w} = 0 \quad & \textit{in} \quad \mathbb{R}^3 \end{aligned}$$

and

$$\begin{aligned}\|\mathbf{w}(t)\|_{L^s(\mathbb{R}^3)} &\le c_s \|\tilde{\mathbf{u}}(t)\|_{L^a(\mathcal{B}_0)}\\ \|\nabla\mathbf{w}(t)\|_{L^s(\mathbb{R}^3)} &\le c_s \|\tilde{\mathbf{u}}(t)\|_{L^s(\mathcal{B}_0)}\\ \|\nabla^2\mathbf{w}(t)\|_{L^s(\mathbb{R}^3)} &\le c_s \|\nabla\tilde{\mathbf{u}}(t)\|_{L^s(\mathcal{B}_0)},\end{aligned}$$

for $a = \max\{1, \frac{3s}{3+s}\}$, $t \in I_0$ *and* $s \in (1, \infty)$.

Let us derive from (6.0.1) the equation for $\mathbf{w}$. For $\boldsymbol{\psi} \in C_0^\infty(\frac{1}{2}Q_0)$ we have

$$\int_{Q_0} \partial_t \mathbf{u} \cdot \operatorname{curl} \boldsymbol{\psi} \,\mathrm{d}x\,\mathrm{d}t = \int_{Q_0} \mathbf{G} : \nabla \operatorname{curl} \boldsymbol{\psi} \,\mathrm{d}x\,\mathrm{d}t.$$

We use $\mathbf{u} = \operatorname{curl}\mathbf{w}$ and partial integration to show that

$$\int_{Q_0} \partial_t \mathbf{w} \cdot \operatorname{curl}\operatorname{curl} \boldsymbol{\psi} \,\mathrm{d}x\,\mathrm{d}t = \int_{Q_0} \mathbf{G} : \nabla \operatorname{curl} \boldsymbol{\psi} \,\mathrm{d}x\,\mathrm{d}t.$$

Now, because

$$\int_{Q_0} \mathbf{w} \cdot \partial_t \nabla \operatorname{div} \boldsymbol{\psi} \,\mathrm{d}x\,\mathrm{d}t = \int_{Q_0} \operatorname{div}\mathbf{w}\, \partial_t \operatorname{div} \boldsymbol{\psi} \,\mathrm{d}x\,\mathrm{d}t = 0$$

and $\operatorname{curl}\operatorname{curl}\boldsymbol{\psi} = -\Delta\boldsymbol{\psi} + \nabla\operatorname{div}\boldsymbol{\psi}$ we obtain

$$\int_{Q_0} \mathbf{w} \cdot \partial_t \Delta \boldsymbol{\psi} \,\mathrm{d}x\,\mathrm{d}t = -\int_{Q_0} \mathbf{G} : \nabla \operatorname{curl} \boldsymbol{\psi} \,\mathrm{d}x\,\mathrm{d}t \tag{6.1.4}$$

for every $\boldsymbol{\psi} \in C_0^\infty(\frac{1}{2}Q_0)$. We can rewrite this as

$$\int_{Q_0} \mathbf{w} \cdot \partial_t \Delta \boldsymbol{\psi} \,\mathrm{d}x\,\mathrm{d}t = -\int_{Q_0} \mathbf{H} : \nabla^2 \boldsymbol{\psi} \,\mathrm{d}x\,\mathrm{d}t, \tag{6.1.5}$$

with $|\mathbf{G}| \sim |\mathbf{H}|$ pointwise. In particular, in the sense of distributions we have

$$\partial_t \Delta \mathbf{w} = -\operatorname{curl}\operatorname{div}\mathbf{G} = -\operatorname{div}\operatorname{div}\mathbf{H}.$$

So in passing from $\mathbf{u}$ to $\mathbf{w}$ we got a system valid for all test functions $\boldsymbol{\psi} \in C_0^\infty(Q_0)$. However, we only have control of $\partial_t\Delta\mathbf{w}$, so that the time derivative of the harmonic part of $\mathbf{w}$ cannot be seen. Hence, a parabolic Poincaré inequality for $\mathbf{w}$ still does not hold; i.e. $\partial_t\mathbf{w}$ is not controlled! In order to remove this harmonic invariance we will replace $\mathbf{w}$ by some function $\mathbf{z}$ such that $\partial_t\Delta\mathbf{w} = \partial_t\Delta\mathbf{z}$. This will imply that $\partial_t\mathbf{z}$ can be controlled by $\mathbf{H}$. To define $\mathbf{z}$ conveniently we need some auxiliary results.

For a ball $\mathcal{B}' \subset \mathbb{R}^3$ and a function $f \in L^s(B')$ we define $\Delta_{\mathcal{B}'}^{-2}\Delta f$ as the weak solution $F \in W_0^{2,s}(B')$ of

$$\int_{\mathcal{B}'} \Delta F \Delta\varphi \,\mathrm{d}x = \int_{\mathcal{B}'} f \Delta\varphi \,\mathrm{d}x \quad \text{for all} \quad \varphi \in C_0^\infty(\mathcal{B}'). \tag{6.1.6}$$

Then $f - \Delta(\Delta_{\mathcal{B}'}^{-2}\Delta f)$ is harmonic on $\mathcal{B}'$.

According to [117] and Lemma 2.1 of [140] we have the following variational estimate.

Lemma 6.1.2. *Let $s \in (1, \infty)$. Then for all $g \in W_0^{2,s}(\mathcal{B}')$ we have*

$$\|\nabla^2 g\|_s \le c_s \sup_{\substack{\varphi \in C_0^\infty(\mathcal{B}') \\ \|\nabla^2 \varphi\|_{s'} \le 1}} \int_{\mathcal{B}'} \Delta g \, \Delta \varphi \, \mathrm{d}x.$$

This implies the following two corollaries.

Corollary 6.1.1. *Let $s \in (1, \infty)$. Then*

$$\int_{\mathcal{B}'} \left|\nabla^2(\Delta_{\mathcal{B}'}^{-2} \Delta f)\right|^s \mathrm{d}x \le c_s \sup_{\substack{\varphi \in C_0^\infty(\mathcal{B}') \\ \|\nabla^2 \varphi\|_{s'} \le 1}} \int_{\mathcal{B}'} f \, \Delta \varphi \, \mathrm{d}x \le c_s \int_{\mathcal{B}'} |f|^s \, \mathrm{d}x$$

for $f \in L^s(\mathcal{B}')$, where c_s is independent of the ball $\mathcal{B}'$.

Proof. The claim follows by Lemma 6.1.2,

$$\int_{\mathcal{B}'} \Delta(\Delta_{\mathcal{B}'}^{-2} \Delta f) \, \Delta \varphi \, \mathrm{d}x = \int_{\mathcal{B}'} f \, \Delta \varphi \, \mathrm{d}x,$$

and Hölder's inequality. □

Corollary 6.1.2. *Let $s \in (1, \infty)$. Then*

$$\int_{\frac{2}{3}\mathcal{B}'} \left|\nabla^3(\Delta_{\mathcal{B}'}^{-2} \Delta f)\right|^s \mathrm{d}x \le c_s \int_{\mathcal{B}'} |\nabla f|^s \, \mathrm{d}x \quad \textit{for} \quad f \in W^{1,s}(\mathcal{B}'),$$

$$\int_{\frac{2}{3}\mathcal{B}'} \left|\nabla^4(\Delta_{\mathcal{B}'}^{-2} \Delta f)\right|^s \mathrm{d}x \le c_s \int_{\mathcal{B}'} |\nabla^2 f|^s \, \mathrm{d}x \quad \textit{for} \quad f \in W^{2,s}(\mathcal{B}'),$$

where c_s is independent of the ball $\mathcal{B}'$.

Proof. The claim follows from Corollary 6.1.1 by standard interior regularity theory (difference quotients, localization and Poincaré's inequality). □

For $V \in L^s(\mathcal{B}')$ we define $\Delta_{\mathcal{B}'}^{-2} \operatorname{div} \operatorname{div} V$ as the weak solution $F \in W_0^{2,s}(\mathcal{B}')$ of

$$\int_{\mathcal{B}'} \Delta F \, \Delta \varphi \, \mathrm{d}x = \int_{\mathcal{B}'} V \, \nabla^2 \varphi \, \mathrm{d}x \qquad \text{for all} \quad \varphi \in C_0^\infty(\mathcal{B}').$$

Similar to Corollary 6.1.1 we get the following result.

Corollary 6.1.3. *Let $s \in (1, \infty)$. Then*

$$\int_{\mathcal{B}'} \left|\nabla^2(\Delta_{\mathcal{B}'}^{-2} \operatorname{div} \operatorname{div} V)\right|^s \mathrm{d}x \le c_s \int_{\mathcal{B}'} |V|^s \, \mathrm{d}x$$

for $V \in L^s(\mathcal{B}')$, where c_s is independent of the ball $\mathcal{B}'$.

The next lemma shows the wanted control of the time derivative.

Lemma 6.1.3. *For a cube* $Q' = I' \times B' \subset Q_0$ *let* $\mathbf{z}_{Q'} := \Delta\Delta_{B'}^{-2}\Delta\mathbf{w}$. *Then for* $s \in (1,\infty)$ *we have*

$$\fint_{Q'} |\mathbf{z}_{Q'}|^s \,\mathrm{d}x \le c_s \fint_{Q'} |\mathbf{w}|^s \,\mathrm{d}x$$

$$\fint_{Q'} \left|\frac{\mathbf{z}_{Q'}}{r'}\right|^s \mathrm{d}x + \fint_{\frac{2}{3}Q'} |\nabla\mathbf{z}_{Q'}|^s \,\mathrm{d}x \le c_s \fint_{Q'} |\nabla\mathbf{w}|^s \,\mathrm{d}x$$

$$\fint_{Q'} \left|\frac{\mathbf{z}_{Q'}}{(r')^2}\right|^s \mathrm{d}x + \fint_{\frac{2}{3}Q'} \left|\frac{\nabla\mathbf{z}_{Q'}}{r'}\right|^s \mathrm{d}x + \fint_{\frac{2}{3}Q'} |\nabla^2\mathbf{z}_{Q'}|^s \,\mathrm{d}x \le c_s \fint_{Q'} |\nabla^2\mathbf{w}|^s \,\mathrm{d}x,$$

$$\fint_{Q'} |\partial_t\mathbf{z}_{Q'}|^s \,\mathrm{d}x\,\mathrm{d}t \le c_s \fint_{Q'} |\mathbf{H}|^s \,\mathrm{d}x\,\mathrm{d}t,$$

where $r' := r_{B'}$.

Proof. The estimate of $\mathbf{z}_{Q'}$ in terms of $\mathbf{w}$ follows directly from Corollary 6.1.1 and integration over time. The estimate of $\mathbf{z}_{Q'}$ in terms of $\nabla\mathbf{w}$ and $\nabla^2\mathbf{w}$ follows from this by Poincaré's inequality, using the fact that we can subtract a linear polynomial from $\mathbf{w}$ without changing the definition of $\mathbf{z}_{Q'}$. The other estimate for $\nabla\mathbf{z}_{Q'}$ and $\nabla^2\mathbf{z}_{Q'}$ follow analogously from Corollary 6.1.2.

For all $\rho \in C_0^\infty(I')$ and $\boldsymbol{\varphi} \in C_0^\infty(B')$ it follows from (6.1.5) that

$$\int_{I'}\int_{B'} \mathbf{w}\cdot\Delta\boldsymbol{\varphi}\,\mathrm{d}x\,\partial_t\rho\,\mathrm{d}t = -\int_{I'}\int_{B'} \mathbf{H} : \nabla^2\boldsymbol{\varphi}\,\mathrm{d}x\,\rho\,\mathrm{d}t.$$

Let d_t^h denote the forward difference quotient in time with step size h. We use $\rho(t) := \fint_t^{t-h} \tilde{\rho}(\tau)\,\mathrm{d}\tau$ with $\tilde{\rho} \in C_0^\infty(I')$ and h sufficiently small. Then $\partial_t\rho = d_t^{-h}\tilde{\rho}$ and

$$\int_{I'}\int_{B'} \mathbf{w}\cdot\Delta\boldsymbol{\varphi}\,\mathrm{d}x\,d_t^{-h}\tilde{\rho}\,\mathrm{d}t = -\int_{I'}\int_{B'} \mathbf{H} : \nabla^2\boldsymbol{\varphi}\,\mathrm{d}x \fint_t^{t-h} \tilde{\rho}(\tau)\,\mathrm{d}\tau\,\mathrm{d}t.$$

This implies that

$$\int_{I'}\int_{B'} d_t^h\mathbf{w}\cdot\Delta\boldsymbol{\varphi}\,\mathrm{d}x\,\tilde{\rho}\,\mathrm{d}t = -\int_{I'}\int_{B'}\fint_t^{t+h} \mathbf{H}(\tau)\,\mathrm{d}\tau : \nabla^2\boldsymbol{\varphi}\,\mathrm{d}x\,\tilde{\rho}\,\mathrm{d}t.$$

Since this is valid for all choices of $\tilde{\rho}$ we have

$$\int_{B'} d_t^h\mathbf{w}\cdot\Delta\boldsymbol{\varphi}\,\mathrm{d}x = -\int_{B'}\fint_t^{t+h} \mathbf{H}(\tau)\,\mathrm{d}\tau : \nabla^2\boldsymbol{\varphi}\,\mathrm{d}x.$$

Since $d_t^h \mathbf{z}_{Q'} = d_t^h(\Delta\Delta_{\mathcal{B}'}^{-2}\Delta\mathbf{w}) = \Delta\Delta_{\mathcal{B}'}^{-2}\Delta(d_t^h\mathbf{w})$, it follows by Corollary 6.1.1 that

$$\begin{aligned}\left(\int_{\mathcal{B}'} |d_t^h \mathbf{z}_{Q'}|^s \,\mathrm{d}x\right)^{\frac{1}{s}} &\leq c \sup_{\substack{\varphi\in C_0^\infty(\mathcal{B})\\ \|\nabla^2\varphi\|_{s'}\leq 1}} \int_{\mathcal{B}'} d_t^h \mathbf{w}\Delta\varphi \,\mathrm{d}x \\ &= \sup_{\substack{\varphi\in C_0^\infty(\mathcal{B})\\ \|\nabla^2\varphi\|_{s'}\leq 1}} \left(-\int_{\mathcal{B}'} \fint_t^{t+h} \mathbf{H}(\tau)\,\mathrm{d}\tau : \nabla^2\varphi \,\mathrm{d}x\right) \\ &\leq c\left(\int_{\mathcal{B}'} \fint_t^{t+h} |\mathbf{H}(\tau)|^s \,\mathrm{d}\tau \,\mathrm{d}x\right)^{\frac{1}{s}}.\end{aligned}$$

Integrating over time and passing to the limit $h \to 0$ yields

$$\left(\int_{Q'} |\partial_t \mathbf{z}_{Q'}|^s \,\mathrm{d}x\,\mathrm{d}t\right)^{\frac{1}{s}} \leq c\left(\int_{Q'} |\mathbf{H}|^s \,\mathrm{d}x\,\mathrm{d}t\right)^{\frac{1}{s}}$$

which finishes the proof. □

Defining $\mathbf{z}(t) := \mathbf{z}_{\frac{1}{2}Q_0}(t) = \Delta\Delta_{\frac{1}{2}\mathcal{B}_0}^{-2}\Delta\mathbf{w}(t)$ for $t \in \frac{1}{2}I_0$, we then have

$$\int_{Q_0} \mathbf{z}\cdot\partial_t\Delta\boldsymbol{\psi}\,\mathrm{d}x\,\mathrm{d}t = \int_{Q_0} \mathbf{w}\cdot\partial_t\Delta\boldsymbol{\psi}\,\mathrm{d}x\,\mathrm{d}t = -\int_{Q_0} \mathbf{H} : \nabla^2\boldsymbol{\psi}\,\mathrm{d}x\,\mathrm{d}t, \tag{6.1.7}$$

for all $\boldsymbol{\psi} \in C_0^\infty(\frac{1}{2}Q_0)$. Since the function $\Delta_{\frac{1}{2}\mathcal{B}_0}^{-2}\mathbf{w}(t) \in W_0^{2,s}(\frac{1}{2}\mathcal{B}_0)$, we can extend it by zero to a function in $W^{2,s}(\mathbb{R}^3)$. In this sense it is natural to extend $\mathbf{z}(t)$ by zero to a function in $L^s(\mathbb{R}^3)$.

Note that Lemma 6.1.3 enables us to control $\partial_t\mathbf{z}$ by $\mathbf{H}$ in $L^s(\frac{1}{2}Q_0)$.

Lemma 6.1.4. *We have*

$$\begin{aligned}\|\mathbf{z}(t)\|_{L^s(\frac{1}{3}\mathcal{B}_0)} &\leq c_s \|\tilde{\mathbf{u}}(t)\|_{L^{\frac{3s}{s+3}}(\mathcal{B}_0)} \\ \|\nabla\mathbf{z}(t)\|_{L^s(\frac{1}{3}\mathcal{B}_0)} &\leq c_s \|\tilde{\mathbf{u}}(t)\|_{L^s(\mathcal{B}_0)} \\ \|\nabla^2\mathbf{z}(t)\|_{L^s(\frac{1}{3}\mathcal{B}_0)} &\leq c_s \|\nabla\tilde{\mathbf{u}}(t)\|_{L^s(\mathcal{B}_0)},\end{aligned}$$

for $t \in I$ and $s \in (1,\infty)$.

Proof. This follows from Corollary 6.1.1, Corollary 6.1.2 and Lemma 6.1.1. □

For $\lambda, \alpha > 0$ and $\sigma > 1$ we define

$$\mathcal{O}_\lambda^\alpha := \{\mathcal{M}_\sigma^\alpha(\chi_{\frac{1}{3}Q_0}|\nabla^2\mathbf{z}|) > \lambda\} \cup \{\alpha\mathcal{M}_\sigma^\alpha(\chi_{\frac{1}{3}Q_0}|\partial_t\mathbf{z}|) > \lambda\}. \tag{6.1.8}$$

Later we will choose $\alpha = \lambda^{2-p}$ and σ smaller than the integrability exponent of $\partial_t \mathbf{z}$. We want to redefine $\mathbf{z}$ on $\mathcal{O}_\lambda^\alpha$. The first step is to cover $\mathcal{O}_\lambda^\alpha$ by well selected cubes. By the lower-semi-continuity property of the maximal functions the set $\mathcal{O}_\lambda^\alpha$ is open. We assume in the following that $\mathcal{O}_\lambda^\alpha$ is non-empty. (In the case that $\mathcal{O}_\lambda^\alpha$ is empty, we do not need to truncate at all.) We cover $\mathcal{O}_\lambda^\alpha$ by an α-parabolic Whitney covering $\{Q_i\}$ with partition of unity in accordance with Lemmas 5.2.1 and 5.2.2.

Due to property (PW3) we have that $16Q_j \cap (\mathbb{R}^{d+1} \setminus \mathcal{O}_\lambda^\alpha) \neq \emptyset$. Thus, the definition of $\mathcal{O}_\lambda^\alpha$ implies that

$$\left(\fint_{16Q_j} |\nabla^2 \mathbf{z}|^\sigma \chi_{\frac{1}{3}Q_0} \,dx\,dt \right)^{\frac{1}{\sigma}} \leq \lambda, \tag{6.1.9}$$

$$\alpha \left(\fint_{16Q_j} |\partial_t \mathbf{z}|^\sigma \chi_{\frac{1}{3}Q_0} \,dx\,dt \right)^{\frac{1}{\sigma}} \leq \lambda. \tag{6.1.10}$$

Lemma 6.1.5. *Assume that there exists $c_0 > 0$ such that $\lambda^p |\mathcal{O}_\lambda^\alpha| \leq c_0$ with $p > \frac{2d}{d+2}$. Then the following holds:*

if $\lambda \geq \lambda_0 = \lambda_0(c_0)$, $\alpha = \lambda^{2-p}$ and $Q_i \cap \frac{1}{4}Q_0 \neq \emptyset$, then $Q_i \subset \frac{1}{3}Q_0$ and $Q_j \subset \frac{1}{3}Q_0$ for all $j \in A_i$.

Proof. Let $Q_i \cap \frac{1}{4}Q_0 \neq \emptyset$. We claim that $Q_i \subset \frac{7}{24}Q_0 \subset \frac{1}{3}Q_0$. Let $s_i := \alpha r_i^2$. It suffices to show that $r_i, s_i \to 0$ as $\lambda \to \infty$. Because $Q_i \subset \mathcal{O}_\lambda^\alpha$ and by assumption, we find

$$\lambda^2 r_i^{d+2} = \lambda^p s_i r_i^d \leq c \lambda^p |Q_i| \leq c \lambda^p |\mathcal{O}_\lambda^\alpha| \leq c c_0. \tag{6.1.11}$$

This implies that $r_i \leq (c c_0 \lambda^{-2})^{\frac{1}{d+2}} \to 0$ as $\lambda \to \infty$. Moreover, $r_i = s_i^{\frac{1}{2}} \alpha^{-\frac{1}{2}}$ and (6.1.11) imply

$$c c_0 \geq \lambda^p s_i r_i^d = \lambda^p s_i^{\frac{d+2}{2}} \alpha^{-\frac{3}{2}} = \lambda^{\frac{d+2}{2}p - d} s_i^{\frac{d+2}{2}}.$$

If $p > \frac{2d}{d+2}$, then $\lambda \to \infty$ implies $s_i \to 0$ as desired.

The claim on $j \in A_i$ follows from the fact that Q_i and Q_j have comparable size and that $\frac{7}{24}Q_0$ is strictly contained in $\frac{1}{3}Q_0$. □

Let us show that the assumption $\lambda^p |\mathcal{O}_\lambda^\alpha| \leq c_0$ from Lemma 6.1.5 is satisfied in our situation. To do this we assume from now on that

$$\alpha := \lambda^{2-p} \tag{6.1.12}$$

and that $\sigma < \min\{p, p'\}$.

Lemma 6.1.6. *Let $c_0 := \|\nabla^2 \mathbf{z}\|^p_{L^p(\frac{1}{3}Q_0)} + \|\partial_t \mathbf{z}\|^{p'}_{L^{p'}(\frac{1}{3}Q_0)}$. Then*

$$\lambda^p |\mathcal{O}_\lambda^\alpha| \leq c_0.$$

Proof. It follows from the weak-type estimate of $\mathcal{M}_\sigma^\alpha$ in (5.2.5), if $\sigma < \min\{p, p'\}$, then

$$|\mathcal{O}_\lambda^\alpha| \le c\lambda^{-p}\|\nabla^2\mathbf{z}\|_{L^p(\frac{1}{3}Q_0)}^p + c(\lambda\alpha^{-1})^{-p'}\|\partial_t\mathbf{z}\|_{L^{p'}(\frac{1}{3}Q_0)}^{p'}$$
$$= c\lambda^{-p}\big(\|\nabla^2\mathbf{z}\|_{L^p(\frac{1}{3}Q_0)}^p + \|\partial_t\mathbf{z}\|_{L^{p'}(\frac{1}{3}Q_0)}^{p'}\big). \qquad \square$$

In the following we choose λ_0 such that the conclusion of Lemma 6.1.5 is valid and assume $\lambda \ge \lambda_0$. Without loss of generality we can assume further that

$$\lambda_0 \ge \left(\fint_{\frac{1}{3}Q_0} |\nabla^2\mathbf{z}|^\sigma \,\mathrm{d}x\,\mathrm{d}t\right)^{\frac{1}{\sigma}} + r_0^{-2}\left(\fint_{\frac{1}{3}Q_0} |\mathbf{z}|^\sigma \,\mathrm{d}x\,\mathrm{d}t\right)^{\frac{1}{\sigma}}. \tag{6.1.13}$$

We define

$$\mathcal{I} := \{i : Q_i \cap \tfrac{1}{4}Q_0 \ne \emptyset\}.$$

Then Lemma 6.1.5 implies that $Q_i \subset \frac{1}{3}Q_0$ (and $Q_j \subset \frac{1}{3}Q_0$ for $j \in A_i$) for all $i \in \mathcal{I}$. For each $i \in \mathcal{I}$ we define local approximation $\mathbf{z}_i$ for $\mathbf{z}$ on Q_i by

$$\mathbf{z}_i := \Pi_{I_i}^0 \Pi_{\mathcal{B}_i}^1(\mathbf{z}), \tag{6.1.14}$$

where $\Pi_{\mathcal{B}_i}^1(\mathbf{z})$ is the first order averaged Taylor polynomial [37,63] with respect to space and $\Pi_{I_i}^0$ is the zero order averaged Taylor polynomial in time. Note that this definition implies the Poincaré-type inequality.

Lemma 6.1.7. *For all $j \in \mathbb{N}$ and $1 \le s < \infty$, if $\nabla^2\mathbf{z}, \partial_t\mathbf{z} \in L^s(\frac{1}{4}Q_0)$, then*

$$\fint_{Q_j}\left|\frac{\mathbf{z}-\mathbf{z}_j}{r_j^2}\right|^s \mathrm{d}x\,\mathrm{d}t + \fint_{Q_j}\left|\frac{\nabla(\mathbf{z}-\mathbf{z}_j)}{r_j}\right|^s \mathrm{d}x\,\mathrm{d}t$$
$$\le c\fint_{Q_j} |\nabla^2\mathbf{z}|^s \,\mathrm{d}x\,\mathrm{d}t + c\alpha^s\fint_{Q_j} |\partial_t\mathbf{z}|^s \,\mathrm{d}x\,\mathrm{d}t.$$

Proof. The estimate is a consequence of Fubini's Theorem, Poincaré estimates and the properties of the averaged Taylor polynomials see Lemma 3.1 of [63]. We find

$$\fint_{Q_j}\left|\frac{\mathbf{z}-\mathbf{z}_j}{r_j^2}\right|^s \mathrm{d}x\,\mathrm{d}t$$
$$\le c\fint_{Q_j}\left|\frac{\mathbf{z}-\Pi_{\mathcal{B}_j}^1(\mathbf{z})}{r_j^2}\right|^s \mathrm{d}x\,\mathrm{d}t + c\fint_{B_j}\fint_{I_j}\left|\frac{\Pi_{\mathcal{B}_j}^1(\mathbf{z}) - \Pi_{I_j}^0\Pi_{\mathcal{B}_j}^1(\mathbf{z})}{r_j^2}\right|^s \mathrm{d}x\,\mathrm{d}t$$
$$\le c\fint_{Q_j} |\nabla^2\mathbf{z}|^s \,\mathrm{d}x\,\mathrm{d}t + c\alpha\fint_{I_j}\fint_{\mathcal{B}_j} |\partial_t\Pi_{\mathcal{B}_j}^1(\mathbf{z})|^s \,\mathrm{d}x\,\mathrm{d}t.$$

Now the continuity of $\Pi^1_{B_j}$ on L^s gives the estimate. Similarly we find (since all norms for polynomials are equivalent)

$$\begin{aligned}
&\fint_{Q_j} \left|\frac{\nabla(\mathbf{z}-\mathbf{z}_j)}{r_j}\right|^s \mathrm{d}x\,\mathrm{d}t \\
&\leq c\fint_{Q_j} \left|\frac{\nabla(\mathbf{z}-\Pi^1_{B_j}(\mathbf{z}))}{r_j}\right|^s \mathrm{d}x\,\mathrm{d}t + c\fint_{Q_j} \left|\frac{\nabla\big(\Pi^1_{B_j}(\mathbf{z})-\Pi^0_{I_j}\Pi^1_{B_j}(\mathbf{z})\big)}{r_j}\right|^s \mathrm{d}x\,\mathrm{d}t \\
&\leq c\fint_{Q_j} \left|\frac{\nabla(\mathbf{z}-\Pi^1_{B_j}(\mathbf{z}))}{r_j}\right|^s \mathrm{d}x\,\mathrm{d}t + c\fint_{Q_j} \left|\frac{\Pi^1_{B_j}(\mathbf{z})-\Pi^0_{I_j}\Pi^1_{B_j}(\mathbf{z})}{r_j^2}\right|^s \mathrm{d}x\,\mathrm{d}t \\
&\leq c\fint_{Q_j} |\nabla^2\mathbf{z}|^s\,\mathrm{d}x\,\mathrm{d}t + c\alpha\fint_{Q_j} |\partial_t\mathbf{z}|^s\,\mathrm{d}x\,\mathrm{d}t. \qquad \square
\end{aligned}$$

We can now define our truncation $\mathbf{z}^\alpha_\lambda$ for $\lambda \geq \lambda_0$ on $\frac14 Q_0$ by

$$\mathbf{z}^\alpha_\lambda := \mathbf{z} - \sum_{i\in\mathcal{I}} \varphi_i(\mathbf{z}-\mathbf{z}_i). \tag{6.1.15}$$

It suffices to sum over i with $Q_i \cap \frac14 Q_0 \neq \emptyset$.

Since the φ_i are locally finite, this sum is pointwise well-defined. We will see later that the sum converges also in other topologies. Using $\sum_{i\in\mathcal{I}}\varphi_i = 1$ on $\frac14 Q_0$, we can also write $\mathbf{z}^\alpha_\lambda$ in the form

$$\mathbf{z}^\alpha_\lambda = \begin{cases} \mathbf{z} & \text{in } \tfrac14 Q_0 \setminus \mathcal{O}^\alpha_\lambda, \\ \sum_{i\in\mathcal{I}}\varphi_i\mathbf{z}_i & \text{in } \tfrac14 Q_0 \cap \mathcal{O}^\alpha_\lambda. \end{cases} \tag{6.1.16}$$

In the following we describe some properties of the truncation (e.g. $\nabla^2\mathbf{z}^\alpha_\lambda \in L^\infty(\frac14 Q_0)$).

Lemma 6.1.8. *For all $j \in \mathbb{N}$ and all $k \in \mathbb{N}$ with $Q_j \cap Q_k \neq \emptyset$ we have*
(a) $\fint_{Q_j} |\nabla^2\mathbf{z}|\,\mathrm{d}x\,\mathrm{d}t + \alpha\fint_{Q_j}|\partial_t\mathbf{z}|\,\mathrm{d}x\,\mathrm{d}t \leq c\,\lambda$.
(b) $\|\mathbf{z}_j - \mathbf{z}_k\|_{L^\infty(Q_j)} \leq c\fint_{Q_j}|\mathbf{z}-\mathbf{z}_j|\,\mathrm{d}x\,\mathrm{d}t + c\fint_{Q_k}|\mathbf{z}-\mathbf{z}_k|\,\mathrm{d}x\,\mathrm{d}t$.
(c) $\|\mathbf{z}_j - \mathbf{z}_k\|_{L^\infty(Q_j)} \leq c\,r_j^2\,\lambda$.

Proof. Part (a) follows from $Q_j \subset 16Q_j$ and $16Q_j \cap \mathcal{O}^{\complement}_\lambda \neq \emptyset$, so

$$\fint_{16Q_j}(|\nabla^2\mathbf{z}| + \alpha|\partial_t\mathbf{z}|)\chi_{\frac13 Q_0}\,\mathrm{d}x\,\mathrm{d}t \leq \left(\fint_{16Q_j}(|\nabla^2\mathbf{z}| + \alpha|\partial_t\mathbf{z}|)^\sigma\chi_{\frac13 Q_0}\,\mathrm{d}x\,\mathrm{d}t\right)^{\frac1\sigma} \leq c\lambda.$$

Part (b) follows from the geometric property of the Q_j. If $Q_j \cap Q_k \neq \emptyset$, then $|Q_j \cap Q_k| \geq c\max\{|Q_j|, |Q_k|\}$. This and the norm equivalence for linear polynomials imply

$$\|\mathbf{z}_j - \mathbf{z}_k\|_{L^\infty(Q_j)} \leq c\fint_{Q_j\cap Q_k}|\mathbf{z}_j - \mathbf{z}_k|\,\mathrm{d}x\,\mathrm{d}t$$

$$\le c\fint_{Q_j} |\mathbf{z}_j - \mathbf{z}|\ \mathrm{d}x + c\fint_{Q_k} |\mathbf{z} - \mathbf{z}_k|\ \mathrm{d}x.$$

Finally, (c) is a consequence of Lemma 6.1.7, (a) and (b). □

Next, we prove the stability of the truncation.

Lemma 6.1.9. *Let* $1 < s < \infty$ *and* $\mathbf{z} \in L^s(\mathbb{R}; W^{2,s}(\mathbb{R}^3))$. *Then*

$$\|\mathbf{z}^\alpha_\lambda\|_{L^s(\frac{1}{4}Q_0)} \le c\,\|\mathbf{z}\|_{L^s(\frac{1}{3}Q_0)},$$

$$\|\nabla\mathbf{z}^\alpha_\lambda\|_{L^s(\frac{1}{4}Q_0)} \le c\,\|\nabla\mathbf{z}\|_{L^s(\frac{1}{3}Q_0)} + c\alpha r_0\|\partial_t\mathbf{z}\|_{L^s(\frac{1}{3}Q_0)},$$

$$\|\nabla^2\mathbf{z}^\alpha_\lambda\|_{L^s(\frac{1}{4}Q_0)} + \alpha\|\partial_t\mathbf{z}^\alpha_\lambda\|_{L^s(\frac{1}{3}Q_0)} \le c\,\|\nabla^2\mathbf{z}\|_{L^s(\frac{1}{3}Q_0)} + c\alpha\|\partial_t\mathbf{z}\|_{L^s(\frac{1}{3}Q_0)}.$$

Moreover, the sum in (6.1.15) *converges in* $L^s(\frac{1}{4}I_0, W^{2,s}(\frac{1}{4}\mathcal{B}_0))$.

Proof. We first show that the sum in (6.1.15) converges absolutely in $L^s(\frac{1}{4}Q_0)$:

$$\int_{\frac{1}{4}Q_0} |\mathbf{z} - \mathbf{z}^\alpha_\lambda|^s\,\mathrm{d}x \le \sum_{i\in\mathcal{I}}\int_{Q_i} |\mathbf{z} - \mathbf{z}_i|^s\,\mathrm{d}x\,\mathrm{d}t \le c\sum_{i\in\mathcal{I}}\int_{Q_i} |\mathbf{z}|^s\,\mathrm{d}x\,\mathrm{d}t \le c\int_{\frac{1}{3}Q_0} |\mathbf{z}|^s\,\mathrm{d}t,$$

where we used continuity of the mapping $\mathbf{z} \mapsto \mathbf{z}_i$ in $L^s(Q_i)$, (PP1) and the finite intersection property of Q_i (PW4). We start by showing the estimate for the second derivatives

$$\int_{\mathcal{O}^\alpha_\lambda} |\nabla^2(\mathbf{z} - \mathbf{z}^\alpha_\lambda)|^s\,\mathrm{d}x\,\mathrm{d}t = \left|\sum_{i\in\mathcal{I}}\int_{Q_i} \nabla^2(\varphi_i(\mathbf{z} - \mathbf{z}_i))\,\mathrm{d}x\,\mathrm{d}t\right|$$
$$\le c\sum_{i\in\mathcal{I}}\int_{Q_i} |\nabla^2\mathbf{z}|^s + \left|\frac{\nabla(\mathbf{z} - \mathbf{z}_i)}{r_i}\right|^s + \left|\frac{\mathbf{z} - \mathbf{z}_i}{r_i^2}\right|^s\,\mathrm{d}x\,\mathrm{d}t.$$

For the time derivative we find (since $\mathbf{z}_i$ is constant in time), that

$$\begin{aligned}\int_{\mathcal{O}^\alpha_\lambda} |\partial_t(\mathbf{z} - \mathbf{z}^\alpha_\lambda)|^s\,\mathrm{d}x\,\mathrm{d}t &= \left|\sum_{i\in\mathcal{I}}\int_{Q_i} \partial_t(\varphi_i(\mathbf{z} - \mathbf{z}_i))\,\mathrm{d}x\,\mathrm{d}t\right| \\ &\le c\sum_{i\in\mathcal{I}}\int_{Q_i} |\partial_t\mathbf{z}|^s + \left|\frac{\mathbf{z} - \mathbf{z}_i}{\alpha r_i^2}\right|^s\,\mathrm{d}x\,\mathrm{d}t.\end{aligned} \tag{6.1.17}$$

Using Lemma 6.1.7 and the finite intersection of the Q_i shows that

$$\int_{\frac{1}{4}Q_0} |\nabla^2(\mathbf{z} - \mathbf{z}^\alpha_\lambda)|^s + \alpha^s|\partial_t(\mathbf{z} - \mathbf{z}^\alpha_\lambda)|^s\,\mathrm{d}x\,\mathrm{d}t \le c\sum_{i\in\mathcal{I}}\int_{Q_i} |\nabla^2\mathbf{z}|^s + \alpha^s|\partial_t\mathbf{z}|^s\,\mathrm{d}x\,\mathrm{d}t$$
$$\le c\int_{\frac{1}{3}Q_0} |\nabla^2\mathbf{z}|^s + \alpha^s|\partial_t\mathbf{z}|^s\,\mathrm{d}x\,\mathrm{d}t.$$

The estimate of the gradient is analogous, since

$$\int_{\mathcal{O}^\alpha_\lambda} |\nabla(\mathbf{z} - \mathbf{z}^\alpha_\lambda)|^s\,\mathrm{d}x\,\mathrm{d}t \le \sum_{i\in\mathcal{I}} |\nabla(\mathbf{z} - \mathbf{z}_i)|^s + \left|\frac{\mathbf{z} - \mathbf{z}_i}{r_i}\right|^s\,\mathrm{d}x\,\mathrm{d}t.$$ □

The truncation $\mathbf{z}_\lambda^\alpha$ has better regularity properties than $\mathbf{z}$. Indeed, $\nabla\mathbf{z}$ is Lipschitz.

Lemma 6.1.10. *For $\lambda > \lambda_0$ we have*

$$\|\nabla^2\mathbf{z}_\lambda^\alpha\|_{L^\infty(\frac{1}{4}Q_0)} + r_0^{-1}\|\nabla\mathbf{z}_\lambda^\alpha\|_{L^\infty(\frac{1}{4}Q_0)} + r_0^{-2}\|\mathbf{z}_\lambda^\alpha\|_{L^\infty(\frac{1}{4}Q_0)} + \alpha\|\partial_t\mathbf{z}_\lambda^\alpha\|_{L^\infty(\frac{1}{4}Q_0)} \le c\lambda.$$

Proof. If $(t,x) \in Q_i$, then

$$|\nabla^2\mathbf{z}_\lambda^\alpha(t,x)| = \Big|\sum_{j\in A_i}\nabla^2(\varphi_j\mathbf{z}_j)(t,x)\Big| \le \sum_{j\in A_i}|\nabla^2(\varphi_j(\mathbf{z}_j-\mathbf{z}_i))(t,x)|$$

because $\{\varphi_j\}$ is a partition of unity. Now we find (since all norms on polynomials are equivalent, $\#A_j \le c$ and Lemma 6.1.8) that

$$|\nabla^2\mathbf{z}_\lambda^\alpha(t,x)| \le c\sum_{j\in A_i}\frac{\|\mathbf{z}_i-\mathbf{z}_j\|_{L^\infty(Q_i)}}{r_i^2} \le c\lambda.$$

Concerning the time derivative for $(t,x)\in Q_i$ since $\mathbf{z}_i$ is constant in time we find that

$$\begin{aligned}|\partial_t\mathbf{z}_\lambda^\alpha(t,x)| &= \Big|\partial_t\sum_{j\in A_i}(\varphi_j\mathbf{z}_j)(t,x)\Big| \le \sum_{j\in A_i}|\partial_t(\varphi_j)(\mathbf{z}_j-\mathbf{z}_i)(t,x)| \\ &\le \sum_{j\in A_i}\frac{\|\mathbf{z}_i-\mathbf{z}_j\|_{L^\infty(Q_i)}}{\alpha r_i^2} \le \frac{c\lambda}{\alpha}.\end{aligned}$$

The zero order term is estimated by Poincaré's inequality; first in time and then in space

$$\begin{aligned}r_0^{-2}\|\mathbf{z}_\lambda^\alpha\|_{L^\infty(\frac{1}{4}I_0;L^\infty(\frac{1}{4}B_0))} &\le c\alpha\|\partial_t\mathbf{z}_\lambda^\alpha\|_{L^\infty(\frac{1}{4}Q_0)} + cr_0^{-2}\|\mathbf{z}_\lambda^\alpha\|_{L^1(\frac{1}{4}I_0;L^\infty(\frac{1}{4}B_0))} \\ &\le c\lambda + c\|\nabla^2\mathbf{z}_\lambda^\alpha\|_{L^\infty(\frac{1}{4}Q_0)} + cr_0^{-2}\|\mathbf{z}_\lambda^\alpha\|_{L^1(\frac{1}{4}Q_0)}.\end{aligned}$$

This implies, by the norm equivalence of polynomials, Jensen's inequality, Lemma 6.1.9 and (6.1.13),

$$r_0^{-1}\|\nabla\mathbf{z}_\lambda^\alpha\|_{L^\infty(\frac{1}{4}Q_0)} + r_0^{-2}\|\mathbf{z}_\lambda^\alpha\|_{L^\infty(\frac{1}{4}Q_0)} \le c\lambda + r_0^{-2}\|\mathbf{z}\|_{L^\sigma(\frac{1}{3}Q_0)} \le c\lambda. \qquad \square$$

The next lemma will control the time error we get when we apply the truncation as a test function.

Lemma 6.1.11. *For all $\zeta \in C_0^\infty(\frac{1}{4}Q_0)$ with $\|\nabla^2\zeta\|_\infty \le c$ and $\lambda \ge \lambda_0$,*

$$\left|\int_{\frac{1}{4}Q_0}\partial_t(\mathbf{z}-\mathbf{z}_\lambda^\alpha)\,\Delta(\zeta\mathbf{z}_\lambda^\alpha)\,\mathrm{d}x\,\mathrm{d}t\right| \le c\alpha^{-1}\lambda^2\,|\mathcal{O}_\lambda^\alpha|,$$

where the constant c is independent of α and λ.

Proof. We use Hölder's inequality and Lemma 6.1.10 to derive

$$
\begin{aligned}
(I) &:= \left| \int_{\frac{1}{4}Q_0} \partial_t(\mathbf{z} - \mathbf{z}_\lambda^\alpha)\, \Delta(\zeta \mathbf{z}_\lambda^\alpha)\, \mathrm{d}x\, \mathrm{d}t \right| \\
&\leq \sum_{i\in\mathcal{I}} \left(\int_{Q_i} |\partial_t(\varphi_i(\mathbf{z} - \mathbf{z}_i))|^\sigma \,\mathrm{d}x\,\mathrm{d}t \right)^{\frac{1}{\sigma}} \left(\int_{Q_i} |\Delta(\zeta \mathbf{z}_\lambda^\alpha)|^{\sigma'} \,\mathrm{d}x\,\mathrm{d}t \right)^{\frac{1}{\sigma'}} \\
&\leq c\lambda \sum_{i\in\mathcal{I}} |Q_i| \left(\fint_{Q_i} |\partial_t(\varphi_i(\mathbf{z} - \mathbf{z}_i))|^\sigma \,\mathrm{d}x\,\mathrm{d}t \right)^{\frac{1}{\sigma}}.
\end{aligned}
$$

Combining this with (6.1.17), (6.1.9) and (6.1.10) yields

$$
\begin{aligned}
(I) &\leq c\lambda \sum_{i\in\mathcal{I}} |Q_i| \left(\alpha^{-1} \left(\fint_{Q_j} |\nabla^2 \mathbf{z}|^\sigma \,\mathrm{d}x\,\mathrm{d}t \right)^{\frac{1}{\sigma}} + \left(\fint_{Q_j} |\partial_t \mathbf{z}|^\sigma \,\mathrm{d}x\,\mathrm{d}t \right)^{\frac{1}{\alpha}} \right) \\
&\leq c\alpha^{-1}\lambda^2 \sum_{i\in\mathcal{I}} |Q_i| \leq c\alpha^{-1}\lambda^2 |\mathcal{O}_\lambda^\alpha|,
\end{aligned}
$$

using the local finiteness of the $\{Q_i\}$ in the final step. □

Theorem 6.1.24. *Let $1 < p < \infty$ with $p, p' > \sigma$. Let $\mathbf{w}_m$ and $\mathbf{H}_m$ satisfy $\partial_t \Delta \mathbf{w}_m = -\operatorname{div}\operatorname{div}\mathbf{H}_m$ in the sense of distributions $\mathcal{D}'(\frac{1}{2}Q_0)$, see* (6.1.5)*. Further assume that $\mathbf{w}_m$ is a weak null sequence in $L^p(\frac{1}{2}I_0; W^{2,p}(\frac{1}{2}\mathcal{B}_0))$ and a strong null sequence in $L^\sigma(\frac{1}{2}Q_0)$. Further, assume that $\mathbf{H}_m = \mathbf{H}_m^1 + \mathbf{H}_m^2$ such that $\mathbf{H}_m^1$ is a weak null sequence in $L^{p'}(Q_0)$ and $\mathbf{H}_m^2$ converges strongly to zero in $L^\sigma(Q_0)$. Define $\mathbf{z}_m := \Delta\Delta_{\frac{1}{2}\mathcal{B}_0}^{-2}\Delta\mathbf{w}_m$ pointwise in time on $\frac{1}{2}I_0$. Then there is a double sequence $(\lambda_{m,k}) \subset \mathbb{R}^+$ and $k_0 \in \mathbb{N}$ such that*

(a) $2^{2^k} \leq \lambda_{m,k} \leq 2^{2^{k+1}}$

such that the double sequence $\mathbf{z}_{m,k} := \mathbf{z}_{\lambda_{m,k}}^{\alpha_{m,k}}$ with $\alpha_{m,k} := \lambda_{m,k}^{2-p}$ satisfies the following properties for all $k \geq k_0$

(b) $\{\mathbf{z}_{m,k} \neq \mathbf{z}\} \subset \mathcal{O}_{m,k} := \mathcal{O}_{\lambda_{m,k}}^{\alpha_{m,k}}$,

(c) $\|\nabla^2 \mathbf{z}_{m,k}\|_{L^\infty(\frac{1}{4}Q_0)} \leq c\lambda_{m,k}$,

(d) *$\mathbf{z}_{m,k} \to 0$ and $\nabla \mathbf{z}_{m,k} \to 0$ in $L^\infty(\frac{1}{4}Q_0)$ for $m \to \infty$ and k fixed,*

(e) *$\nabla^2 \mathbf{z}_{m,k} \rightharpoonup^* 0$ in $L^\infty(\frac{1}{4}Q_0)$ for $m \to \infty$ and k fixed,*

(f) *We have for all $\zeta \in C_0^\infty(\frac{1}{4}Q_0)$*

$$
\left| \int \big(\partial_t(\mathbf{z}_m - \mathbf{z}_{m,k})\big) \cdot \Delta(\zeta \mathbf{z}_{m,k})\,\mathrm{d}x\,\mathrm{d}t \right| \leq c\lambda_{m,k}^p |\mathcal{O}_{m,k}|,
$$

(g) $\limsup_{m\to\infty} \lambda_{m,k}^p |\mathcal{O}_{m,k}| \leq c\,2^{-k} \sup_m (\|\nabla^2 \mathbf{z}_m\|_p + c\,\|\mathbf{H}_m^1\|_{p'}^{\frac{1}{p-1}})$.

Proof. Let us assume that $\lambda_{m,k}$ satisfies (a). We will choose the precise values of $\lambda_{m,k}$ later. Due to Lemma 6.1.3 we have $\mathbf{z}_m \rightharpoonup 0$ in $L^p(\frac{1}{4}I_0; W^{2,p}(\frac{1}{4}\mathcal{B}_0))$;

this is due to the fact that the operator $\mathbf{w} \mapsto \Delta\Delta_{\frac{1}{2}\mathcal{B}_0}^{-2}\Delta\mathbf{w} = \mathbf{z}$ is linear and continuous in $L^p(\frac{1}{4}I_0; W^{2,p}(\frac{1}{4}\mathcal{B}_0))$. Then the properties (b) and (c) follow from Lemma 6.1.10. Moreover, Corollary 6.1.1 ensures that the strong convergence in $L^\sigma(\frac{1}{2}Q_0)$ transfers from $\mathbf{w}_m$ to $\mathbf{z}_m$. By Lemma 6.1.9 we get the same for $\mathbf{z}_{m,k}$ and that the sequence $\nabla^2\mathbf{z}_{m,k}$ is, for fixed k and s, bounded in $L^s(\frac{1}{4}Q_0)$. The combination of these convergence properties implies (by interpolation) (d). Moreover, the boundedness of $\nabla^2\mathbf{z}_{m,k}$ in $L^s(\frac{1}{4}Q_0)$ implies the weak convergence of a subsequence. Since (d) ensures that the limit is zero, we get, by the usual arguments, weak convergence of the whole sequence. This proves (e). Moreover, (f) follows by Lemma 6.1.11 and the choice of $\alpha_{m,k}$.

It remains to choose $2^{2^k} \le \lambda_{m,k} \le 2^{2^{k+1}}$ such that (g) holds. We use the decomposition

$$\begin{aligned}\partial_t\mathbf{z}_m &= \Delta\Delta_{\frac{1}{2}\mathcal{B}_0}^{-2}\operatorname{div}\operatorname{div}\mathbf{H}_m = \Delta\Delta_{\frac{1}{2}\mathcal{B}_0}^{-2}\operatorname{div}\operatorname{div}\mathbf{H}_m^1 + \Delta\Delta_{\frac{1}{2}\mathcal{B}_0}^{-2}\operatorname{div}\operatorname{div}\mathbf{H}_m^2\\ &=: \mathbf{h}_m^1 + \mathbf{h}_m^2.\end{aligned}$$

We decompose

$$\begin{aligned}\mathcal{O}_{m,k} &= \{\mathcal{M}_\sigma^{\alpha_{m,k}}(\chi_{\frac{1}{3}Q_0}|\nabla^2\mathbf{z}_{m,k}|) > \lambda_{m,k}\} \cup \{\alpha_{m,k}\mathcal{M}_\sigma^{\alpha_{m,k}}(\chi_{\frac{1}{3}Q_0}|\partial_t\mathbf{z}_m|) > \lambda_{m,k}\}\\ &\subset \{\mathcal{M}_\sigma^{\alpha_{m,k}}(\chi_{\frac{1}{3}Q_0}|\nabla^2\mathbf{z}_{m,k}|) > \lambda_{m,k}\} \cup \{\alpha_{m,k}\mathcal{M}_\sigma^{\alpha_{m,k}}(\chi_{\frac{1}{3}Q_0}|\mathbf{h}_m^1|) > \tfrac{1}{2}\lambda_{m,k}\}\\ &\quad \cup \{\alpha_{m,k}\mathcal{M}_\sigma^{\alpha_{m,k}}(\chi_{\frac{1}{3}Q_0}|\mathbf{h}_m^2|) > \tfrac{1}{2}\lambda_{m,k}\}\\ &=: I \cup II \cup III.\end{aligned}$$

Define

$$g_m := 2\mathcal{M}_\sigma^{\alpha_{m,k}}(\chi_{\frac{1}{3}Q_0}|\nabla^2\mathbf{z}_m|) + \left(2\mathcal{M}_\sigma^{\alpha_{m,k}}(\chi_{\frac{1}{3}Q_0}|\mathbf{h}_m^1|)\right)^{\frac{1}{p-1}}.$$

Then by the boundedness of $\mathcal{M}_\sigma$ on L^p and $L^{p'}$ (using $p, p' > \sigma$), as well as Corollary 6.1.3, we have

$$\begin{aligned}\|g_m\|_p &\le \left\|2\mathcal{M}_\sigma^{\alpha_{m,k}}(\chi_{\frac{1}{3}Q_0}|\nabla^2\mathbf{z}_m|)\right\|_p + \left\|(2\mathcal{M}_\sigma^{\alpha_{m,k}}(\chi_{\frac{1}{3}Q_0}|\mathbf{h}_m^1|))^{\frac{1}{p-1}}\right\|_p\\ &= \left\|2\mathcal{M}_\sigma^{\alpha_{m,k}}(\chi_{\frac{1}{3}Q_0}|\nabla^2\mathbf{z}_m|)\right\|_p + \left\|2\mathcal{M}_\sigma^{\alpha_{m,k}}(\chi_{\frac{1}{3}Q_0}|\mathbf{h}_m^1|)\right\|_{p'}^{\frac{1}{p-1}}\\ &\le c\left\|\nabla^2\mathbf{z}_m\right\|_{L^p(\frac{1}{3}Q_0)} + c\left\|\mathbf{h}_m^1\right\|_{L^{p'}(\frac{1}{2}Q_0)}^{\frac{1}{p-1}}\\ &\le c\left\|\nabla^2\mathbf{z}_m\right\|_{L^p(\frac{1}{3}Q_0)} + c\left\|\mathbf{H}_m^1\right\|_{L^{p'}(\frac{1}{2}Q_0)}^{\frac{1}{p-1}}.\end{aligned}$$

Let $K := \sup_m(\|\nabla^2\mathbf{z}_m\|_p + c\|\mathbf{h}_m^1\|_{p'}^{\frac{1}{p-1}})$. In particular, $\|g_m\|_p \le K$ uniformly in k. Note that

$$I \cup II = \{\mathcal{M}_\sigma^{\alpha_{m,k}}(\chi_{\frac{1}{3}Q_0}|\nabla^2\mathbf{z}_{m,k}|) > \lambda_{m,k}\} \cup \{(\mathcal{M}_\sigma^{\alpha_{m,k}}(\chi_{\frac{1}{3}Q_0}|\mathbf{h}_m^1|))^{\frac{1}{p-1}} > \lambda_{m,k}\}$$

$$\begin{aligned}
&\subset \{2\mathcal{M}_\sigma^{\alpha_{m,k}}(\chi_{\frac{1}{3}Q_0}|\nabla^2 \mathbf{z}_{m,k}|) + (2\mathcal{M}_\sigma^{\alpha_{m,k}}(\chi_{\frac{1}{3}Q_0}|\mathbf{h}_m^1|))^{\frac{1}{p-1}} > \lambda_{m,k}\} \\
&= \{g_m > \lambda_{m,k}\}.
\end{aligned}$$

We estimate

$$\begin{aligned}
\int_{\mathbb{R}^{d+1}} |g_m|^p \, dx &= \int_{\mathbb{R}^{d+1}} \int_0^\infty \frac{1}{p} t^{p-1} \chi_{\{|g_m|>t\}} \, dt \, dx \geq \int_{\mathbb{R}^{d+1}} \sum_{k \in \mathbb{N}} \frac{1}{p} 2^k \chi_{\{|g_m|>2^{k+1}\}} \, dx \\
&\geq \sum_{j \in \mathbb{N}} \sum_{k=2^j}^{2^{j+1}-1} \frac{1}{p} 2^{kp} |\{|g_m| > 2^{k+1}\}|.
\end{aligned}$$

For fixed m, j the sum over k involves 2^j summands and not all of them can be large. Consequently there exists $\lambda_{m,k} \in \{2^{2^k+1}, \dots, 2^{2^{k+1}}\}$, such that

$$\lambda_{m,k}^p \big|\{|g_m| > \lambda_{m,k}\}\big| \leq c\, 2^{-k} K^p$$

uniformly in m and k, and hence

$$\lambda_{m,k}^p \big|I \cup II\big| \leq \lambda_{m,k}^p \big|\{g_m > \lambda_{m,k}\}\big| \leq c\, 2^{-k} K^p. \tag{6.1.18}$$

On the other hand, from the weak-L^σ estimate for $\mathcal{M}_\sigma^{\alpha_{m,k}}$ we see that

$$\begin{aligned}
\limsup_{m\to\infty} \left(\lambda_{m,k}^p |III|\right) &= \limsup_{m\to\infty} \left(\lambda_{m,k}^p \big|\{\alpha_{m,k} \mathcal{M}_\sigma^{\alpha_{m,k}}(\chi_{\frac{1}{3}Q_0}|\mathbf{h}_m^2|) > \tfrac{1}{2}\lambda_{m,k}\}\big|\right) \\
&\leq \limsup_{m\to\infty} \left(c\lambda_{m,k}^p \big\|\mathbf{h}_m^2\big\|_{L^\sigma(\frac{1}{3}Q_0)}^\sigma (\alpha_{m,k}/\lambda_{m,k})^\sigma\right).
\end{aligned}$$

Since $2^{2^k+1} \leq \lambda_{m,k} \leq 2^{2^{k+1}}$, $\alpha_{m,k} = \lambda_{m,k}^{2-p}$ and $\mathbf{h}_m^2 \to 0$ in $L^\sigma(\frac{1}{2}Q_0)$ (which is a consequence of $\mathbf{H}_m^2 \to 0$ in $L^\sigma(\frac{1}{2}Q_0)$ and Corollary 6.1.3), it follows that

$$\limsup_{m\to\infty} \left(\lambda_{m,k}^p |III|\right) = 0.$$

This and (6.1.18) prove (g). □

Theorem 6.1.25. *Let $1 < p < \infty$ with $p, p' > \sigma$. Let $\mathbf{u}_m$ and $\mathbf{G}_m$ satisfy $\partial_t \mathbf{u}_m = -\operatorname{div} \mathbf{G}_m$ in the sense of distributions $\mathcal{D}'_{\text{div}}(Q_0)$. Assume that $\mathbf{u}_m$ is a weak null sequence in $L^p(I_0; W^{1,p}(\mathcal{B}_0))$ and a strong null sequence in $L^\sigma(Q_0)$ and bounded in $L^\infty(I_0, T; L^\sigma(\mathcal{B}_0))$. Further assume that $\mathbf{G}_m = \mathbf{G}_m^1 + \mathbf{G}_m^2$ such that $\mathbf{G}_m^1$ is a weak null sequence in $L^{p'}(Q_0)$ and $\mathbf{G}_m^2$ converges strongly to zero in $L^\sigma(Q_0)$. Then there is a double sequence $(\lambda_{m,k}) \subset \mathbb{R}^+$ and $k_0 \in \mathbb{N}$ with*

(a) $2^{2^k} \leq \lambda_{m,k} \leq 2^{2^{k+1}}$

such that the double sequences $\mathbf{u}_{m,k} := \mathbf{u}_{\lambda_{m,k}}^{\alpha_{m,k}} \in L^1(Q_0)$, $\alpha_{m,k} := \lambda_{m,k}^{2-p}$ and $\mathcal{O}_{m,k} := \mathcal{O}_{\lambda_{m,k}}^{\alpha_{m,k}}$ (defined in Theorem 6.1.24) satisfy the following properties for all $k \geq k_0$.

(b) $\mathbf{u}_{m,k} \in L^s(\frac{1}{4}I_0; W^{1,s}_{0,\mathrm{div}}(\frac{1}{6}\mathcal{B}_0))$ *for all* $s < \infty$ *and* $\operatorname{supp}(\mathbf{u}_{m,k}) \subset \frac{1}{6}Q_0$.

(c) $\mathbf{u}_{m,k} = \mathbf{u}_m$ *a.e. on* $\frac{1}{8}Q_0 \setminus \mathcal{O}_{m,k}$.

(d) $\|\nabla \mathbf{u}_{m,k}\|_{L^\infty(\frac{1}{4}Q_0)} \leq c\,\lambda_{m,k}$.

(e) $\mathbf{u}_{m,k} \to 0$ *in* $L^\infty(\frac{1}{4}Q_0)$ *for* $m \to \infty$ *and k fixed.*

(f) $\nabla \mathbf{u}_{m,k} \rightharpoonup^* 0$ *in* $L^\infty(\frac{1}{4}Q_0)$ *for* $m \to \infty$ *and k fixed.*

(g) $\limsup_{m\to\infty} \lambda^p_{m,k}|\mathcal{O}_{m,k}| \leq c\,2^{-k}$.

(h) $\limsup_{m\to\infty} \left| \int \mathbf{G}_m : \nabla \mathbf{u}_{m,k}\,\mathrm{d}x\,\mathrm{d}t \right| \leq c\,\lambda^p_{m,k}|\mathcal{O}_{m,k}|$.

Proof. We define, pointwise in time on I_0,

$$\tilde{\mathbf{u}}_m := \gamma \mathbf{u}_m - \mathrm{Bog}_{\mathcal{B}_0 \setminus \frac{1}{2}\mathcal{B}_0}(\nabla\gamma \cdot \mathbf{u}_m),$$
$$\mathbf{w}_m := \operatorname{curl}^{-1} \tilde{\mathbf{u}}_m,$$
$$\mathbf{z}_m := \Delta \Delta^{-2}_{\frac{1}{2}Q_0} \Delta \mathbf{w}_m,$$

where $\gamma \in C_0^\infty(Q_0)$ with $\chi_{\frac{1}{2}Q_0} \leq \gamma \leq \chi_{Q_0}$. Then we apply Theorem 6.1.24 to the sequence $\mathbf{z}_m$. Finally, let

$$\mathbf{u}_{m,k} := \operatorname{curl}(\zeta \mathbf{z}_{m,k}) + \operatorname{curl}(\zeta(\mathbf{w}_m - \mathbf{z}_m)), \tag{6.1.19}$$

where $\zeta \in C_0^\infty(\frac{1}{6}Q_0)$ with $\chi_{\frac{1}{8}Q_0} \leq \zeta \leq \chi_{\frac{1}{6}Q_0}$. This means on $\frac{1}{8}Q_0$ we have

$$\mathbf{u}_{m,k} = \mathbf{u}_m + \operatorname{curl}(\mathbf{z}_{m,k} - \mathbf{z}_m).$$

Note that $\operatorname{curl}(\mathbf{w}_m - \mathbf{z}_m)$ is harmonic (in space) on $\frac{1}{2}Q_0$ and bounded in time, due to the assumption that $\mathbf{u}_m$ is bounded uniformly in $L^\infty(I_0; L^\sigma(\mathcal{B}_0))$, which transfers to $\mathbf{w}_m$ and $\mathbf{z}_m$ by Lemma 6.1.1 and 6.1.4. This allows us to estimate the higher order spaces derivatives on $\frac{1}{4}Q_0$ by lower order ones on $\frac{1}{2}Q_0$. This, (6.1.19) and Theorem 6.1.24 immediately imply all the claimed properties except (h).

The claim of (g) follows exactly as (g) of Theorem 6.1.24.

Let us prove (h). It follows by simple density arguments that $\mathbf{u}_{m,k}$ is an admissible test function for the equation $\partial_t \mathbf{u}_m = -\operatorname{div} \mathbf{G}_m$. We thus obtain

$$\begin{aligned}
&\int \mathbf{G}_m : \nabla \mathbf{u}_{m,k}\,\mathrm{d}x\,\mathrm{d}t \\
&\quad = \int \partial_t \mathbf{u}_m \cdot \mathbf{u}_{m,k}\,\mathrm{d}x\,\mathrm{d}t \\
&\quad = \int \left(\partial_t \operatorname{curl} \mathbf{w}_m\right) \cdot \operatorname{curl}\left(\zeta \mathbf{z}_{m,k}\right) \mathrm{d}x\,\mathrm{d}t \\
&\quad + \int \left(\partial_t \operatorname{curl} \mathbf{w}_m\right) \cdot \operatorname{curl}\left(\zeta\left(\mathbf{w}_m - \mathbf{z}_m\right)\right) \mathrm{d}x\,\mathrm{d}t
\end{aligned}$$

$$= -\int (\partial_t \mathbf{z}_m)\cdot \Delta(\zeta \mathbf{z}_{m,k})\,dx\,dt - \int (\partial_t \mathbf{z}_m)\cdot \Delta(\zeta(\mathbf{w}_m - \mathbf{z}_m))\,dx\,dt$$
$$=: T_1 + T_2.$$

Here we took into account $\operatorname{curl}\operatorname{curl}\mathbf{w}_m = -\Delta\mathbf{w}_m$ (due to $\operatorname{div}\mathbf{w}_m = 0$) and $\Delta\mathbf{w}_m = \Delta\mathbf{z}_m$. By assumption $\mathbf{G}_m$ is bounded in $L^\sigma(Q_0)$. Using regularity properties of harmonic functions (for $\mathbf{w}_m - \mathbf{z}_m$) as well as Lemma 6.1.3 and Lemma 6.1.1 we gain (after choosing a subsequence)

$$\Big(\fint_{Q_0} |\Delta(\zeta(\mathbf{w}_m - \mathbf{z}_m))|^{\sigma'}\,dx\,dt\Big)^{\frac{1}{\sigma'}}$$
$$\le c\, r_0^{-2}\Big(\fint_{\frac14 Q_0} |\mathbf{w}_m - \mathbf{z}_m|^{\frac{3\sigma}{3+\sigma}}\,dx\,dt\Big)^{\frac{3+\sigma}{3\sigma}}$$
$$\le c\, r_0^{-2}\Big(\fint_{\frac12 Q_0} |\mathbf{w}_m|^{\frac{3\sigma}{3+\sigma}}\,dx\,dt\Big)^{\frac{3+\sigma}{3\sigma}}$$
$$\le c\, r_0^{-3}\Big(\fint_{Q_0} |\tilde{\mathbf{u}}_m|^{\sigma}\,dx\,dt\Big)^{\frac{1}{\sigma}} \longrightarrow 0 \quad \text{as} \quad m\to\infty.$$

Since, additionally, $\partial_t\mathbf{z}_m$ is uniformly bounded in $L^\sigma(\frac12 Q_0)$ by Lemma 6.1.3, we have $T_2 \to 0$ as $m\to\infty$. Furthermore, there holds

$$T_1 = \int (\partial_t(\mathbf{z}_m - \mathbf{z}_{m,k}))\cdot\Delta(\zeta\mathbf{z}_{m,k})\,dx\,dt + \int (\partial_t\mathbf{z}_{m,k})\cdot\Delta(\zeta\mathbf{z}_{m,k})\,dx\,dt$$
$$=: T_{1,1} + T_{1,2},$$

where the first term can be bounded using Theorem 6.1.24 (f). So it remains to show that

$$T_{1,2} := \int (\partial_t\mathbf{z}_{m,k})\cdot\Delta(\zeta\mathbf{z}_{m,k})\,dx\,dt \longrightarrow 0 \qquad \text{as} \quad m\to\infty.$$

We have

$$T_{1,2} = -\int \frac12 \partial_t(|\nabla\mathbf{z}_{m,k}|^2)\zeta\,dx\,dt + \int (\partial_t\mathbf{z}_{m,k})\cdot \operatorname{div}(\nabla\zeta\otimes\mathbf{z}_{m,k})\,dx\,dt$$
$$= \int \frac12 |\nabla\mathbf{z}_{m,k}|^2\partial_t\zeta\,dx\,dt + \int (\partial_t\mathbf{z}_{m,k})\cdot\operatorname{div}(\nabla\zeta\otimes\mathbf{z}_{m,k})\,dx\,dt.$$

The first term is estimated by Theorem 6.1.24(d). For the second we use Lemma 6.1.9 and Lemma 6.1.3 ($s=\sigma$) to find

$$\int |\partial_t\mathbf{z}_{m,k}||\operatorname{div}(\nabla\zeta\otimes\mathbf{z}_{m,k})|\,dx\,dt$$
$$\le c\Big(\int_{\frac13 Q_0} |\mathbf{G}_m|^\sigma + |\nabla^2\mathbf{z}_m|^\sigma\,dx\,dt\Big)^{\frac1\sigma}\Big(\int_{\frac13 Q_0} |\nabla\mathbf{z}_{m,k}|^{\sigma'} + |\mathbf{z}_{m,k}|^{\sigma'}\,dx\,dt\Big)^{\frac{1}{\sigma'}}.$$

Now because $\mathbf{G}_m$ and $\nabla^2 \mathbf{z}_m$ are uniformly bounded in $L^\sigma(\frac{1}{2}Q_0)$ we find by Theorem 6.1.24 (d), that

$$\lim_{m\to\infty} T_{1,2} = 0,$$

which proves the claim of (h). □

The following corollary is useful in the application of the solenoidal Lipschitz truncation.

Corollary 6.1.4. *Let all assumptions of Theorem 6.1.25 be satisfied with* $\zeta \in C_0^\infty(\frac{1}{6}Q_0)$ *with* $\chi_{\frac{1}{8}Q_0} \le \zeta \le \chi_{\frac{1}{6}Q_0}$ *as in the proof of Theorem 6.1.25. If additionally* $\mathbf{u}_m$ *is uniformly bounded in* $L^\infty(I_0, L^\sigma(\mathcal{B}_0))$*, then for every* $\mathbf{K} \in L^{p'}(\frac{1}{6}Q_0)$

$$\limsup_{m\to\infty} \left| \int ((\mathbf{G}_m^1 + \mathbf{K}) : \nabla \mathbf{u}_m)\zeta \chi_{\mathcal{O}_{m,k}^\complement} \,\mathrm{d}x\,\mathrm{d}t \right| \le c\,2^{-k/p}.$$

Proof. It follows from (f), (g) and (h) of Theorem 6.1.25 that

$$\limsup_{m\to\infty} \left| \int (\mathbf{G}_m + \mathbf{K}) : \nabla \mathbf{u}_{m,k} \,\mathrm{d}x\,\mathrm{d}t \right| \le c\lambda_{m,k}^p |\mathcal{O}_{m,k}| \le c\,2^{-k}. \tag{6.1.20}$$

Recall that $\mathbf{u}_{m,k} = \operatorname{curl}(\zeta \mathbf{z}_{m,k}) + \operatorname{curl}(\zeta(\mathbf{w}_m - \mathbf{z}_m))$. So, by Theorem 6.1.24, we have $\mathbf{z}_{m,k}, \nabla \mathbf{z}_{m,k} \to 0$ in $L^\infty(\frac{1}{4}Q_0)$ as $m \to \infty$ with k fixed. Since $\mathbf{u}_m$ is a strong null sequence in $L^\sigma(Q_0)$ and is bounded in $L^\infty(I_0, L^\sigma(\mathcal{B}_0))$ we see that $\mathbf{u}_m \to 0$ strongly in $L^s(I_0, L^\sigma(\mathcal{B}_0))$ for any $s \in (1, \infty)$. By continuity of the Bogovskiĭ operator (see Theorem 2.1.6 with $A(t) = B(t) = t^\sigma$) we have the same convergence for $\tilde{\mathbf{u}}_m$. Now, Lemma 6.1.1 implies $\mathbf{w}_m = \operatorname{curl}^{-1} \tilde{\mathbf{u}}_m \to 0$ in $L^s(I_0, W^{1,\sigma}(\mathbb{R}^3))$. Using $\mathbf{z}_m := \Delta\Delta_{\frac{1}{2}Q_0}^{-2} \Delta \mathbf{w}_m$ and Corollary 6.1.1 we also have $\mathbf{z}_m \to 0$ in $L^s(I_0, W^{1,\sigma}(\mathbb{R}^3))$. Since $\mathbf{z}_m - \mathbf{w}_m$ is harmonic on $\frac{1}{4}Q_0$, we have $\mathbf{z}_m - \mathbf{w}_m \to 0$ in $L^s(I_0, W^{2,s}(\frac{1}{6}\mathcal{B}_0))$. These convergences imply that

$$\nabla \mathbf{u}_{m,k} = \zeta \nabla \operatorname{curl} \mathbf{z}_{m,k} + \mathbf{a}_{m,k},$$

with $\mathbf{a}_{m,k} \to 0$ in $L^s(\frac{1}{6}Q_0)$ as $m \to \infty$ with k fixed. This, the boundedness of $\mathbf{G}_m$ in $L^\sigma(\frac{1}{6}Q_0)$, $\mathbf{K} \in L^{p'}(\frac{1}{6}Q_0)$ and (6.1.20) imply (using $s > \sigma'$)

$$\limsup_{m\to\infty} \left| \int ((\mathbf{G}_m + \mathbf{K}) : \nabla \operatorname{curl} \mathbf{z}_{m,k})\zeta \,\mathrm{d}x\,\mathrm{d}t \right| \le c\,2^{-k}.$$

Since $\mathbf{G}_m = \mathbf{G}_m^1 + \mathbf{G}_m^2$, $\mathbf{G}_m^2 \to 0$ in $L^\sigma(\frac{1}{6}Q_0)$ and $\mathbf{z}_{m,k} \rightharpoonup 0$ in $L^{\sigma'}(\frac{1}{6}Q_z)$ for $m \to \infty$ and k fixed, we have

$$\limsup_{m\to\infty} \left| \int ((\mathbf{G}_m^1 + \mathbf{K}) : \nabla \operatorname{curl} \mathbf{z}_{m,k})\zeta \,\mathrm{d}x\,\mathrm{d}t \right| \le c\,2^{-k}. \tag{6.1.21}$$

The boundedness of $\mathbf{G}_m^1$ and $\mathbf{K}$ in $L^{p'}(\frac{1}{6}Q_0)$ and Theorem 6.1.24 and (g) prove

$$\limsup_{m\to\infty}\left|\int((\mathbf{G}_m^1+\mathbf{K}):\nabla\operatorname{curl}\mathbf{z}_{m,k})\zeta\,\chi_{\mathcal{O}_{m,k}}\,\mathrm{d}x\,\mathrm{d}t\right|\le c\,2^{-k/p}.$$

This, together with (6.1.21) and $\mathbf{z}_{m,k}=\mathbf{z}_m$ in $\mathcal{O}_{m,k}^{\complement}$ yield

$$\limsup_{m\to\infty}\left|\int((\mathbf{G}_m^1+\mathbf{K}):\nabla\operatorname{curl}\mathbf{z}_m)\zeta\,\chi_{\mathcal{O}_{m,k}^{\complement}}\,\mathrm{d}x\,\mathrm{d}t\right|\le c\,2^{-k/p}.$$

Recall that $\mathbf{z}_m-\mathbf{w}_m\to 0$ in $L^s(I_0,W^{2,s}(\frac{1}{6}\mathcal{B}_0))$ for any $s\in(1,\infty)$. This and the boundedness of $\mathbf{G}_m^1$ in $L^{p'}(Q_0)$ allows us to replace $\mathbf{z}_m$ in the previous integral by $\mathbf{w}_m$. Now $\operatorname{curl}\mathbf{w}_m=\mathbf{u}_m$ proves the claim. □

The next corollary follows by combining Lemma 6.1.9, Lemma 6.1.10, Theorem 6.1.24 (g) (with $\alpha=1$) and the continuity of curl^{-1} with a scaling procedure.

Corollary 6.1.5. *For some $\sigma>0$ let $\mathbf{u}\in L^\sigma(I_0;W^{1,\sigma}_{\operatorname{div}}(\mathcal{B}_0))\cap L^\infty(I;L^\sigma(\mathcal{B}_0))$ with $\partial_t\mathbf{u}=\operatorname{div}\mathbf{H}$ in $D'_{\operatorname{div}}(Q_0)$ for some $\mathbf{H}\in L^\sigma(Q_0)$. Then for every $m_0\gg 1$ and $\gamma>0$ there exist $\lambda\in[2^{m_0}\gamma,2^{2m_0}\gamma]$ and a function $\mathbf{u}_\lambda$ with the following properties.*

(a) *It holds $\mathbf{u}_\lambda\in L^\infty(I_0,W^{1,\infty}_{0,\operatorname{div}}(B_0))$ with $\|\nabla\mathbf{u}_\lambda\|_\infty\le c\lambda$.*

(b) *We have*

$$\lambda^\sigma\frac{\mathcal{L}^{d+1}\left(\frac{1}{2}Q_0\cap\{\mathbf{u}_\lambda\neq\mathbf{u}\}\right)}{|Q_0|}\le\frac{c}{m_0}\left(\fint_{Q_0}r_0^{-\sigma}|\mathbf{u}|^\sigma+|\nabla\mathbf{u}|^\sigma\,\mathrm{d}x\,\mathrm{d}t+\fint_{Q_0}|\mathbf{H}|^\sigma\,\mathrm{d}x\,\mathrm{d}t\right).$$

(c) *It holds*

$$\fint_{Q_0}|\mathbf{u}_\lambda|^\sigma\,\mathrm{d}x\,\mathrm{d}t\le c\left(\fint_{Q_0}|\mathbf{u}|^\sigma+\fint_{Q_0}r_0^\sigma|\mathbf{H}|^\sigma\,\mathrm{d}x\,\mathrm{d}t\right),$$

$$\fint_{Q_0}|\nabla\mathbf{u}_\lambda|^\sigma\,\mathrm{d}x\,\mathrm{d}t\le c\left(\fint_{Q_0}r_0^{-\sigma}|\mathbf{u}|^\sigma+|\nabla\mathbf{u}|^\sigma\,\mathrm{d}x\,\mathrm{d}t+\fint_{Q_0}|\mathbf{H}|^\sigma\,\mathrm{d}x\,\mathrm{d}t\right).$$

(d) *We have $\partial_t(\mathbf{u}-\mathbf{u}_\lambda)\in L^{\sigma'}(\frac{1}{2}I_0,W^{-1,\sigma'}(\frac{1}{2}\mathcal{B}_0))$ and*

$$-\fint_{\frac{1}{2}Q_0}(\mathbf{u}-\mathbf{u}_\lambda)\cdot\partial_t\boldsymbol{\varphi}\,\mathrm{d}x\,\mathrm{d}t$$

$$\le c(\kappa)\int_{\frac{1}{2}Q_0}\chi_{\{\mathbf{u}_\lambda\neq\mathbf{u}\}}|\nabla\boldsymbol{\varphi}|^{\sigma'}\,\mathrm{d}x\,\mathrm{d}t+\kappa\left(\fint_{Q_0}r_0^{-\sigma}|\mathbf{u}|^\sigma+|\nabla\mathbf{u}|^\sigma+|\mathbf{H}|^\sigma\,\mathrm{d}x\,\mathrm{d}t\right)$$

for all $\boldsymbol{\varphi}\in C_0^\infty(\frac{1}{2}Q_0)$ and all $\kappa>0$.

Proof. We apply the arguments used in the proof of Theorem 6.1.25 to the constant sequence $\mathbf{u}$ with the choice $\alpha = 1$. So we set

$$\mathbf{u}_\lambda := \operatorname{curl}(\zeta \mathbf{z}_\lambda) + \operatorname{curl}(\zeta(\mathbf{w} - \mathbf{z})), \tag{6.1.22}$$

where $\zeta \in C_0^\infty(\frac{1}{6}Q_0)$ with $\chi_{\frac{1}{8}Q_0} \leq \zeta \leq \chi_{\frac{1}{6}Q_0}$. Hence, in $\frac{1}{8}Q_0$ we have

$$\mathbf{u}_\lambda = \mathbf{u} + \operatorname{curl}(\mathbf{z}_\lambda - \mathbf{z}).$$

We immediately obtain the claim of (a). As a consequence of the Lemmas 6.1.1, 6.1.4 and 6.1.9 we obtain the inequalities

$$\fint_Q |\mathbf{u}_\lambda|^\sigma \,\mathrm{d}x\,\mathrm{d}t \leq c\left(\fint_{Q_0} |\mathbf{u}|^\sigma + \fint_Q r_0^\sigma |\partial_t \mathbf{z}|^\sigma \,\mathrm{d}x\,\mathrm{d}t\right),$$

$$\fint_{Q_0} |\nabla\mathbf{u}_\lambda|^\sigma \,\mathrm{d}x\,\mathrm{d}t \leq c\left(\fint_{Q_0} r_0^{-\sigma} |\mathbf{u}|^\sigma + |\nabla\mathbf{u}|^\sigma \,\mathrm{d}x\,\mathrm{d}t + \fint_Q |\partial_t \mathbf{z}|^\sigma \,\mathrm{d}x\,\mathrm{d}t\right),$$

claimed in c). Finally we can replace $\partial_t \mathbf{z}$ by $\mathbf{H}$ on account of Lemma 6.1.3.

It remains to find good levels. We define for some $s \in (1, \max\{\sigma, \sigma'\})$

$$\mathcal{O}_\lambda := \{\mathcal{M}_s(\chi_{\frac{1}{3}Q_0} |\nabla^2 \mathbf{z}|) > \lambda\} \cup \{\mathcal{M}_s(\chi_{\frac{1}{3}Q_0} |\mathbf{H}|) > \lambda\},$$
$$g := \mathcal{M}_s(\chi_{\frac{1}{3}Q_0} |\nabla^2 \mathbf{z}|) + \mathcal{M}_s(\chi_{\frac{1}{3}Q_0} |\mathbf{H}|).$$

It now follows from the continuity of $\mathcal{M}_s$ in (5.2.4), together with Lemmas 6.1.1 and 6.1.4

$$\int_{\mathbb{R}^{d+1}} |g|^\sigma \,\mathrm{d}x \leq c\left(\int_{Q_0} r_0^{-\sigma} |\mathbf{u}|^\sigma + |\nabla\mathbf{u}|^\sigma \,\mathrm{d}x\,\mathrm{d}t + \int_{Q_0} |\mathbf{H}|^\sigma \,\mathrm{d}x\,\mathrm{d}t\right). \tag{6.1.23}$$

Furthermore, the following holds for every $m_0 \in \mathbb{N}$ and every $\gamma > 0$

$$\begin{aligned}\int_{\mathbb{R}^{d+1}} |g|^\sigma \,\mathrm{d}x &= \int_{\mathbb{R}^{d+1}} \int_0^\infty \frac{1}{\sigma} t^{\sigma-1} \chi_{\{|g|>t\}} \,\mathrm{d}t\,\mathrm{d}x \\ &\geq \int_{\mathbb{R}^{d+1}} \sum_{m=m_0}^{2m_0-1} \frac{1}{\sigma} (2^m \gamma)^\sigma \chi_{\{|g|>\gamma 2^{m+1}\}} \,\mathrm{d}x.\end{aligned}$$

So, there is $m_1 \in \{m_0, \ldots, 2m_0 - 1\}$ such that

$$\int_{\mathbb{R}^{d+1}} (2^{m_1}\gamma)^\sigma \chi_{\{|g|>\gamma 2^{m_1+1}\}} \,\mathrm{d}x \leq \frac{c}{m_0} \int_{\mathbb{R}^{d+1}} |g|^\sigma \,\mathrm{d}x.$$

Setting $\lambda = \gamma 2^{m_1+1}$ yields

$$\lambda^\sigma |\tfrac{1}{3} Q_0 \cap \{|g| > \lambda\}| \leq \frac{c}{m_0} \int_{\mathbb{R}^{d+1}} |g|^\sigma \,\mathrm{d}x.$$

Combining this with (6.1.23) gives the estimate in b) due to the definition of $\mathcal{O}_\lambda$.

Finally, we prove d). We have $\mathbf{u}_\lambda - \mathbf{u} = \operatorname{curl}(\mathbf{z}_\lambda - \mathbf{z})$ in $\frac{1}{8}Q_0$ such that Lemma 6.1.3 and 6.1.9 imply $\partial_t(\mathbf{u}_\lambda - \mathbf{u}) \in L^{\sigma'}(\frac{1}{8}I_0, W^{-1,\sigma'}(\frac{1}{8}\mathcal{B}_0))$. Moreover, we have for $\boldsymbol{\varphi} \in C_0^\infty(\frac{1}{8}Q_0)$

$$-\int_{\frac{1}{8}Q_0} (\mathbf{u} - \mathbf{u}_\lambda) \cdot \partial_t \boldsymbol{\varphi} \,\mathrm{d}x\,\mathrm{d}t = \int_{\frac{1}{8}Q_0} \chi_{\mathcal{O}_\lambda} \partial_t(\mathbf{z} - \mathbf{z}_\lambda) \cdot \operatorname{curl} \boldsymbol{\varphi} \,\mathrm{d}x\,\mathrm{d}t$$
$$\leq c(\kappa) \int_{\frac{1}{2}Q_0} \chi_{\mathcal{O}_\lambda} |\nabla \boldsymbol{\varphi}|^{\sigma'} \,\mathrm{d}x\,\mathrm{d}t + \kappa \left(\fint_{Q_0} |\partial_t(\mathbf{z}_\lambda - \mathbf{z})|^\sigma \,\mathrm{d}x\,\mathrm{d}t \right),$$

as a consequence of Young's inequality. Applying Lemmas 6.1.3, 6.1.4 and 6.1.9 yields

$$\fint_{Q_0} |\partial_t(\mathbf{z}_\lambda - \mathbf{z})|^\sigma \,\mathrm{d}x\,\mathrm{d}t \leq c\left(\fint_{Q_0} r_0^{-\sigma} |\mathbf{u}|^\sigma + |\nabla \mathbf{u}|^\sigma \,\mathrm{d}x\,\mathrm{d}t + \fint_{Q_0} |\mathbf{H}|^\sigma \,\mathrm{d}x\,\mathrm{d}t \right).$$

So we have shown the estimate claimed in (d) on $\frac{1}{8}Q_0$.

A simple scaling argument allows us the obtain all the estimates on $\frac{1}{2}Q_0$. □

Remark 6.1.10 (The higher dimensional case). For general dimensions, the solenoidal Lipschitz truncation is best understood in terms of differential forms. We start with $\tilde{\mathbf{u}}$ as given in (6.1.3). Now, we have to find $\mathbf{w}$ such that $\operatorname{curl} \mathbf{w} = \tilde{\mathbf{u}}$ and $\operatorname{div} \mathbf{w} = 0$. Let us define the 1-form α in $\mathbb{R}^d$ associated to the vector field $\tilde{\mathbf{u}}$ by $\alpha := \sum_i \tilde{u}_i dx^i$. Then we need to find a 2-form G such that $d^*G = \alpha$ and $dG = 0$, where d is the outer derivative and d^* its adjoint by the scalar product for k-forms. Similar to $\mathbf{w} = \operatorname{curl}^{-1} \tilde{\mathbf{u}} = \operatorname{curl} \Delta^{-1} \tilde{\mathbf{u}}$ we get G by $G := d\Delta^{-1}\alpha$. Since we are on the whole space, Δ^{-1} can be constructed by mollification with $c\,|x|^{2-d}$. Thus, we have

$$G(x) = (d\Delta^{-1}\alpha)(x) = d\sum_i \left(\int_{\mathbb{R}^d} \frac{\mathbf{u}_i(y)}{|x-y|^{d-2}} dy \right) dx^i.$$

Let us explain how to substitute the equation $\partial_t \Delta \mathbf{w} = -\operatorname{curl} \operatorname{div} \mathbf{G}$, see (6.1.4). Instead of test functions $\boldsymbol{\psi}$ with $\operatorname{div} \boldsymbol{\psi} = 0$ we use the associated 1-forms $\beta = \sum_i \psi_i dx^i$ with $d^*\beta = 0$. Thus there exists a 2-form γ with $d^*\gamma = 0$. Then

$$\langle \partial_t \tilde{\mathbf{u}}, \boldsymbol{\psi} \rangle = \langle \partial_t \alpha, \beta \rangle = \langle \partial_t d^* G, d^* \gamma \rangle = \langle \partial_t d d^* G, \gamma \rangle = \langle -\partial_t \Delta G, \gamma \rangle,$$

where we used $-\Delta = dd^* + d^*d$ and $d\alpha = 0$ in the last step. Note that $-\Delta$ applied to the form G is the same as $-\Delta$ applied to the vector field of all components of G. Now we define $\mathbf{w}$ as the associated vector field (with $\binom{d}{2}$ components) of G and we arrive again at an equation for $\partial_t \Delta \mathbf{w}$. This concludes the construction; the rest can be done exactly as for dimension three. The restriction $p > \frac{6}{5}$ used in this section will change to $\frac{2d}{d+2}$.

6.2 $\mathcal{A}$-STOKES APPROXIMATION – EVOLUTIONARY CASE

By $\mathcal{A}$ we denote a symmetric, elliptic tensor, i.e.

$$c_0|\boldsymbol{\tau}|^2 \le \mathcal{A}(\boldsymbol{\tau},\boldsymbol{\tau}) \le c_1|\boldsymbol{\tau}|^2 \quad \text{for all} \quad \boldsymbol{\tau} \in \mathbb{R}^{d\times d}. \tag{6.2.24}$$

We set $|\mathcal{A}| := c_1/c_0$. Let $\mathcal{B} \subset \mathbb{R}^d$ be a ball and $J = (t_0, t_1)$ a bounded interval. We set $Q = J \times \mathcal{B}$. For a function $\mathbf{w} \in L^1(Q)$ with $\partial_t \mathbf{w} \in L^{q'}(J; W^{-1,q'}(\mathcal{B}))$ we introduce the unique function $\mathbf{H}_\mathbf{w} \in L_0^{q'}(Q)$ with

$$\int_Q \mathbf{w}\cdot\partial_t\boldsymbol{\varphi}\,\mathrm{d}x\,\mathrm{d}t = \int_Q \mathbf{H}_\mathbf{w} : \nabla\boldsymbol{\varphi}\,\mathrm{d}x\,\mathrm{d}t$$

for all $\boldsymbol{\varphi} \in C^\infty_{0,\mathrm{div}}(Q)$. We begin with a variational inequality for the non-stationary $\mathcal{A}$-Stokes system.

Lemma 6.2.1. *Suppose that* (6.2.24) *holds and that* $q > 1$. *There holds for every* $\mathbf{u} \in C_w([t_0,t_1]; L^1(\mathcal{B})) \cap L^q(J; W^{1,q}(\mathcal{B}))$ *with* $\mathbf{u}(t_0,\cdot) = 0$ *a.e.*

$$\begin{aligned}\fint_Q |\nabla\mathbf{u}|^q\,\mathrm{d}x\,\mathrm{d}t \le c \sup_{\boldsymbol{\xi}\in C^\infty_{0,\mathrm{div}}(Q)} \bigg[&\fint_Q \Big(\mathcal{A}(\boldsymbol{\varepsilon}(\mathbf{u}),\boldsymbol{\varepsilon}(\boldsymbol{\xi})) - \mathbf{u}\cdot\partial_t\boldsymbol{\xi}\Big)\,\mathrm{d}x\,\mathrm{d}t \\ &- \fint_Q \Big(|\nabla\boldsymbol{\xi}|^{q'} + |\mathbf{H}_{\boldsymbol{\xi}}|^{q'}\Big)\,\mathrm{d}x\,\mathrm{d}t\bigg],\end{aligned}$$

where c *only depends on* $\mathcal{A}$, q *and* d.

Proof. Duality arguments show that

$$\frac{1}{q}\fint_Q |\nabla\mathbf{u}|^q\,\mathrm{d}x\,\mathrm{d}t = \sup_{\mathbf{G}\in L^{q'}(Q)} \bigg[\fint_Q \nabla\mathbf{u} : \mathbf{G}\,\mathrm{d}x\,\mathrm{d}t - \frac{1}{q'}\fint_Q |\mathbf{G}|^{q'}\,\mathrm{d}x\,\mathrm{d}t\bigg].$$

For a given $\mathbf{G} \in L^{q'}(Q)$ let $\mathbf{z}_\mathbf{G}$ be the unique $L^{q'}(J; W^{1,q'}_{0,\mathrm{div}}(\mathcal{B}))$-solution to

$$\fint_Q \mathbf{z}\cdot\partial_t\boldsymbol{\xi}\,\mathrm{d}x\,\mathrm{d}t + \int_Q \mathcal{A}(\boldsymbol{\varepsilon}(\mathbf{z}),\boldsymbol{\varepsilon}(\boldsymbol{\xi}))\,\mathrm{d}x\,\mathrm{d}t = \int_Q \mathbf{G} : \nabla\boldsymbol{\xi}\,\mathrm{d}x\,\mathrm{d}t \tag{6.2.25}$$

for all $\boldsymbol{\xi} \in C^\infty_{0,\mathrm{div}}((t_0,t_1] \times B)$. This is a backward parabolic equation with end datum zero. We have that $\partial_t \mathbf{z}_\mathbf{G} \in L^{q'}(J; W^{-1,q'}_{\mathrm{div}}(\mathcal{B}))$, so test-functions can be chosen from the space $L^q(J; W^{1,q}_{0,\mathrm{div}}(\mathcal{B}))$. Due to Theorem B.3.50 (which can be applied to $\tilde{\mathbf{z}}_{\tilde{\mathbf{G}}}(t,\cdot) = \mathbf{z}_\mathbf{G}(t_1 - t,\cdot)$, where $\tilde{\mathbf{G}}(t,\cdot) = \mathbf{G}(t_1 - t,\cdot)$) this solution satisfies

$$\fint_Q |\nabla\mathbf{z}_\mathbf{G}|^{q'}\,\mathrm{d}x\,\mathrm{d}t + \fint_Q |\mathbf{H}_{\mathbf{z}_\mathbf{G}}|^{q'}\,\mathrm{d}x\,\mathrm{d}t \le c\fint_Q |\mathbf{G}|^{q'}\,\mathrm{d}x\,\mathrm{d}t.$$

In other words, the mapping $L^{q'}(\mathcal{B}) \ni \mathbf{G} \mapsto \mathbf{z_G} \in L^{q'}(J; W^{1,q'}_{0,\mathrm{div}}(\mathcal{B}))$ is continuous. This and $\mathbf{u}(t_0, \cdot) = 0$ yield (using $\mathbf{u}$ as a test-function in (6.2.25))

$$\begin{aligned}
&\fint_Q |\nabla \mathbf{u}|^q \,\mathrm{d}x\,\mathrm{d}t \\
&\quad\le c \sup_{\mathbf{G}\in L^{q'}(Q)} \bigg[\fint_Q \mathcal{A}(\boldsymbol{\varepsilon}(\mathbf{u}), \boldsymbol{\varepsilon}(\mathbf{z_G}))\,\mathrm{d}x\,\mathrm{d}t - \fint_Q \partial_t \mathbf{z_G}\cdot \mathbf{u}\,\mathrm{d}x\,\mathrm{d}t \\
&\quad - \fint_Q \Big(|\nabla \mathbf{z_H}|^{q'} + |\mathbf{H}_{\mathbf{z_G}}|^{q'}\Big)\,\mathrm{d}x\,\mathrm{d}t\bigg] \\
&\quad\le c \sup_{\boldsymbol{\xi}\in C^\infty_{0,\mathrm{div}}(Q)} \bigg[\fint_Q \mathcal{A}(\boldsymbol{\varepsilon}(\mathbf{u}), \boldsymbol{\varepsilon}(\boldsymbol{\xi}))\,\mathrm{d}x\,\mathrm{d}t - \fint_Q \mathbf{u}\cdot\partial_t \boldsymbol{\xi}\,\mathrm{d}x\,\mathrm{d}t \\
&\quad - \fint_Q \Big(|\nabla \boldsymbol{\xi}|^{q'} + |\mathbf{H}_{\boldsymbol{\xi}}|^{q'}\Big)\,\mathrm{d}x\,\mathrm{d}t\bigg],
\end{aligned}$$

which yields the claim. □

Let us now state the $\mathcal{A}$-Stokes approximation. In the following let $\mathcal{B}$ be a ball with radius r and J an interval with length $2r^2$. Let $\tilde{Q}$ denote either $Q = J \times \mathcal{B}$ or $2Q$. We use similar notations for $\tilde{J}$ and $\tilde{\mathcal{B}}$.

Theorem 6.2.26. *Suppose that (6.2.24) holds. Let $\mathbf{v} \in L^{qs}(2\tilde{J}; W^{1,qs}_{\mathrm{div}}(2\tilde{\mathcal{B}}))$, $q, s > 1$, be an almost $\mathcal{A}$-Stokes solution in the sense that*

$$\left|\fint_{2Q} \mathbf{v}\cdot\partial_t\boldsymbol{\xi}\,\mathrm{d}x\,\mathrm{d}t - \fint_{2Q} \mathcal{A}(\boldsymbol{\varepsilon}(\mathbf{v}), \boldsymbol{\varepsilon}(\boldsymbol{\xi}))\,\mathrm{d}x\,\mathrm{d}t\right| \le \delta \fint_{2Q} |\boldsymbol{\varepsilon}(\mathbf{v})|\,\mathrm{d}x\,\mathrm{d}t\, \|\nabla\boldsymbol{\xi}\|_\infty \tag{6.2.26}$$

for all $\boldsymbol{\xi} \in C^\infty_{0,\mathrm{div}}(2Q)$ and some small $\delta > 0$. Then the unique solution $\mathbf{w} \in L^q(J; W^{1,q}_{0,\mathrm{div}}(\mathcal{B}))$ to

$$\begin{aligned}
&\int_Q \mathbf{w}\cdot\partial_t\boldsymbol{\xi}\,\mathrm{d}x\,\mathrm{d}t - \int_Q \mathcal{A}(\boldsymbol{\varepsilon}(\mathbf{w}), \boldsymbol{\varepsilon}(\boldsymbol{\xi}))\,\mathrm{d}x\,\mathrm{d}t \\
&\quad = \int_Q \mathbf{v}\cdot\partial_t\boldsymbol{\xi}\,\mathrm{d}x\,\mathrm{d}t - \int_Q \mathcal{A}(\boldsymbol{\varepsilon}(\mathbf{v}), \boldsymbol{\varepsilon}(\boldsymbol{\xi}))\,\mathrm{d}x\,\mathrm{d}t
\end{aligned} \tag{6.2.27}$$

for all $\boldsymbol{\xi} \in C^\infty_{0,\mathrm{div}}([t_0, t_1) \times \mathcal{B})$ satisfies

$$\fint_Q \left|\frac{\mathbf{w}}{r}\right|^q \mathrm{d}x\,\mathrm{d}t + \fint_Q |\nabla \mathbf{w}|^q\,\mathrm{d}x\,\mathrm{d}t \le \kappa \left(\fint_{2\tilde{Q}} |\nabla \mathbf{v}|^{qs}\,\mathrm{d}x\,\mathrm{d}t\right)^{\frac{1}{s}}.$$

It holds $\kappa = \kappa(q, s, \delta)$ and $\lim_{\delta\to 0}\kappa(q, s, \delta) = 0$. The function $\mathbf{h} := \mathbf{v} - \mathbf{w}$ is called the $\mathcal{A}$-Stokes approximation of $\mathbf{v}$.

Remark 6.2.11. From the proof of Theorem 6.2.26 we have the following stability result, choosing $p = qs = q$.

$$\fint_Q \left|\frac{\mathbf{w}}{r}\right|^p \mathrm{d}x\,\mathrm{d}t + \fint_Q |\nabla \mathbf{w}|^p \,\mathrm{d}x\,\mathrm{d}t \le c \fint_{2\tilde{Q}} |\nabla \mathbf{v}|^p \,\mathrm{d}x\,\mathrm{d}t.$$

Indeed κ stays bounded if $s \to 1$.

Proof. Let $\mathbf{w}$ be defined as in (6.2.27). Combining Poincaré's inequality with Lemma 6.2.1 and (6.2.27) shows that

$$\begin{aligned}
&\fint_Q \left|\frac{\mathbf{w}}{r}\right|^q \mathrm{d}x\,\mathrm{d}t + \fint_Q |\nabla \mathbf{w}|^q \,\mathrm{d}x\,\mathrm{d}t \\
&\quad \le c \sup_{\boldsymbol{\xi} \in C^\infty_{0,\mathrm{div}}(Q)} \left[\fint_Q \mathcal{A}(\boldsymbol{\varepsilon}(\mathbf{v}), \boldsymbol{\varepsilon}(\boldsymbol{\xi}))\,\mathrm{d}x\,\mathrm{d}t - \fint_Q \mathbf{v} \cdot \partial_t \boldsymbol{\xi}\,\mathrm{d}x\,\mathrm{d}t \right. \\
&\quad \left. - \fint_Q \left(|\nabla \boldsymbol{\xi}|^{q'} + |\mathbf{H}_{\boldsymbol{\xi}}|^{q'} \right) \mathrm{d}x\,\mathrm{d}t \right].
\end{aligned} \tag{6.2.28}$$

In the following let us fix $\boldsymbol{\xi} \in C^\infty_{0,\mathrm{div}}(Q)$. Let

$$\gamma := \left(\fint_Q |\nabla \boldsymbol{\xi}|^{q'} \,\mathrm{d}x\,\mathrm{d}t + \fint_Q |\mathbf{H}_{\boldsymbol{\xi}}|^{q'} \,\mathrm{d}x\,\mathrm{d}t \right)^{\frac{1}{q'}},$$

and $m_0 \in \mathbb{N}$, $m_0 \gg 1$. Due to Corollary 6.1.5, applied with $\sigma = q'$, we find $\lambda \in [2^{m_0}\gamma, 2^{2m_0}\gamma]$ and $\boldsymbol{\xi}_\lambda \in L^\infty(4J; W^{1,\infty}_{0,\mathrm{div}}(4\mathcal{B}))$ such that

$$\|\nabla \boldsymbol{\xi}_\lambda\|_{L^\infty(4Q)} \le c\lambda, \tag{6.2.29}$$

$$\lambda^{q'} \frac{\mathcal{L}^{d+1}(2Q \cap \{\boldsymbol{\xi}_\lambda \ne \boldsymbol{\xi}\})}{|Q|} \le \frac{c}{m_0} \left(\fint_Q |\nabla \boldsymbol{\xi}|^{q'} \,\mathrm{d}x\,\mathrm{d}t + \fint_Q |\mathbf{H}_{\boldsymbol{\xi}}|^{q'} \,\mathrm{d}x\,\mathrm{d}t \right), \tag{6.2.30}$$

$$\fint_{4Q} |\boldsymbol{\xi}_\lambda|^{q'} \,\mathrm{d}x\,\mathrm{d}t \le c \left(\fint_Q |\boldsymbol{\xi}|^{q'} \,\mathrm{d}x\,\mathrm{d}t + \fint_Q r^{q'} |\mathbf{H}_{\boldsymbol{\xi}}|^{q'} \,\mathrm{d}x\,\mathrm{d}t \right), \tag{6.2.31}$$

$$\fint_{4Q} |\nabla \boldsymbol{\xi}_\lambda|^{q'} \,\mathrm{d}x\,\mathrm{d}t \le c \left(\fint_Q |\nabla \boldsymbol{\xi}|^{q'} \,\mathrm{d}x\,\mathrm{d}t + \fint_Q |\mathbf{H}_{\boldsymbol{\xi}}|^{q'} \,\mathrm{d}x\,\mathrm{d}t \right). \tag{6.2.32}$$

Note that $\boldsymbol{\xi}$ can be extended by 0 to $4Q$ thus the equation

$$\partial_t \boldsymbol{\xi} = \operatorname{div} \operatorname{Bog}_{\mathcal{B}}(\partial_t \boldsymbol{\xi}) =: \operatorname{div} \mathbf{H}_{\boldsymbol{\xi}}$$

holds on $4Q$ by the properties of $\operatorname{Bog}_{\mathcal{B}}$ (since $\mathbf{H}_{\boldsymbol{\xi}}$ can be extended as well). For the properties of the Bogovskiĭoperator we refer to Section 2.1, in particular Theorem 2.1.6. Corollary 6.1.5 (d) implies that $\partial_t(\boldsymbol{\xi} - \boldsymbol{\xi}_\lambda) \in L^{q'}(2J, W^{-1,q'}(2\mathcal{B}))$ and

$$\begin{aligned}
&\int_{2J} \langle \partial_t(\boldsymbol{\xi} - \boldsymbol{\xi}_\lambda), \boldsymbol{\varphi} \rangle \,\mathrm{d}t \\
&\quad \le c(\kappa) \int_{2Q} \chi_{\{\boldsymbol{\xi} \ne \boldsymbol{\xi}_\lambda\}} |\nabla \boldsymbol{\varphi}|^q \,\mathrm{d}x\,\mathrm{d}t + \kappa \left(\int_Q |\nabla \boldsymbol{\xi}|^{q'} + |\mathbf{H}_{\boldsymbol{\xi}}|^{q'} \,\mathrm{d}x\,\mathrm{d}t \right),
\end{aligned} \tag{6.2.33}$$

for all $\varphi \in W_0^{1,q}(2Q)$. For $\eta \in C_0^\infty(2Q)$ with $\eta \equiv 1$ on Q, $|\nabla^k\eta| \leq cr^{-k}$ and $|\partial_t\nabla^{k-1}\eta| \leq cr^{-(k+1)}$ $(k=1,2)$ we see that

$$\begin{aligned}
&\fint_Q \mathcal{A}(\boldsymbol{\varepsilon}(\mathbf{v}), \boldsymbol{\varepsilon}(\boldsymbol{\xi}))\,\mathrm{d}x\,\mathrm{d}t - \fint_Q \mathbf{v}\cdot\partial_t\boldsymbol{\xi}\,\mathrm{d}x\,\mathrm{d}t\\
&= 2^{d+2}\fint_{2Q} \mathcal{A}\big(\boldsymbol{\varepsilon}(\mathbf{v}), \boldsymbol{\varepsilon}(\eta\boldsymbol{\xi} - \mathrm{Bog}_{2\mathcal{B}\setminus\mathcal{B}}(\nabla\eta\boldsymbol{\xi}))\big)\,\mathrm{d}x\,\mathrm{d}t\\
&- \fint_{2Q} \mathbf{v}\cdot\partial_t\big(\eta\boldsymbol{\xi} - \ldots\big)\,\mathrm{d}x\,\mathrm{d}t\\
&= 2^{d+2}\Big(\fint_{2Q} \mathcal{A}\big(\boldsymbol{\varepsilon}(\mathbf{v}), \boldsymbol{\varepsilon}(\eta\boldsymbol{\xi}_\lambda - \mathrm{Bog}_{2\mathcal{B}\setminus\mathcal{B}}(\nabla\eta\boldsymbol{\xi}_\lambda))\big)\,\mathrm{d}x\,\mathrm{d}t\\
&+ \fint_{2Q} \partial_t\mathbf{v}\cdot\big(\eta\boldsymbol{\xi}_\lambda - \ldots\big)\,\mathrm{d}x\,\mathrm{d}t\Big)\\
&+ 2^{d+2}\fint_{2Q} \mathcal{A}\big(\boldsymbol{\varepsilon}(\mathbf{v}), \boldsymbol{\varepsilon}(\eta(\boldsymbol{\xi}-\boldsymbol{\xi}_\lambda) - \mathrm{Bog}_{2\mathcal{B}\setminus\mathcal{B}}(\nabla\eta(\boldsymbol{\xi}-\boldsymbol{\xi}_\lambda)))\big)\,\mathrm{d}x\,\mathrm{d}t\\
&+ 2^{d+2}\fint_{2Q} \partial_t\mathbf{v}\cdot\big(\eta(\boldsymbol{\xi}-\boldsymbol{\xi}_\lambda) - \mathrm{Bog}_{2\mathcal{B}\setminus\mathcal{B}}(\nabla\eta(\boldsymbol{\xi}-\boldsymbol{\xi}_\lambda))\big)\,\mathrm{d}x\,\mathrm{d}t\\
&=: 2^{d+2}(I + II + III).
\end{aligned}$$

Note that the time-derivative of $\mathbf{v}$ exists in the $W_{\mathrm{div}}^{-1,\infty}$-sense as a consequence of (6.2.26), so all terms are well-defined by the properties of $\boldsymbol{\xi}_\lambda$. We have the following inequality on account of the continuity properties of $\nabla\mathrm{Bog}$ on L^p-spaces, (6.2.31), (6.2.32) and Poincaré's inequality (we set $\tilde{\boldsymbol{\xi}}_\lambda := \boldsymbol{\xi} - \boldsymbol{\xi}_\lambda$):

$$\begin{aligned}
\fint_{2Q} |\nabla\boldsymbol{\Psi}_\lambda|^{q'}\,\mathrm{d}x\,\mathrm{d}t &:= \fint_{2Q} |\nabla(\eta\tilde{\boldsymbol{\xi}}_\lambda) - \nabla\mathrm{Bog}_{2\mathcal{B}\setminus\mathcal{B}}(\nabla\eta\tilde{\boldsymbol{\xi}}_\lambda)|^{q'}\,\mathrm{d}x\,\mathrm{d}t\\
&\leq c\fint_{2Q} |\nabla\tilde{\boldsymbol{\xi}}_\lambda|^{q'}\,\mathrm{d}x\,\mathrm{d}t + c\fint_{2Q} \left|\frac{\tilde{\boldsymbol{\xi}}_\lambda}{r}\right|^{q'}\,\mathrm{d}x\,\mathrm{d}t\\
&\leq c\fint_Q |\nabla\boldsymbol{\xi}|^{q'}\,\mathrm{d}x\,\mathrm{d}t + c\fint_Q \left|\frac{\boldsymbol{\xi}}{r}\right|^{q'}\,\mathrm{d}x\,\mathrm{d}t + c\fint_Q |\mathbf{H}_{\boldsymbol{\xi}}|^{q'}\,\mathrm{d}x\,\mathrm{d}t\\
&\leq c\fint_Q |\nabla\boldsymbol{\xi}|^{q'}\,\mathrm{d}x\,\mathrm{d}t + c\fint_Q |\mathbf{H}_{\boldsymbol{\xi}}|^{q'}\,\mathrm{d}x\,\mathrm{d}t. \qquad (6.2.34)
\end{aligned}$$

Young's inequality, with an appropriate choice of $\varepsilon > 0$, together with (6.2.31) and (6.2.32), implies that

$$\begin{aligned}
II &\leq c(\varepsilon)\fint_{2Q} |\boldsymbol{\varepsilon}(\mathbf{v})|^q\chi_{\{\boldsymbol{\xi}\neq\boldsymbol{\xi}_\lambda\}}\,\mathrm{d}x\,\mathrm{d}t + \varepsilon\fint_{2Q} |\nabla\boldsymbol{\Psi}_\lambda|^{q'}\,\mathrm{d}x\,\mathrm{d}t\\
&\leq c\fint_{2Q} |\boldsymbol{\varepsilon}(\mathbf{v})|^q\chi_{\{\boldsymbol{\xi}\neq\boldsymbol{\xi}_\lambda\}}\,\mathrm{d}x\,\mathrm{d}t + \frac{1}{3}\fint_Q |\nabla\boldsymbol{\xi}|^{q'} + |\mathbf{H}_{\boldsymbol{\xi}}|^{q'}\,\mathrm{d}x\,\mathrm{d}t\\
&=: II_1 + II_2,
\end{aligned}$$

where c depends on $\mathcal{A}$, q and q'. Hölder's inequality now yields

$$II_1 \leq c\bigg(\fint_{2Q} |\nabla \mathbf{v}|^{qs}\,\mathrm{d}x\,\mathrm{d}t\bigg)^{\frac{1}{s}}\bigg(\frac{\mathcal{L}^{d+1}\left(2Q\cap\{\boldsymbol{\xi}_\lambda\neq\boldsymbol{\xi}\}\right)}{|Q|}\bigg)^{1-\frac{1}{s}}.$$

If follows from (6.2.30), by the choice of γ and $\lambda \geq \gamma$ that

$$\frac{\mathcal{L}^{d+1}\left(2Q\cap\{\boldsymbol{\xi}_\lambda\neq\boldsymbol{\xi}\}\right)}{|Q|} \leq \frac{c\gamma^{q'}}{m_0\lambda^{q'}} \leq \frac{c}{m_0}. \tag{6.2.35}$$

Thus

$$II_1 \leq c\bigg(\fint_{2Q} |\nabla \mathbf{v}|^{qs}\,\mathrm{d}x\,\mathrm{d}t\bigg)^{\frac{1}{s}}\bigg(\frac{c}{m_0}\bigg)^{1-\frac{1}{s}}.$$

We choose m_0 sufficiently large that

$$II_1 \leq \frac{\kappa}{3}\bigg(\fint_{2Q} |\nabla \mathbf{v}|^{qs}\,\mathrm{d}x\,\mathrm{d}t\bigg)^{\frac{1}{s}}.$$

Since $\partial_t(\boldsymbol{\xi}-\boldsymbol{\xi}_\lambda) \in L^{q'}(2J, W^{-1,q'}(2\mathcal{B}))$ we can write III as

$$\begin{aligned}III &= \fint_{2Q} \mathbf{v}\cdot\partial_t\eta(\boldsymbol{\xi}-\boldsymbol{\xi}_\lambda)\,\mathrm{d}x\,\mathrm{d}t + \fint_{2Q} \eta\mathbf{v}\cdot\partial_t(\boldsymbol{\xi}-\boldsymbol{\xi}_\lambda)\,\mathrm{d}x\,\mathrm{d}t\\ &\quad - \fint_{2Q} \mathbf{v}\cdot\mathrm{Bog}_{2\mathcal{B}\setminus\mathcal{B}}(\partial_t\nabla\eta(\boldsymbol{\xi}-\boldsymbol{\xi}_\lambda))\,\mathrm{d}x\,\mathrm{d}t\\ &\quad - \fint_{2Q} \mathrm{Bog}^*_{2\mathcal{B}\setminus\mathcal{B}}(\mathbf{v})\nabla\eta\cdot\partial_t(\boldsymbol{\xi}-\boldsymbol{\xi}_\lambda)\,\mathrm{d}x\,\mathrm{d}t\\ &=: III_1 + III_2 + III_3 + III_4.\end{aligned}$$

The Bogovskiĭ operator is continuous from $L^2_\perp \to L^2$. Hence its dual (in the sense of L^2-duality) is continuous from $L^2 \to L^2_\perp$. Therefore $\mathrm{Bog}^*_{2\mathcal{B}\setminus\mathcal{B}}(\mathbf{v})$ is well-defined. We consider the four terms separately. For the first one we have

$$\begin{aligned}III_1 &\leq c\fint_{2I}\fint_{2\mathcal{B}\setminus\mathcal{B}} \chi_{\{\boldsymbol{\xi}_\lambda\neq\boldsymbol{\xi}\}}\left|\frac{\mathbf{v}}{r}\right|\left|\frac{\boldsymbol{\xi}-\boldsymbol{\xi}_\lambda}{r}\right|\mathrm{d}x\,\mathrm{d}t\\ &\leq c(\varepsilon)\fint_{2I}\fint_{2\mathcal{B}\setminus\mathcal{B}}\left|\frac{\mathbf{v}}{r}\right|^q \chi_{\{\boldsymbol{\xi}_\lambda\neq\boldsymbol{\xi}\}}\,\mathrm{d}x\,\mathrm{d}t + \varepsilon\fint_{2Q}\left|\frac{\boldsymbol{\xi}-\boldsymbol{\xi}_\lambda}{r}\right|^{q'}\mathrm{d}x\,\mathrm{d}t\\ &=: c(\varepsilon)III_{11} + \varepsilon III_{12}.\end{aligned}$$

Poincaré's inequality and Young's inequality yield

$$\begin{aligned}III_{11} &\leq c\bigg(\fint_{2I}\fint_{2\mathcal{B}\setminus\mathcal{B}}\left|\frac{\mathbf{v}}{r}\right|^{qs}\mathrm{d}x\,\mathrm{d}t\bigg)^{\frac{1}{s}}\bigg(\frac{\mathcal{L}^{d+1}\left(2Q\cap\{\boldsymbol{\xi}_\lambda\neq\boldsymbol{\xi}\}\right)}{|Q|}\bigg)^{1-\frac{1}{s}}\\ &\leq c\bigg(\fint_{2Q} |\nabla \mathbf{v}|^{qs}\,\mathrm{d}x\,\mathrm{d}t\bigg)^{\frac{1}{s}}\bigg(\frac{\mathcal{L}^{d+1}\left(2Q\cap\{\boldsymbol{\xi}_\lambda\neq\boldsymbol{\xi}\}\right)}{|Q|}\bigg)^{1-\frac{1}{s}}.\end{aligned}$$

Arguing as for the term II_1 implies that

$$III_{11} \leq \frac{\kappa}{12}\left(\fint_{2Q} |\nabla \mathbf{v}|^{qs}\, \mathrm{d}x\, \mathrm{d}t\right)^{\frac{1}{s}}.$$

Moreover, we gain from (6.2.31) and Poincaré's inequality

$$\begin{aligned}
III_{12} &\leq c\fint_{Q} \left|\frac{\boldsymbol{\xi}}{r}\right|^{q'} \mathrm{d}x\, \mathrm{d}t + c\fint_{Q} |\mathbf{H}_{\boldsymbol{\xi}}|^{q'}\, \mathrm{d}x\, \mathrm{d}t \\
&\leq c\fint_{Q} |\nabla \boldsymbol{\xi}|^{q'}\, \mathrm{d}x\, \mathrm{d}t + c\fint_{Q} |\mathbf{H}_{\boldsymbol{\xi}}|^{q'}\, \mathrm{d}x\, \mathrm{d}t,
\end{aligned}$$

and finally

$$III_1 \leq \frac{\kappa}{12}\left(\fint_{2Q} |\nabla \mathbf{v}|^{qs}\, \mathrm{d}x\, \mathrm{d}t\right)^{\frac{1}{s}} + \frac{1}{12}\left(\int_{Q} |\nabla \boldsymbol{\xi}|^{q'} + |\mathbf{H}_{\boldsymbol{\xi}}|^{q'}\, \mathrm{d}x\, \mathrm{d}t\right).$$

The formulation in (6.2.26) does not change if we subtract terms which are constant in space from $\mathbf{v}$ (note that $(\partial_t \boldsymbol{\xi}) = 0$ for every t due to $\partial_t \boldsymbol{\xi}(t, \cdot) \in C^\infty_{0,\mathrm{div}}(\mathcal{B})$). So we can assume that

$$\fint_{2\mathcal{B}\setminus\mathcal{B}} \mathbf{v}(t)\, \mathrm{d}x = 0 \quad \text{for a.e.} \quad t \in 2J. \tag{6.2.36}$$

As a consequence of (6.2.33), (6.2.36) and Poincaré's inequality we obtain similarly as for III_1

$$\begin{aligned}
III_2 &\leq c(\varepsilon)\int_{2Q} \chi_{\{\boldsymbol{\xi}\neq\boldsymbol{\xi}_\lambda\}} |\nabla(\eta \mathbf{v})|^q\, \mathrm{d}x\, \mathrm{d}t + \varepsilon\left(\int_{Q} |\nabla \boldsymbol{\xi}|^{q'} + |\mathbf{H}_{\boldsymbol{\xi}}|^{q'}\, \mathrm{d}x\, \mathrm{d}t\right) \\
&\leq \frac{\kappa}{12}\left(\fint_{2Q} |\nabla \mathbf{v}|^{qs}\, \mathrm{d}x\, \mathrm{d}t\right)^{\frac{1}{s}} + \frac{1}{12}\left(\int_{Q} |\nabla \boldsymbol{\xi}|^{q'} + |\mathbf{H}_{\boldsymbol{\xi}}|^{q'}\, \mathrm{d}x\, \mathrm{d}t\right).
\end{aligned}$$

Taking into account continuity properties of the Bogovskiĭ operator from $L^q_\perp \to W^{1,q}_0$ we can estimate III_3 via (we use again (6.2.36) and Poincaré's inequality)

$$\begin{aligned}
III_3 &\leq \left(\fint_{2Q}\fint_{2\mathcal{B}\setminus\mathcal{B}} \left|\frac{\mathbf{v}}{r}\right|^{qs} \mathrm{d}x\, \mathrm{d}t\right)^{\frac{1}{qs}} \\
&\quad \times \left(\fint_{2Q} r^{(qs)'} \left|\mathrm{Bog}_{2\mathcal{B}\setminus\mathcal{B}}(\partial_t \nabla \eta(\boldsymbol{\xi} - \boldsymbol{\xi}_\lambda))\right|^{(qs)'} \mathrm{d}x\, \mathrm{d}t\right)^{\frac{1}{(qs)'}} \\
&\leq c\left(\fint_{2Q} |\nabla \mathbf{v}|^{qs}\, \mathrm{d}x\, \mathrm{d}t\right)^{\frac{1}{qs}} \left(\fint_{2Q} r^{2(qs)'} \left|\partial_t \nabla \eta(\boldsymbol{\xi} - \boldsymbol{\xi}_\lambda)\right|^{(qs)'} \mathrm{d}x\, \mathrm{d}t\right)^{\frac{1}{(qs)'}} \\
&\leq c\left(\fint_{2Q} |\nabla \mathbf{v}|^{qs}\, \mathrm{d}x\, \mathrm{d}t\right)^{\frac{1}{qs}} \left(\fint_{2Q} \chi_{\{\boldsymbol{\xi}\neq\boldsymbol{\xi}_\lambda\}} \left|\frac{\boldsymbol{\xi} - \boldsymbol{\xi}_\lambda}{r}\right|^{(qs)'} \mathrm{d}x\, \mathrm{d}t\right)^{\frac{1}{(qs)'}}.
\end{aligned}$$

Hence, from Young's inequality for every $\varepsilon > 0$, we deduce that

$$
\begin{aligned}
III_3 &\le \frac{\varepsilon^q}{12}\left(\fint_{2Q} |\nabla \mathbf{v}|^{qs}\,\mathrm{d}x\,\mathrm{d}t\right)^{\frac{1}{s}} + c\,\varepsilon^{-q'}\left(\fint_{2Q} \chi_{\{\boldsymbol{\xi}\neq\boldsymbol{\xi}_\lambda\}}\left|\frac{\boldsymbol{\xi}-\boldsymbol{\xi}_\lambda}{r}\right|^{(qs)'}\mathrm{d}x\,\mathrm{d}t\right)^{\frac{q'}{(qs)'}}\\
&=: \frac{\varepsilon^q}{12} III_{31} + c\varepsilon^{-q'} III_{32}.
\end{aligned}
$$

It follows due to Hölder's inequality, Poincaré's inequality, (6.2.30), (6.2.32) and (6.2.35) for m_0 large enough

$$
\begin{aligned}
III_{32} &\le \left(\frac{\mathcal{L}^{d+1}\left(2Q\cap\{\boldsymbol{\xi}_\lambda\neq\boldsymbol{\xi}\}\right)}{|Q|}\right)^{1-\frac{1}{s}}\left(\fint_{2Q}\left|\frac{\boldsymbol{\xi}-\boldsymbol{\xi}_\lambda}{r}\right|^{q'}\mathrm{d}x\,\mathrm{d}t\right)\\
&\le c\left(\frac{\mathcal{L}^{d+1}\left(2Q\cap\{\boldsymbol{\xi}_\lambda\neq\boldsymbol{\xi}\}\right)}{|Q|}\right)^{1-\frac{1}{s}}\left(\fint_{4Q}\left|\frac{\boldsymbol{\xi}-\boldsymbol{\xi}_\lambda}{4r}\right|^{q'}\mathrm{d}x\,\mathrm{d}t\right)\\
&\le c\left(\frac{\mathcal{L}^{d+1}\left(2Q\cap\{\boldsymbol{\xi}_\lambda\neq\boldsymbol{\xi}\}\right)}{|Q|}\right)^{1-\frac{1}{s}}\left(\fint_{4Q}\left(|\nabla\boldsymbol{\xi}|^{q'}+|\nabla\boldsymbol{\xi}_\lambda|^{q}\right)\mathrm{d}x\,\mathrm{d}t\right)\\
&\le \frac{\kappa}{12c}\left(\int_{Q}|\nabla\boldsymbol{\xi}|^{q'}+|\mathbf{H}_{\boldsymbol{\xi}}|^{q'}\,\mathrm{d}x\,\mathrm{d}t\right).
\end{aligned}
$$

Choosing $\varepsilon := \kappa^{1/q'}$ implies

$$
III_3 \le \frac{\kappa}{12}\left(\fint_{2Q}|\nabla\mathbf{v}|^{qs}\,\mathrm{d}x\,\mathrm{d}t\right)^{\frac{1}{s}} + \frac{1}{12}\left(\int_Q|\nabla\boldsymbol{\xi}|^{q'}+|\mathbf{H}_{\boldsymbol{\xi}}|^{q'}\,\mathrm{d}x\,\mathrm{d}t\right).
$$

By (6.2.33) and (6.2.35) we have for m_0 large enough

$$
\begin{aligned}
III_4 &\le c\fint_{2Q}\chi_{\{\boldsymbol{\xi}\neq\boldsymbol{\xi}_\lambda\}}|\nabla(\nabla\eta\,\mathrm{Bog}^*_{2\mathcal{B}\setminus\mathcal{B}}(\mathbf{v}))|^q\,\mathrm{d}x\,\mathrm{d}t + \frac{1}{12}\left(\int_Q|\nabla\boldsymbol{\xi}|^{q'}+|\mathbf{H}_{\boldsymbol{\xi}}|^{q'}\,\mathrm{d}x\,\mathrm{d}t\right)\\
&\le \varepsilon\left(\fint_{2Q}|\nabla(\nabla\eta\,\mathrm{Bog}^*_{2\mathcal{B}\setminus\mathcal{B}}(\mathbf{v}))|^{sq}\,\mathrm{d}x\,\mathrm{d}t\right)^{\frac{1}{s}} + \frac{1}{12}\left(\int_Q|\nabla\boldsymbol{\xi}|^{q'}+|\mathbf{H}_{\boldsymbol{\xi}}|^{q'}\,\mathrm{d}x\,\mathrm{d}t\right)\\
&=: \varepsilon III_{41} + \frac{1}{12} III_{42}.
\end{aligned}
$$

Due to the continuity of Bog(div(·)) on L^p for any $1 < p < \infty$ (see [85, III.3, Theorem 3.3] and Theorem 2.1.7 for the Bogovskiĭ operator and negative norms) we have continuity of $\nabla\mathrm{Bog}^*$ as well. This, Poincaré's inequality (note that $\mathrm{Bog}^*_{2\mathcal{B}\setminus\mathcal{B}}(\mathbf{v}) \in L^p_0(2\mathcal{B}\setminus\mathcal{B})$) and (6.2.36) yield

$$
\begin{aligned}
III_{41} &\le c\left(\fint_{2Q}\left|\frac{\mathrm{Bog}^*_{2\mathcal{B}\setminus\mathcal{B}}(\mathbf{v})}{r^2}\right|^{sq}\mathrm{d}x\,\mathrm{d}t + \fint_{2Q}\left|\frac{\nabla\mathrm{Bog}^*_{2\mathcal{B}\setminus\mathcal{B}}(\mathbf{v})}{r}\right|^{sq}\mathrm{d}x\,\mathrm{d}t\right)^{\frac{1}{s}}\\
&\le c\left(\fint_{2Q}\left|\frac{\nabla\mathrm{Bog}^*_{2\mathcal{B}\setminus\mathcal{B}}(\mathbf{v})}{r}\right|^{sq}\mathrm{d}x\,\mathrm{d}t\right)^{\frac{1}{s}} \le c\left(\fint_{2I}\fint_{2\mathcal{B}\setminus\mathcal{B}}\left|\frac{\mathbf{v}}{r}\right|^{sq}\mathrm{d}x\,\mathrm{d}t\right)^{\frac{1}{s}}
\end{aligned}
$$

$$\leq c\left(\fint_{2Q} |\nabla \mathbf{v}|^{sq}\, dx\, dt\right)^{\frac{1}{s}},$$

and hence for $\varepsilon := \kappa/12c$,

$$III_4 \leq \frac{\kappa}{12}\left(\fint_{2Q} |\nabla \mathbf{v}|^{qs}\, dx\, dt\right)^{\frac{1}{s}} + \frac{1}{12}\left(\int_{Q} |\nabla \boldsymbol{\xi}|^{q'} + |\mathbf{H}_{\xi}|^{q'}\, dx\, dt\right).$$

Combining the estimates for III_1–III_4, we see that

$$III \leq \frac{\kappa}{3}\left(\fint_{2Q} |\nabla \mathbf{v}|^{qs}\, dx\, dt\right)^{\frac{1}{s}} + \frac{1}{3}\left(\int_{Q} |\nabla \boldsymbol{\xi}|^{q'} + |\mathbf{H}_{\xi}|^{q'}\, dx\, dt\right).$$

Since $\mathbf{v}$ is an almost $\mathcal{A}$-Stokes solution and $\|\nabla \boldsymbol{\xi}_\lambda\|_\infty \leq c\lambda \leq c2^{m_0}\gamma$ we have

$$\begin{aligned} |I| &\leq \delta \fint_{2\tilde{Q}} |\nabla \mathbf{v}|\, dx\, dt\, \|\nabla \boldsymbol{\xi}_\lambda\|_{\infty, 2Q} \\ &\leq \delta \left(\fint_{2\tilde{Q}} |\nabla \mathbf{v}|^{qs}\, dx\, dt\right)^{\frac{1}{qs}} c2^{m_0}\gamma. \end{aligned}$$

We apply Young's inequality and Jensen's inequality to give

$$\begin{aligned} |I| &\leq \delta 2^{m_0} c\left(\fint_{2\tilde{Q}} |\nabla \mathbf{v}|^{qs}\, dx\, dt\right)^{\frac{1}{s}} + \delta 2^{m_0} c\gamma^{q'} \\ &\leq \delta 2^{m_0} c\left(\fint_{2\tilde{Q}} |\nabla \mathbf{v}|^{qs}\, dx\, dt\right)^{\frac{1}{s}} + \delta 2^{m_0} c\left(\fint_{Q} |\nabla \boldsymbol{\xi}|^{q'}\, dx\, dt + \fint_{Q} |\mathbf{H}_{\xi}|^{q'}\, dx\, dt\right). \end{aligned}$$

Now, we choose $\delta > 0$ so small such that $\delta 2^{m_0} c \leq \kappa/3$. Thus

$$|I| \leq \frac{\kappa}{3}\left(\fint_{2\tilde{Q}} |\nabla \mathbf{v}|^{qs}\, dx\, dt\right)^{\frac{1}{s}} + \frac{1}{3}\left(\fint_{Q} |\nabla \boldsymbol{\xi}|^{q'}\, dx\, dt + \fint_{Q} |\mathbf{H}_{\xi}|^{q'}\, dx\, dt\right).$$

Combining the estimates for I, II and III we have established

$$\begin{aligned} &\fint_{2Q} \mathcal{A}(\boldsymbol{\varepsilon}(\mathbf{v}), \boldsymbol{\varepsilon}(\boldsymbol{\xi}))\, dx\, dt - \fint_{Q} \mathbf{v} \cdot \partial_t \boldsymbol{\xi}\, dx\, dt \\ &\quad \leq \kappa\left(\fint_{2\tilde{Q}} |\nabla \mathbf{v}|^{qs}\, dx\, dt\right)^{\frac{1}{s}} + \fint_{Q} |\nabla \boldsymbol{\xi}|^{q'}\, dx\, dt + \fint_{Q} |\mathbf{H}_{\xi}|^{q'}\, dx\, dt. \end{aligned}$$

Inserting this in (6.2.28) shows the claim. □

CHAPTER 7

Power law fluids

Contents

Abstract

We study non-stationary motions of power law fluids in a bounded Lipschitz domain. Based on the solenoidal Lipschitz truncation from Chapter 6 we show the existence of weak solutions to the generalized Navier–Stokes system for $p > \frac{2d}{d+2}$. Our approach completely avoids the appearance of the pressure function.

The flow of a homogeneous incompressible fluid in a bounded body $G \subset \mathbb{R}^d$ $(d = 2, 3)$ during the time interval $(0, T)$ is described by the following set of equations

$$\begin{cases} \rho\partial_t \mathbf{v} + \rho(\nabla\mathbf{v})\mathbf{v} = \operatorname{div}\mathbf{S} - \nabla\pi + \rho\mathbf{f} & \text{in} \quad Q, \\ \operatorname{div}\mathbf{v} = 0 & \text{in} \quad Q, \\ \mathbf{v} = 0 & \text{on} \quad \partial G, \\ \mathbf{v}(0,\cdot) = \mathbf{v}_0 & \text{in} \quad G, \end{cases} \tag{7.0.1}$$

see for instance [23]. Here the unknown quantities are the velocity field $\mathbf{v} : Q \to \mathbb{R}^d$ and the pressure $\pi : Q \to \mathbb{R}$. The function $\mathbf{f} : Q \to \mathbb{R}^d$ represents a system of volume forces and $\mathbf{v}_0 : G \to \mathbb{R}^d$ the initial datum, while $\mathbf{S} : Q \to \mathbb{R}^{d\times d}_{sym}$ is the stress deviator and $\rho > 0$ is the density of the fluid. Equation $(7.0.1)_1$ and $(7.0.1)_2$ describe the conservation of balance and the conservation of mass respectively. Both are valid for all homogeneous incompressible liquids and gases. Very popular among rheologists is the power law model

$$\mathbf{S}(\boldsymbol{\varepsilon}(\mathbf{v})) = \nu_0\big(1 + |\boldsymbol{\varepsilon}(\mathbf{v})|\big)^{p-2}\boldsymbol{\varepsilon}(\mathbf{v}) \tag{7.0.2}$$

where $\nu_0 > 0$ and $p \in (1, \infty)$, cf. [13,23]. We recall that the case $p \in [\frac{3}{2}, 2]$ covers many interesting applications. For further comments on the physical background we refer to Section 1.4.

The first mathematical results concerning (7.0.1), (7.0.2) were achieved by Ladyshenskaya and Lions for $p \geq \frac{3d+2}{d+2}$ (see [106] and [109]). They show

Existence Theory for Generalized Newtonian Fluids.
DOI: http://dx.doi.org/10.1016/B978-0-12-811044-7.00009-4

the existence of a weak solution in the space

$$L^p(0,T;W^{1,p}_{0,\mathrm{div}}(G))\cap L^\infty(0,T;L^2(G)).$$

The weak formulation reads as

$$\begin{aligned}\int_Q \mathbf{S}(\boldsymbol{\varepsilon}(\mathbf{v})):\boldsymbol{\varepsilon}(\boldsymbol{\varphi})\,\mathrm{d}x\,\mathrm{d}t &= \int_Q \mathbf{f}\cdot\boldsymbol{\varphi}\,\mathrm{d}x\,\mathrm{d}t-\int_Q(\nabla\mathbf{v})\mathbf{v}\cdot\boldsymbol{\varphi}\,\mathrm{d}x\,\mathrm{d}t\\ &\quad+\int_Q \mathbf{v}\cdot\partial_t\boldsymbol{\varphi}\,\mathrm{d}x\,\mathrm{d}t+\int_G \mathbf{v}_0\cdot\boldsymbol{\varphi}(0)\,\mathrm{d}x\end{aligned}$$

for all $\boldsymbol{\varphi}\in C^\infty_{0,\mathrm{div}}([0,T)\times G)$ with $\mathbf{S}$ given by (7.0.2). In this case it follows from parabolic interpolation that $(\nabla\mathbf{v})\mathbf{v}\cdot\mathbf{v}\in L^1(Q)$. So the solution is also a test-function and the existence proof is based on monotone operator theory and compactness arguments. These results have been improved by Wolf to the case $p>\frac{2d+2}{d+2}$ via L^∞-truncation. Wolf's result was improved to $p>\frac{2d}{d+2}$ in [65] by the Lipschitz truncation method. Under this restriction on p we have $\mathbf{v}\otimes\mathbf{v}\in L^1(Q)$. Hence we can test by Lipschitz continuous functions. So we have to approximate $\mathbf{v}$ by a Lipschitz continuous function $\mathbf{v}_\lambda$ which is quite challenging in the parabolic situation. For further historical comments we refer to Section 5.3.

We will revise the existence proof from [65]. Using the solenoidal Lipschitz truncation constructed in section 6.1 we can completely avoid the appearance of the pressure and therefore highly simplify the method. For simplicity we only consider the case $d=3$ but all results of this chapter extend to the general case. The main result is as follows.

Theorem 7.0.27. *Let $p>\frac{6}{5}$, $\mathbf{f}\in L^{p'}(Q)$ and $\mathbf{v}_0\in L^2(G)$. Then there is a solution $\mathbf{v}\in L^\infty(0,T;L^2(G))\cap L^p(0,T;W^{1,p}_{0,\mathrm{div}}(G))$ to*

$$\begin{aligned}\int_Q \mathbf{S}(\boldsymbol{\varepsilon}(\mathbf{v})):\boldsymbol{\varepsilon}(\boldsymbol{\varphi})\,\mathrm{d}x\,\mathrm{d}t &= \int_Q \mathbf{f}\cdot\boldsymbol{\varphi}\,\mathrm{d}x\,\mathrm{d}t+\int_Q \mathbf{v}\otimes\mathbf{v}:\boldsymbol{\varepsilon}(\boldsymbol{\varphi})\,\mathrm{d}x\,\mathrm{d}t\\ &\quad+\int_Q \mathbf{v}\cdot\partial_t\boldsymbol{\varphi}\,\mathrm{d}x\,\mathrm{d}t+\int_G \mathbf{v}_0\cdot\boldsymbol{\varphi}(0)\,\mathrm{d}x\end{aligned}\tag{7.0.3}$$

for all $\boldsymbol{\varphi}\in C^\infty_{0,\mathrm{div}}([0,T)\times G)$.

Remark 7.0.12. It is still open whether there exists a weak solution in the case $1<p\le\frac{6}{5}$. Unlike the stationary case the convective term $\mathbf{v}\otimes\mathbf{v}$ is always well-defined independent of the dimension. However, it is not clear how to obtain the compactness of the approximated velocity $\mathbf{v}_m$ in $L^2(Q)$. This seems to be necessary to identify the limit of $\mathbf{v}_m\otimes\mathbf{v}_m$ with $\mathbf{v}\otimes\mathbf{v}$.

In the next section we show the existence of weak solutions to (7.0.3) in case $p > \frac{11}{5}$. Due to this bound on q the space of test functions coincides with the space where the solution is constructed and the convective term becomes a compact perturbation. The result of Section 7.1 will later be used to obtain an approximate solution in the proof of Theorem 7.0.27, see Section 7.2.

7.1 THE APPROXIMATED SYSTEM

Throughout this section we assume that $\mathbf{S} \in C^0(\mathbb{R}^{d\times d}_{sym}) \cap C^1(\mathbb{R}^{d\times d}_{sym} \setminus \{0\})$ and for some $\kappa \geq 0$

$$\lambda\big(\kappa + |\boldsymbol{\varepsilon}|\big)^{q-2}|\boldsymbol{\sigma}|^2 \leq D\mathbf{S}(\boldsymbol{\varepsilon})(\boldsymbol{\sigma},\boldsymbol{\sigma}) \leq \Lambda\big(\kappa + |\boldsymbol{\varepsilon}|\big)^{q-2}|\boldsymbol{\sigma}|^2 \tag{7.1.4}$$

for all $\boldsymbol{\varepsilon}, \boldsymbol{\sigma} \in \mathbb{R}^{d\times d}_{sym} \setminus \{0\}$ with some positive constants λ, Λ.

Theorem 7.1.28. *Assume (7.1.4) with $q > \frac{11}{5}$, $\mathbf{f} \in L^{q'}(Q)$ and $\mathbf{v}_0 \in L^2(G)$. Then there is a solution $\mathbf{v} \in L^\infty(0, T; L^2(G)) \cap L^q(0, T; W^{1,q}_{0,\mathrm{div}}(G))$ to*

$$\begin{aligned}\int_Q \mathbf{S}(\boldsymbol{\varepsilon}(\mathbf{v})) : \boldsymbol{\varepsilon}(\boldsymbol{\varphi})\,\mathrm{d}x\,\mathrm{d}t &= \int_Q \mathbf{f}\cdot\boldsymbol{\varphi}\,\mathrm{d}x\,\mathrm{d}t + \int_Q \mathbf{v}\otimes\mathbf{v} : \boldsymbol{\varepsilon}(\boldsymbol{\varphi})\,\mathrm{d}x\,\mathrm{d}t \\ &+ \int_Q \mathbf{v}\cdot\partial_t\boldsymbol{\varphi}\,\mathrm{d}x\,\mathrm{d}t + \int_G \mathbf{v}_0\cdot\boldsymbol{\varphi}(0)\,\mathrm{d}x\end{aligned} \tag{7.1.5}$$

for all $\boldsymbol{\varphi} \in C^\infty_{0,\mathrm{div}}([0, T) \times G)$.

Proof. We mainly follow the ideas of [111], chapter 5. We separate space and time and approximate the corresponding Sobolev space by a finite dimensional subspace. From [111] we infer the existence of a sequence $(\lambda_k) \subset \mathbb{R}$ and a sequence of functions $(\mathbf{w}_k) \subset W^{l,2}_{0,\mathrm{div}}(G)$, $l \in \mathbb{N}$, such that

i) $\mathbf{w}_k$ is an eigenvector to the eigenvalue λ_k of the Stokes-operator in the sense that

$$\langle \mathbf{w}_k, \boldsymbol{\varphi}\rangle_{W^{l,2}_0} = \lambda_k \int_G \mathbf{w}_k\cdot\boldsymbol{\varphi}\,\mathrm{d}x \quad \text{for all} \quad \boldsymbol{\varphi} \in W^{l,2}_{0,\mathrm{div}}(G),$$

ii) $\int_G \mathbf{w}_k\mathbf{w}_m\,\mathrm{d}x = \delta_{km}$ for all $k, m \in \mathbb{N}$,

iii) $1 \leq \lambda_1 \leq \lambda_2 \leq \ldots$ and $\lambda_k \to \infty$,

iv) $\langle \frac{\mathbf{w}_k}{\sqrt{\lambda_k}}, \frac{\mathbf{w}_m}{\sqrt{\lambda_m}}\rangle_{W^{l,2}_0} = \delta_{km}$ for all $k, m \in \mathbb{N}$,

v) $(\mathbf{w}_k)$ is a basis of $W^{l,2}_{0,\mathrm{div}}(G)$.

We choose $l > 1 + \frac{d}{2}$ so that $W^{l,2}_0(G) \hookrightarrow W^{1,\infty}(G)$. We are looking for an approximate solution $\mathbf{v}_N$ of the form

$$\mathbf{v}_N = \sum_{k=1}^{N} c^k_N \mathbf{w}_k$$

where $\mathbf{C}_N = (c_N^k)_{k=1}^N : (0, T) \to \mathbb{R}^N$. We will construct $\mathbf{C}_N$ so that $\mathbf{v}_N$ is a solution to

$$\begin{aligned}\int_G \mathbf{S}(\boldsymbol{\varepsilon}(\mathbf{v}_N)) : \boldsymbol{\varepsilon}(\mathbf{w}_k)\,\mathrm{d}x &= -\int_G \partial_t \mathbf{v}_N \cdot \mathbf{w}_k\,\mathrm{d}x + \int_G \mathbf{v}_N \otimes \mathbf{v}_N : \boldsymbol{\varepsilon}(\mathbf{w}_k)\,\mathrm{d}x \\ &\quad + \int_G \mathbf{f}\cdot \mathbf{w}_k\,\mathrm{d}x, \quad k = 1, \dots, N \qquad (7.1.6)\\ \mathbf{v}_N(0,\cdot) &= \mathcal{P}_N \mathbf{v}_0.\end{aligned}$$

Here $\mathcal{P}_N$ is the $L^2(G)$-orthogonal projection into $\mathcal{X}_N := \operatorname{span}\{\mathbf{w}_1, \dots, \mathbf{w}_N\}$, i.e.

$$\mathcal{P}_N(\mathbf{u}) := \sum_{k=1}^N \left(\int_G \mathbf{w}_k \cdot \mathbf{u}\,\mathrm{d}x\right)\mathbf{w}_k.$$

On account of the properties of $(\mathbf{w}_k)$ equation (7.1.6) is equivalent to

$$\begin{aligned}\frac{\mathrm{d}c_N^k}{\mathrm{d}t} &= -\int_G \mathbf{S}(\mathbf{C}_N \cdot \boldsymbol{\varepsilon}(\mathbf{w}_N)) : \boldsymbol{\varepsilon}(\mathbf{w}_k)\,\mathrm{d}x + \sum_{l,j} c_N^l c_N^j \int_G \mathbf{w}_l \otimes \mathbf{w}_j : \nabla \mathbf{w}_k\,\mathrm{d}x \\ &\quad + \int_G \mathbf{f}\cdot \mathbf{w}_k\,\mathrm{d}x, \quad k = 1, \dots, N \qquad (7.1.7)\\ c_N^k(0) &= \int_G \mathbf{w}_k \cdot \mathbf{v}_0\,\mathrm{d}x, \quad k = 1, \dots, N.\end{aligned}$$

Since the right-hand-side is not globally Lipschitz continuous in $\mathbf{C}_N$ the Picard–Lindelöff Theorem only gives a local solution which does not suffice for our purpose. The following lemma helps (see [143], chapter 30).

Lemma 7.1.1. *Consider the ODE*

$$y' = F(t, y), \quad y(0) = y_0,$$

where F is continuous in t and locally Lipschitz continuous in y. Assume that every possible solution satisfies

$$|y(t)| \le C \quad \textit{for all} \quad t \in [0, T]. \qquad (7.1.8)$$

Then there is a global solution on $[0, T]$.

So we need to show boundedness of $\mathbf{C}_N$ in t (assuming its existence). Therefore we multiply the k-th equation of (7.1.6) by c_N^k and sum with respect to k. Using $\int_G \mathbf{v}_N \otimes \mathbf{v}_N : \nabla \mathbf{v}_N\,\mathrm{d}x$ we obtain

$$\frac{1}{2}\frac{\mathrm{d}}{\mathrm{d}t}\int_G |\mathbf{v}_N|^2\,\mathrm{d}x + \int_G |\nabla \mathbf{v}_N|^q\,\mathrm{d}x = \int_G \mathbf{f}\cdot \mathbf{v}_N\,\mathrm{d}x.$$

Integration over $[0, s]$ with $0 < s \leq T$ implies for all $\kappa > 0$

$$\begin{aligned}\frac{1}{2}\int_G |\mathbf{v}_N(s,\cdot)|^2\,\mathrm{d}x + \lambda\int_{Q_s}\big(|\nabla\mathbf{v}_N|^q - 1\big)\,\mathrm{d}x &\leq \int_{Q_s}\mathbf{f}\cdot\mathbf{v}^N\,\mathrm{d}x\,\mathrm{d}t\\ &\leq c(\kappa)\int_{Q_s}|\mathbf{f}|^{q'}\,\mathrm{d}x\,\mathrm{d}t + \kappa\int_0^s\int_G|\mathbf{v}_N|^q\,\mathrm{d}x\,\mathrm{d}t\\ &\leq c(\kappa)\int_{Q_s}|\mathbf{f}|^{q'}\,\mathrm{d}x\,\mathrm{d}t + \kappa\int_0^s\int_G|\nabla\mathbf{v}_N|^q\,\mathrm{d}x\,\mathrm{d}t,\end{aligned}$$

where we used the inequalities of Korn, Young and Poincaré as well as (7.1.4). Choosing κ small enough leads to

$$\sup_{t\in(0,T)}\int_G|\mathbf{v}_N(t,\cdot)|^2\,\mathrm{d}x + \int_Q|\nabla\mathbf{v}_N|^q\,\mathrm{d}x\,\mathrm{d}t \leq c\int_Q\big(|\mathbf{f}|^{q'}+1\big)\,\mathrm{d}x\,\mathrm{d}t \tag{7.1.9}$$

which also implies (7.1.8). Lemma 7.1.1 shows the existence of a solution $\mathbf{v}_N$ to (7.1.6). Moreover we established with (7.1.9) a useful a priori estimate. Passing to a subsequence implies

$$\mathbf{v}_N \rightharpoonup: \mathbf{v} \quad \text{in} \quad L^q(0,T;W^{1,q}_{0,div}(G)), \tag{7.1.10}$$

$$\mathbf{v}_N \rightharpoonup^* \mathbf{v} \quad \text{in} \quad L^\infty(0,T;L^2(G)). \tag{7.1.11}$$

In order to pass to the limit in the convective term we need compactness of $\mathbf{v}_N$. We obtain from (7.1.6)

$$\begin{aligned}&\int_G\partial_t\mathbf{v}_N\cdot\boldsymbol{\varphi}\,\mathrm{d}x = \int_G\partial_t\mathbf{v}_N\cdot\mathcal{P}^l_N\boldsymbol{\varphi}\,\mathrm{d}x\\ &= -\int_G\mathbf{S}(\boldsymbol{\varepsilon}(\mathbf{v}^N)):\boldsymbol{\varepsilon}(\mathcal{P}^l_N\boldsymbol{\varphi})\,\mathrm{d}x + \int_G\mathbf{v}_N\otimes\mathbf{v}_N:\boldsymbol{\varepsilon}(\mathcal{P}^l_N\boldsymbol{\varphi})\,\mathrm{d}x + \int_G\mathbf{F}:\nabla\mathcal{P}^l_N\boldsymbol{\varphi}\,\mathrm{d}x\\ &=: \int_G\mathbf{H}_N:\nabla\mathcal{P}^l_N\boldsymbol{\varphi}\,\mathrm{d}x\end{aligned}$$

for all $\boldsymbol{\varphi}\in W^{l,2}_{0,div}(G)$ (setting $\mathbf{F} := \nabla\Delta^{-1}\mathbf{f}$). Here $\mathcal{P}^l_N$ denotes the orthogonal projection into $\mathcal{X}_N$ with respect to the $W^{l,2}_0(G)$ inner product. We have uniformly in N

$$\mathbf{H}_N\in L^{q_0}(Q), \quad q_0 =: \min\left\{\tfrac{5}{6}q, q'\right\} > 1, \tag{7.1.12}$$

as a consequence of (7.1.9), (7.1.4) and $\mathbf{F}\in L^{q'}(Q)$ (which follows from $\mathbf{f}\in L^{q'}(Q)$). On account of (7.1.12) and Sobolev's embedding (recall the choice of l) we obtain

$$\begin{aligned}\|\partial_t\mathbf{v}_N\|_{L^{q_0}(0,T;W^{-l,2}_{div}(G))} &= \|\partial_t\mathbf{v}_N\|_{L^{q_0'}(0,T;W^{l,2}_{0,div}(G))'}\\ &= \sup_{\|\boldsymbol{\varphi}\|_{L^{q_0'}(0,T;W^{l,2}_{0,div}(G))}\leq 1}\int_0^T\int_G\partial_t\mathbf{v}_N\cdot\boldsymbol{\varphi}\,\mathrm{d}x\,\mathrm{d}t\end{aligned}$$

$$= \sup_{\|\varphi\|_{L^{q_0'}(0,T;W^{l,2}_{0,div}(G))}\leq 1} \int_0^T\int_G \mathbf{H}_N : \nabla P^l_N\varphi \,\mathrm{d}x\,\mathrm{d}t$$

$$\leq \sup_{\|\varphi\|_{L^{q_0'}(0,T;W^{l,2}_{0,div}(G))}\leq 1} \left(\int_Q |\mathbf{H}_N|^{q_0}\,\mathrm{d}x\,\mathrm{d}t\right)^{\frac{1}{q_0}}\left(\int_Q |\nabla P^N_l\varphi|^{q_0'}\,\mathrm{d}x\,\mathrm{d}t\right)^{\frac{1}{q_0'}}$$

and finally

$$\begin{aligned}
&\|\partial_t\mathbf{v}_N\|_{L^{q_0}(0,T;W^{-l,2}_{div}(G))}\\
&\quad\leq c \sup_{\|\varphi\|_{L^{q_0'}(0,T;W^{l,2}_{0,div}(G))}\leq 1} \|\nabla\mathcal{P}^l_N\varphi\|_{L^{q_0}(0,T;L^\infty(G))}\\
&\quad\leq c \sup_{\|\varphi\|_{L^{q_0'}(0,T;W^{l,2}_{0,div}(G))}\leq 1} \|\mathcal{P}^l_N\varphi\|_{L^{q_0}(0,T;W^{l,2}_{0,div}(G))}\leq c.
\end{aligned}\tag{7.1.13}$$

Combining (7.1.9) and (7.1.13) with the Aubin–Lions compactness Theorem (see Theorem 5.1.23) shows $\mathbf{v}_N\to\mathbf{v}$ in $L^2(0,T;L^2_{\mathrm{div}}(G))$ (recall that $q>\frac{11}{5}$). This and a parabolic interpolation imply

$$\mathbf{v}_N\otimes\mathbf{v}_N\to\mathbf{v}\otimes\mathbf{v}\quad\text{in}\quad L^s(Q)\tag{7.1.14}$$

for all $s<\frac{5}{6}p$. Due to (7.1.9) and (7.1.4) we know that $\mathbf{S}(\boldsymbol{\varepsilon}(\mathbf{v}_N))$ is bounded in $L^{q'}(G)$ thus

$$\mathbf{S}(\boldsymbol{\varepsilon}(\mathbf{v}_N))\rightharpoonup\tilde{\mathbf{S}}\quad\text{in}\quad L^{q'}(Q).\tag{7.1.15}$$

Passing to the limit in (7.1.6) leads to

$$\begin{aligned}
\int_Q \tilde{\mathbf{S}}:\boldsymbol{\varepsilon}(\varphi)\,\mathrm{d}x\,\mathrm{d}t &= \int_Q \mathbf{f}\cdot\varphi\,\mathrm{d}x\,\mathrm{d}t+\int_Q \mathbf{v}\otimes\mathbf{v}:\boldsymbol{\varepsilon}(\varphi)\,\mathrm{d}x\,\mathrm{d}t\\
&\quad+\int_Q \mathbf{v}\cdot\partial_t\varphi\,\mathrm{d}x\,\mathrm{d}t+\int_G \mathbf{v}_0\cdot\varphi(0)\,\mathrm{d}x
\end{aligned}\tag{7.1.16}$$

for all $\varphi\in C^\infty_{0,\mathrm{div}}([0,T)\times G)$. Note that the class of test-functions which factorize in space and time is dense, see Lemma 5.1.2. In (7.1.16) we also used (for $k\in\mathbb{N}$ and $g\in C^\infty_0[0,T)$)

$$\int_Q \partial_t\mathbf{v}_N\cdot g\mathbf{w}_k\,\mathrm{d}x\,\mathrm{d}t=-\int_G \mathcal{P}_N\mathbf{v}_0\cdot\mathbf{w}_k\,\mathrm{d}x\,g(0)-\int_0^T\int_G \mathbf{v}_N\cdot\mathbf{w}_k\partial_t g\,\mathrm{d}x\,\mathrm{d}t$$

and $\mathcal{P}_N\mathbf{v}_0\to\mathbf{v}_0$ in $L^2(G)$.

Finally we need to show that

$$\tilde{\mathbf{S}}=\mathbf{S}(\boldsymbol{\varepsilon}(\mathbf{v}))\tag{7.1.17}$$

holds. We first investigate the time derivative of $\mathbf{v}$. On account of $q > \frac{11}{5}$ the mapping

$$\varphi \mapsto \int_Q \mathbf{v} \otimes \mathbf{v} : \boldsymbol{\varepsilon}(\varphi)\,\mathrm{d}x\,\mathrm{d}t$$

belongs to $L^{q'}(0, T; W_{div}^{-1,q'}(G))$. The same is true for

$$\varphi \mapsto \int_Q \tilde{\mathbf{S}} : \boldsymbol{\varepsilon}(\varphi)\,\mathrm{d}x\,\mathrm{d}t, \quad \varphi \mapsto \int_Q \mathbf{f} \cdot \varphi\,\mathrm{d}x\,\mathrm{d}t.$$

Thus we have $\partial_t \mathbf{v} \in L^{q'}(0, T; W_{div}^{-1,q'}(G))$ by (7.1.16) and

$$\begin{aligned}\int_Q \tilde{\mathbf{S}} : \boldsymbol{\varepsilon}(\varphi)\,\mathrm{d}x\,\mathrm{d}t = &\int_Q \mathbf{f} \cdot \varphi\,\mathrm{d}x\,\mathrm{d}t + \int_Q \mathbf{v} \otimes \mathbf{v} : \boldsymbol{\varepsilon}(\varphi)\,\mathrm{d}x\,\mathrm{d}t \\ &- \int_0^T \langle \partial_t \mathbf{v}, \mathbf{v} \rangle\,\mathrm{d}t\end{aligned} \tag{7.1.18}$$

for all $\varphi \in L^q(0, T; W_{0,div}^{1,q}(G))$. Especially $\mathbf{v}$ is an admissible test function. We claim that

$$\mathbf{v} \in C_w([0, T]; L^2(G)). \tag{7.1.19}$$

Hence we have $\mathbf{v}(0) = \mathbf{v}_0$ and $\mathbf{v}(t)$ is uniquely determined for every $t \in [0, T]$. Sobolev's Theorem for Bochner spaces leads to

$$\mathbf{v} \in C([0, T]; W_{div}^{-1,q'}(G)). \tag{7.1.20}$$

Let $(t_n) \subset [0, T]$ for which $t_n \to t_0$ and $\mathbf{v}(t_n, \cdot) \in L^2(G)$ for all $n \in \mathbb{N}$. Then $\mathbf{v}(t_n, \cdot)$ is bounded in $L^2(G)$ by (7.1.9) thus

$$\mathbf{v}(t_n, \cdot) \rightharpoonup: \mathbf{w} \quad \text{in} \quad L^2(G). \tag{7.1.21}$$

We have as a consequence of (7.1.20)

$$\mathbf{v}(t_n, \cdot) \to \mathbf{v}(t_0, \cdot) \quad \text{in} \quad W_0^{-1,q'}(G). \tag{7.1.22}$$

This leads to $\mathbf{w} = \mathbf{v}(t_0, \cdot)$ and (7.1.21) implies (7.1.19).

We apply monotone operator theory to show (7.1.17). On account of

$$\begin{aligned}&\int_Q \big(\mathbf{S}(\boldsymbol{\varepsilon}(\mathbf{v}_N)) - \mathbf{S}(\boldsymbol{\varepsilon}(\mathbf{v}))\big) : \boldsymbol{\varepsilon}\big(\mathbf{v}_N - \mathbf{v}\big)\,\mathrm{d}x\,\mathrm{d}t \\ &= \int_Q \big(\tilde{\mathbf{S}} - \mathbf{S}(\boldsymbol{\varepsilon}(\mathbf{v}_N))\big) : \boldsymbol{\varepsilon}(\mathbf{v})\,\mathrm{d}x\,\mathrm{d}t - \int_Q \mathbf{S}(\boldsymbol{\varepsilon}(\mathbf{v})) : \boldsymbol{\varepsilon}\big(\mathbf{v}_N - \mathbf{v}\big)\,\mathrm{d}x\,\mathrm{d}t \\ &+ \int_Q \big(\mathbf{S}(\boldsymbol{\varepsilon}(\mathbf{v}_N)) : \boldsymbol{\varepsilon}(\mathbf{v}_N)\,\mathrm{d}x\,\mathrm{d}t - \int_Q \tilde{\mathbf{S}} : \boldsymbol{\varepsilon}(\mathbf{v})\,\mathrm{d}x\,\mathrm{d}t\end{aligned}$$

we obtain from (7.1.10) and (7.1.15)

$$\int_Q \Big(\tilde{\mathbf{S}} - \mathbf{S}(\boldsymbol{\varepsilon}(\mathbf{v}_N))\Big) : \boldsymbol{\varepsilon}(\mathbf{v})\,\mathrm{d}x\,\mathrm{d}t \longrightarrow 0, \quad N \to \infty,$$

$$\int_Q \mathbf{S}(\boldsymbol{\varepsilon}(\mathbf{v})) : \boldsymbol{\varepsilon}\big(\mathbf{v}_N - \mathbf{v}\big)\,\mathrm{d}x\,\mathrm{d}t \longrightarrow 0, \quad N \to \infty.$$

Due to (7.1.6) and (7.1.18) we have

$$\begin{aligned}
&\int_Q \Big(\mathbf{S}(\boldsymbol{\varepsilon}(\mathbf{v}_N)) : \boldsymbol{\varepsilon}(\mathbf{v}_N)\,\mathrm{d}x\,\mathrm{d}t - \int_Q \tilde{\mathbf{S}} : \boldsymbol{\varepsilon}(\mathbf{v})\,\mathrm{d}x\,\mathrm{d}t \\
&= \int_Q \mathbf{f}\cdot\big(\mathbf{v}_N - \mathbf{v}\big)\,\mathrm{d}x\,\mathrm{d}t + \int_Q \Big(\mathbf{v}_N \otimes \mathbf{v}_N : \boldsymbol{\varepsilon}(\mathbf{v}_N) - \mathbf{v}\otimes\mathbf{v} : \boldsymbol{\varepsilon}(\mathbf{v})\Big)\,\mathrm{d}x\,\mathrm{d}t \\
&- \int_Q \partial_t \mathbf{v}_N \cdot \mathbf{v}_N\,\mathrm{d}x\,\mathrm{d}t + \int_0^T \big\langle \partial_t \mathbf{v}, \mathbf{v}\big\rangle\,\mathrm{d}t \\
&=: (I) + (II) + (III).
\end{aligned}$$

We deduce from (7.1.10), (7.1.14) and (7.1.15) that

$$\lim_{N\to\infty} (I) = \lim_{N\to\infty} (II) = 0.$$

For the integral involving the convective term we used the assumption $q > \frac{11}{5}$. Finally, we obtain

$$\begin{aligned}
(III) &= \frac{1}{2}\int_G |\mathbf{v}_N(0)|^2\,\mathrm{d}x - \frac{1}{2}\int_G |\mathbf{v}(0)|^2\,\mathrm{d}x + \frac{1}{2}\int_G |\mathbf{v}(T)|^2\,\mathrm{d}x \\
&- \frac{1}{2}\int_G |\mathbf{v}_N(T)|^2\,\mathrm{d}x \\
&= -\frac{1}{2}\int_G |\mathbf{v}_N(T) - \mathbf{v}(T)|^2\,\mathrm{d}x - \int_G \mathbf{v}_N(T)\cdot\mathbf{v}(T)\,\mathrm{d}x + \int_G |\mathbf{v}(T)|^2\,\mathrm{d}x \\
&+ \frac{1}{2}\int_G |\mathcal{P}_N\mathbf{v}_0|^2\,\mathrm{d}x - \frac{1}{2}\int_G |\mathbf{v}_0|^2\,\mathrm{d}x.
\end{aligned}$$

We infer from (7.1.10) and the continuity of $\mathcal{P}_N$ that $\limsup_{N\to\infty}(III) \le 0$ (here we took into account (7.1.19) and used $\mathbf{v}_N(T) \rightharpoonup \mathbf{v}(T)$ in $L^2(G)$ as a consequence of (7.1.9) and passing to a subsequence) thus

$$\int_Q \Big(\mathbf{S}(\boldsymbol{\varepsilon}(\mathbf{v}_N)) - \mathbf{S}(\boldsymbol{\varepsilon}(\mathbf{v}))\Big) : \boldsymbol{\varepsilon}\big(\mathbf{v}_N - \mathbf{v}\big)\,\mathrm{d}x\,\mathrm{d}t \longrightarrow 0, \quad N \to \infty.$$

Monotonicity of **S** (which follows from (7.1.4)) implies (7.1.17). □

Corollary 7.1.1. *Under the assumptions of Theorem 7.0.27 there is a function* $\tilde{\pi} \in C^0_w([0,T]; L^{q'}_0(G))$ *for which*

$$\int_Q \mathbf{S}(\boldsymbol{\varepsilon}(\mathbf{v})) : \boldsymbol{\varepsilon}(\boldsymbol{\varphi})\,\mathrm{d}x\,\mathrm{d}t = \int_Q \mathbf{f}\cdot\boldsymbol{\varphi}\,\mathrm{d}x\,\mathrm{d}t + \int_Q \mathbf{v}\otimes\mathbf{v} : \boldsymbol{\varepsilon}(\boldsymbol{\varphi})\,\mathrm{d}x\,\mathrm{d}t$$

$$+\int_Q \mathbf{v}\cdot\partial_t\boldsymbol{\varphi}\,\mathrm{d}x\,\mathrm{d}t+\int_G \mathbf{v}_0\,\boldsymbol{\varphi}(0)\,\mathrm{d}x+\int_Q \tilde{\pi}\,\partial_t\operatorname{div}\boldsymbol{\varphi}\,\mathrm{d}x\,\mathrm{d}t$$

for all $\boldsymbol{\varphi}\in C_0^\infty([0,T)\times G)$.

Proof. We follow [140], Thm. 2.6. Let $\boldsymbol{\varphi}(t,x)=g(t)\boldsymbol{\psi}(x)$ with $g\in C_0^\infty(0,T)$ and $\boldsymbol{\psi}\in C_{0,\mathrm{div}}^\infty(G)$. Setting $\mathbf{F}:=\nabla\Delta^{-1}\mathbf{f}\in L^{q'}(Q)$ we have

$$\begin{aligned}\int_Q \mathbf{v}\cdot\boldsymbol{\psi}\,g'\,\mathrm{d}x\,\mathrm{d}t&=\int_Q \mathbf{S}(\boldsymbol{\varepsilon}(\mathbf{v})):\nabla\boldsymbol{\psi}\,g\,\mathrm{d}x\,\mathrm{d}t-\int_Q \mathbf{v}\otimes\mathbf{v}:\nabla\boldsymbol{\psi}\,g\,\mathrm{d}x\,\mathrm{d}t\\&\quad-\int_Q \mathbf{F}:\nabla\boldsymbol{\psi}\,g\,\mathrm{d}x\,\mathrm{d}t\\&=:-\int_Q \mathbf{Q}:\nabla\boldsymbol{\psi}\,g\,\mathrm{d}x\,\mathrm{d}t\end{aligned}$$

which is equivalent to

$$\int_0^T\left(\int_G \mathbf{v}\cdot\boldsymbol{\psi}\,\mathrm{d}x\right)g'\,\mathrm{d}t=-\int_0^T\left(\int_G \mathbf{Q}:\nabla\boldsymbol{\psi}\,\mathrm{d}x\right)g\,\mathrm{d}t. \tag{7.1.23}$$

If we define

$$\begin{aligned}\alpha(t)&:=\int_G \mathbf{v}(t,\cdot)\cdot\boldsymbol{\psi}\,\mathrm{d}x,\\ \beta(t)&:=\int_G \mathbf{Q}:\nabla\boldsymbol{\psi}\,\mathrm{d}x,\end{aligned}$$

we obtain from (7.1.23)

$$\int_0^T \alpha g'\,\mathrm{d}t=-\int_0^T \beta g\,\mathrm{d}t \tag{7.1.24}$$

for all $g\in C_0^\infty(0,T)$. Since α and β belong to $L^1(0,T)$ this implies $\alpha'=\beta$. Hence the following holds

$$\alpha(t)=\alpha(0)+\int_0^t \beta(s)\,\mathrm{d}s,\quad t\in(0,T). \tag{7.1.25}$$

Now we define

$$\tilde{\mathbf{Q}}(t):=\int_0^t \mathbf{Q}(s)\,ds$$

and follow from (7.1.25)

$$\int_G\Big((\mathbf{v}(t,\cdot)-\mathbf{v}(0,\cdot))\cdot\boldsymbol{\psi}+\tilde{\mathbf{Q}}(t):\nabla\boldsymbol{\psi}\Big)\,\mathrm{d}x=0$$

for all $\boldsymbol{\psi}\in W_{0,\mathrm{div}}^{1,q}(G)$. Here we took into account $\tilde{\mathbf{Q}}(t)\in L^{q'}(G)$ which is a consequence of $q>\frac{11}{5}$ and (7.1.4). De Rahm's Theorem implies the existence of $\tilde{\pi}(t) \in L_0^{q'}(G)$ for which

$$\int_G \Big((\mathbf{v}(t,\cdot) - \mathbf{v}(0,\cdot)) \cdot \boldsymbol{\psi} + \tilde{\mathbf{Q}} : \nabla \boldsymbol{\psi} \Big) \, \mathrm{d}x = \int_G \tilde{\pi}(t) \operatorname{div} \boldsymbol{\psi} \, \mathrm{d}x \tag{7.1.26}$$

for all $\boldsymbol{\psi} \in W_0^{1,q}(G)$. For $u \in L^q(G)$ we set $\boldsymbol{\psi} = \mathrm{Bog}_G\big(u - (u)_G\big) \in W_{0,\mathrm{div}}^{1,q}(G)$ so that

$$\begin{aligned}\int_G \tilde{\pi}(t)\, u \, \mathrm{d}x &= \int_G \tilde{\pi}(t)\, (\operatorname{div} \boldsymbol{\psi} + (u)_G) \, \mathrm{d}x = \int_G \tilde{\pi}(t) \operatorname{div} \boldsymbol{\psi} \, \mathrm{d}x \\ &= \int_G \Big((\mathbf{v}(t,\cdot) - \mathbf{v}(0,\cdot)) \cdot \boldsymbol{\psi} + \tilde{\mathbf{Q}}(t) : \nabla \boldsymbol{\psi} \Big) \, \mathrm{d}x.\end{aligned}$$

Due to $\mathbf{v} \in C_w([0,T]; L^2(G))$ (see (7.1.19)) and $\tilde{\mathbf{Q}} \in C([0,T]; L^{q'}(G))$ we obtain

$$\lim_{t \to t_0} \int_G \tilde{\pi}(t)\, u \, \mathrm{d}x = \int_G \tilde{\pi}(t_0)\, u \, \mathrm{d}x$$

for all $u \in L^q(G)$, hence $\tilde{\pi} \in C_w([0,T]; L^{q'}(G))$. Equation (7.1.25) finally implies

$$\begin{aligned}&\int_Q \mathbf{S}(\boldsymbol{\varepsilon}(\mathbf{v})) : \nabla \boldsymbol{\varphi} \, \mathrm{d}x \, \mathrm{d}t \\ &\quad = \int_Q \mathbf{v} \otimes \mathbf{v} : \nabla \boldsymbol{\varphi} \, \mathrm{d}x \, \mathrm{d}t + \int_Q \mathbf{f} \cdot \boldsymbol{\varphi} \, \mathrm{d}x \, \mathrm{d}t \\ &\quad + \int_Q \mathbf{v} \cdot \partial_t \boldsymbol{\varphi} \, \mathrm{d}x \, \mathrm{d}t + \int_G \mathbf{v}_0 \cdot \boldsymbol{\varphi}(0,\cdot) \, \mathrm{d}x + \int_Q \tilde{\pi}\, \partial_t \operatorname{div} \boldsymbol{\varphi} \, \mathrm{d}x \, \mathrm{d}t\end{aligned}$$

for all $\boldsymbol{\varphi}$ of the class

$$Y := \operatorname{span}\big\{ g \boldsymbol{\psi},\ g \in C_0^\infty[0,T),\ \boldsymbol{\psi} \in C_0^\infty(G) \big\}$$

which is dense (see Lemma 5.1.1). □

Remark 7.1.13. The "original" pressure term π can be obtained by setting $\pi := \partial_t \tilde{\pi}$. But without further information about the regularity with respect to time of the quantities involved in the equation it only exists in the sense of distributions.

7.2 NON-STATIONARY FLOWS

In this section we show how the solenoidal Lipschitz truncation can be used to simplify the existence proof for weak solutions to the power law model for non-Newtonian fluids. We are able to work completely in the pressure free formulation and establish the existence of a solution $\mathbf{v} \in L^\infty(0, T; L^2(G)) \cap L^p(0, T; W^{1,p}_{0,\mathrm{div}}(G))$ to

$$\begin{aligned}\int_Q \mathbf{S}(\boldsymbol{\varepsilon}(\mathbf{v})) : \boldsymbol{\varepsilon}(\boldsymbol{\varphi})\,\mathrm{d}x\,\mathrm{d}t = \int_Q \mathbf{f}\cdot\boldsymbol{\varphi}\,\mathrm{d}x\,\mathrm{d}t + \int_Q \mathbf{v}\otimes\mathbf{v} : \boldsymbol{\varepsilon}(\boldsymbol{\varphi})\,\mathrm{d}x\,\mathrm{d}t \\ + \int_Q \mathbf{v}\,\partial_t\boldsymbol{\varphi}\,\mathrm{d}x\,\mathrm{d}t + \int_G \mathbf{v}_0\,\boldsymbol{\varphi}(0)\,\mathrm{d}x\end{aligned} \tag{7.2.27}$$

for all $\boldsymbol{\varphi} \in C^\infty_{0,\mathrm{div}}([0, T) \times G)$.

Proof. We start with an approximate system whose solution is known to exist. Let $\mathbf{v}_m \in L^q(0, T; W^{1,q}_{0,\mathrm{div}}(G)) \cap L^\infty(0, T; L^2(G))$ be a solution to

$$\begin{aligned}&\int_Q \mathbf{S}(\boldsymbol{\varepsilon}(\mathbf{v})) : \boldsymbol{\varepsilon}(\boldsymbol{\varphi})\,\mathrm{d}x\,\mathrm{d}t + \frac{1}{m}\int_Q |\boldsymbol{\varepsilon}(\mathbf{v})|^{q-2}\boldsymbol{\varepsilon}(\mathbf{v}) : \boldsymbol{\varepsilon}(\boldsymbol{\varphi})\,\mathrm{d}x\,\mathrm{d}t \qquad (7.2.28)\\ &= \int_Q \mathbf{f}\cdot\boldsymbol{\varphi}\,\mathrm{d}x\,\mathrm{d}t + \int_Q \mathbf{v}\otimes\mathbf{v} : \boldsymbol{\varepsilon}(\boldsymbol{\varphi})\,\mathrm{d}x\,\mathrm{d}t + \int_Q \mathbf{v}\,\partial_t\boldsymbol{\varphi}\,\mathrm{d}x\,\mathrm{d}t + \int_G \mathbf{v}_0\,\boldsymbol{\varphi}(0)\,\mathrm{d}x\end{aligned}$$

for all $\boldsymbol{\varphi} \in C^\infty_{0,\mathrm{div}}([0, T) \times G)$, where $q > \max\{\frac{11}{5}, p\}$.

The existence of $\mathbf{v}_m$ follows from Theorem 7.1.28. Since we are allowed to test with $\mathbf{v}_m$, we find

$$\begin{aligned}&\frac{1}{2}\|\mathbf{v}_m(t)\|^2_{L^2} + \int_0^t\int_G \mathbf{S}(\boldsymbol{\varepsilon}(\mathbf{v}_m)) : \boldsymbol{\varepsilon}(\mathbf{v}_m)\,\mathrm{d}x\,\mathrm{d}\sigma + \frac{1}{m}\int_0^t\int_G |\boldsymbol{\varepsilon}(\mathbf{v}_m)|^q\,\mathrm{d}x\,\mathrm{d}\sigma\\ &= \frac{1}{2}\|\mathbf{v}_0\|^2_{L^2} + \int_0^t\int_G \mathbf{f} : \mathbf{v}_m\,dx\,\mathrm{d}\sigma,\end{aligned} \tag{7.2.29}$$

for all $t \in (0, T)$. By coercivity and Korn's inequality we obtain

$$\int_Q \mathbf{S}(\boldsymbol{\varepsilon}(\mathbf{v}_m)) : \boldsymbol{\varepsilon}(\mathbf{v}_m)\,\mathrm{d}x\,\mathrm{d}t \geq c\Big(\int_Q |\nabla\mathbf{v}_m|^p\,\mathrm{d}x\,\mathrm{d}t - 1\Big)$$

thus

$$\|m^{-1/q}\boldsymbol{\varepsilon}(\mathbf{v}_m)\|_{q,Q} + \|\mathbf{v}_m\|^2_{L^\infty(0,T;L^2)} + \|\nabla\mathbf{v}_m\|_{p,Q} \leq c. \tag{7.2.30}$$

Hence we find a function $\mathbf{v} \in L^p(0, T; W^{1,p}_{0,\mathrm{div}}(G)) \cap L^\infty(0, T; L^2(G))$ for which (passing to a subsequence)

$$\begin{aligned}\nabla\mathbf{v}_m &\rightharpoonup \nabla\mathbf{v} &&\text{in}\quad L^p(Q),\\ \mathbf{v}_m &\overset{*}{\rightharpoonup} \mathbf{v} &&\text{in}\quad L^\infty(0, T; L^2(G)),\\ \tfrac{1}{m}|\boldsymbol{\varepsilon}(\mathbf{v}_m)|^{q-2}\boldsymbol{\varepsilon}(\mathbf{v}_m) &\to 0 &&\text{in}\quad L^{q'}(Q).\end{aligned} \tag{7.2.31}$$

Since $\mathbf{S}(\boldsymbol{\varepsilon}(\mathbf{v}_m))$ is bounded in $L^{p'}(Q)$ by (7.2.30), there exist $\tilde{\mathbf{S}} \in L^{p'}(Q)$ with

$$\mathbf{S}(\boldsymbol{\varepsilon}(\mathbf{v}_m)) \rightharpoonup \tilde{\mathbf{S}} \quad\text{in}\quad L^{p'}(Q). \tag{7.2.32}$$

Let us have a look at the time derivative. From equation (7.2.28) we get the uniform boundedness of $\partial_t \mathbf{v}_m$ in $L^{p'}(0,T;W^{-3,2}_{\text{div}}(G))$ and weak convergence of $\partial_t \mathbf{v}_m$ to $\partial_t \mathbf{v}$ in the same space (for a subsequence). This shows by using the compactness of the embedding $W^{1,p}_{0,\text{div}}(G) \hookrightarrow L^{2\sigma_2}_{\text{div}}(G)$ for some $\sigma_2 > 1$ (which follows from our assumption $p > \frac{6}{5}$, resp. $p > \frac{2n}{n+2}$) and the Aubin–Lions theorem (see Theorem 5.1.23) that $\mathbf{v}_m \to \mathbf{v}$ in $L^{\sigma}(0,T;L^{2\sigma_2}_{\text{div}}(G))$. This and the boundedness in $L^{\infty}(0,T;L^2(G))$ imply that for some $\sigma > 1$

$$\mathbf{v}_m \to \mathbf{v} \quad \text{in} \quad L^s(0,T;L^{2\sigma}(G)) \quad \text{for all} \quad s < \infty. \tag{7.2.33}$$

As a consequence we have

$$\mathbf{v}_m \otimes \mathbf{v}_m \to \mathbf{v} \otimes \mathbf{v} \quad \text{in} \quad L^s(0,T;L^{\sigma}(G)) \quad \text{for all} \quad s < \infty. \tag{7.2.34}$$

Overall, we get our limit equation

$$\begin{aligned}\int_Q \tilde{\mathbf{S}} : \boldsymbol{\varepsilon}(\boldsymbol{\varphi})\,\mathrm{d}x\,\mathrm{d}t &= \int_Q \mathbf{f}\cdot\boldsymbol{\varphi}\,\mathrm{d}x\,\mathrm{d}t + \int_Q \mathbf{v}\otimes\mathbf{v} : \boldsymbol{\varepsilon}(\boldsymbol{\varphi})\,\mathrm{d}x\,\mathrm{d}t \\ &\quad + \int_Q \mathbf{v}\,\partial_t\boldsymbol{\varphi}\,\mathrm{d}x\,\mathrm{d}t + \int_G \mathbf{v}_0\,\boldsymbol{\varphi}(0)\,\mathrm{d}x\end{aligned} \tag{7.2.35}$$

for all $\boldsymbol{\varphi} \in C^{\infty}_{0,\text{div}}([0,T) \times G)$.

The entire forthcoming effort is to prove $\tilde{\mathbf{S}} = \mathbf{S}(\boldsymbol{\varepsilon}(\mathbf{v}))$ almost everywhere. We start with the difference of the equation of $\mathbf{v}_m$ and the limit equation which is

$$\begin{aligned}&-\int_Q (\mathbf{v}_m - \mathbf{v})\cdot\partial_t\boldsymbol{\varphi}\,\mathrm{d}x\,\mathrm{d}t + \int_Q (\mathbf{S}(\boldsymbol{\varepsilon}(\mathbf{v}_m)) - \tilde{\mathbf{S}}) : \nabla\boldsymbol{\varphi}\,\mathrm{d}x\,\mathrm{d}t \\ &\quad = \int_Q \left(\mathbf{v}_m\otimes\mathbf{v}_m - \mathbf{v}\otimes\mathbf{v} + m^{-1}|\boldsymbol{\varepsilon}(\mathbf{v}_m)|^{q-2}\boldsymbol{\varepsilon}(\mathbf{v}_m)\right) : \nabla\boldsymbol{\varphi}\,\mathrm{d}x\,\mathrm{d}t\end{aligned} \tag{7.2.36}$$

for all $\boldsymbol{\varphi} \in C^{\infty}_{0,\text{div}}([0,T) \times G)$. We define $\mathbf{u}_m := \mathbf{v}_m - \mathbf{v}$. Then by (7.2.33)

$$\begin{aligned}\mathbf{u}_m &\rightharpoonup \mathbf{0} \quad &&\text{in} \quad L^p(0,T;W^{1,p}_{0,\text{div}}(G)),\\ \mathbf{u}_m &\to \mathbf{0} \quad &&\text{in} \quad L^{2\sigma}(Q),\\ \mathbf{u}_m &\rightharpoonup^* \mathbf{0} \quad &&\text{in} \quad L^{\infty}(0,T;L^2(G)).\end{aligned} \tag{7.2.37}$$

Thus, we can write (7.2.36) as

$$\int_Q \mathbf{u}_m\cdot\partial_t\boldsymbol{\varphi}\,\mathrm{d}x\,\mathrm{d}t = \int_Q \mathbf{H}_m : \nabla\boldsymbol{\varphi}\,\mathrm{d}x\,\mathrm{d}t \tag{7.2.38}$$

for all $\boldsymbol{\varphi} \in C^{\infty}_{0,\text{div}}(Q)$, where $\mathbf{H}_m := \mathbf{H}^1_m + \mathbf{H}^2_m$ with

$$\begin{aligned}\mathbf{H}^1_m &:= \mathbf{S}(\boldsymbol{\varepsilon}(\mathbf{v}_m)) - \tilde{\mathbf{S}},\\ \mathbf{H}^2_m &:= \mathbf{v}_m\otimes\mathbf{v}_m - \mathbf{v}\otimes\mathbf{v} + m^{-1}|\boldsymbol{\varepsilon}(\mathbf{v}_m)|^{q-2}\boldsymbol{\varepsilon}(\mathbf{v}_m).\end{aligned}$$

Moreover, (7.2.31) and (7.2.33) imply

$$\|\mathbf{H}_m^1\|_{p'} \le c \tag{7.2.39}$$

as well as

$$\mathbf{H}_m^2 \to 0 \quad \text{in} \quad L^\sigma(Q). \tag{7.2.40}$$

Now take any cylinder $Q_0 \Subset (0, T) \times G$. Now, (7.2.37), (7.2.38), (7.2.39) and (7.2.40) ensure that we can apply Corollary 6.1.4. In particular, for suitable $\zeta \in C_0^\infty(\frac{1}{6}Q_0)$ with $\chi_{\frac{1}{8}Q_0} \le \zeta \le \chi_{\frac{1}{6}Q_0}$ Corollary 6.1.4 implies

$$\limsup_{m\to\infty} \left| \int (\mathbf{H}_m^1 : \nabla(\mathbf{v}_m - \mathbf{v}))\zeta\,\chi_{\mathcal{O}_{m,k}^\complement}\,\mathrm{d}x\,\mathrm{d}t \right| \le c\,2^{-k/p}.$$

In other words

$$\limsup_{m\to\infty} \left| \int \Big(\big(\mathbf{S}(\boldsymbol{\varepsilon}(\mathbf{v}_m)) - \tilde{\mathbf{S}}\big) : \nabla(\mathbf{v}_m - \mathbf{v}) \Big)\zeta\,\chi_{\mathcal{O}_{m,k}^\complement}\,\mathrm{d}x\,\mathrm{d}t \right| \le c\,2^{-k/p}.$$

Now, the boundedness of $\mathbf{S}(\boldsymbol{\varepsilon}(\mathbf{v}))$ and $\tilde{\mathbf{S}}$ in $L^{p'}(\frac{1}{6}Q_0)$ and Theorem 6.1.25 (h) and (g) give

$$\limsup_{m\to\infty} \left| \int \Big(\big(\tilde{\mathbf{S}} - \mathbf{S}(\boldsymbol{\varepsilon}(\mathbf{v}))\big) : \nabla(\mathbf{v}_m - \mathbf{v}) \Big)\zeta\,\chi_{\mathcal{O}_{m,k}^\complement}\,\mathrm{d}x\,\mathrm{d}t \right| \le c\,2^{-k/p}.$$

This and the previous estimate imply

$$\limsup_{m\to\infty} \left| \int \Big(\big(\mathbf{S}(\boldsymbol{\varepsilon}(\mathbf{v}_m)) - \mathbf{S}(\boldsymbol{\varepsilon}(\mathbf{v}))\big) : \nabla(\mathbf{v}_m - \mathbf{v}) \Big)\zeta\,\chi_{\mathcal{O}_{m,k}^\complement}\,\mathrm{d}x\,\mathrm{d}t \right| \le c\,2^{-k/p}.$$

Let $\theta \in (0, 1)$. Then by Hölder's inequality and Theorem 6.1.25 (g)

$$\begin{aligned}
&\limsup_{m\to\infty} \int \Big(\big(\mathbf{S}(\boldsymbol{\varepsilon}(\mathbf{v}_m)) - \mathbf{S}(\boldsymbol{\varepsilon}(\mathbf{v}))\big) : \nabla(\mathbf{v}_m - \mathbf{v}) \Big)^\theta \zeta\,\chi_{\mathcal{O}_{m,k}}\,\mathrm{d}x\,\mathrm{d}t \\
&\quad \le c\,|\mathcal{O}_{m,k}|^{1-\theta} \le c\,2^{-(1-\theta)\frac{k}{p}}.
\end{aligned}$$

This, the previous estimate and Hölder's inequality lead to

$$\limsup_{m\to\infty} \int \Big(\big(\mathbf{S}(\boldsymbol{\varepsilon}(\mathbf{v}_m)) - \mathbf{S}(\boldsymbol{\varepsilon}(\mathbf{v}))\big) : \nabla(\mathbf{v}_m - \mathbf{v}) \Big)^\theta \zeta\,\mathrm{d}x\,\mathrm{d}t \le c\,2^{-(1-\theta)\frac{k}{p}}.$$

For $k \to \infty$ the right-hand-side converges to zero. Now, the monotonicity of $\mathbf{S}$ implies that $\mathbf{S}(\boldsymbol{\varepsilon}(\mathbf{v}_m)) \to \mathbf{S}(\boldsymbol{\varepsilon}(\mathbf{v}))$ a.e. in $\frac{1}{8}Q_0$. This concludes the proof of Theorem 7.0.27. □

Remark 7.2.14. As done in Corollary 7.1.1 the pressure can be reconstructed. Here we have $\tilde{\pi} \in L^\sigma(Q)$.

PART 3

Stochastic problems

CHAPTER 8

Preliminaries

Contents

Abstract

In this chapter we revise some basic concepts from stochastic analysis. We begin with the properties of stochastic processes before defining the stochastic integral in the sense of Itô. The third section is concerned with the chain rule in stochastic integration which is known as Itô's formula. Finally we present classical as well as more recent results on stochastic ordinary differential equations. These will be used in the finite dimensional approximation of stochastic PDEs in chapter 10.

8.1 STOCHASTIC PROCESSES

We consider random variables on a probability space $(\Omega, \mathcal{F}, \mathbb{P})$. Let $(\mathcal{F}_t)_{t\geq 0}$ be a filtration such that $\mathcal{F}_s \subseteq \mathcal{F}_t \subseteq \mathcal{F}$ for $0 \leq s \leq t < \infty$. A real-valued *stochastic process* is a set of random variables $X = (X_t)_{t\geq 0}$ on $(\Omega, \mathcal{F})$ with values in $(\mathbb{R}, \mathfrak{B}(\mathbb{R}))$. A stochastic process can be interpreted as a function of t and ω, where t can be interpreted as time. For fixed $\omega \in \Omega$ the mapping $t \mapsto X_t(\omega)$ is called *path* or *trajectory* of X. We follow the presentation from [100] where the interested reader may also find details of the proofs.

Definition 8.1.1. A stochastic process is called *measurable*, if the mapping

$$(t, \omega) \mapsto X_t(\omega) : \ ([0, \infty) \times \Omega, \mathfrak{B}([0, \infty)) \otimes \mathcal{F}) \to (\mathbb{R}, \mathfrak{B}(\mathbb{R}))$$

is measurable.

Definition 8.1.2. A stochastic process is called *adapted* to the filtration $(\mathcal{F}_t)_{t\geq 0}$, if the mapping

$$\omega \mapsto X_t(\omega) : \ (\Omega, \mathcal{F}_t) \to (\mathbb{R}, \mathfrak{B}(\mathbb{R}))$$

is measurable for all $t \geq 0$.

Existence Theory for Generalized Newtonian Fluids.
DOI: http://dx.doi.org/10.1016/B978-0-12-811044-7.00011-2

Definition 8.1.3. A stochastic process is called *progressively measurable*, if the mapping

$$(s,\omega)\mapsto X_s(\omega):\ ([0,t]\times\Omega,\mathfrak{B}([0,t])\otimes\mathcal{F}_t)\to(\mathbb{R},\mathfrak{B}(\mathbb{R}))$$

is measurable for all $t\geq 0$.

Remark 8.1.15. Progressive measurability implies adaptedness.

Theorem 8.1.29. *If a stochastic process X is adapted to the filtration $(\mathcal{F}_t)_{t\geq 0}$ and a.e. path is left-continuous or right-continuous, then X is progressively measurable.*

The most important process is the Wiener process.

Definition 8.1.4 (Wiener process). A Wiener process is a real valued stochastic process $W=(W_t)_{t\geq 0}$ with the following properties.

i) The increments of B are independent, i.e. for arbitrary $0\leq t_0<t_1<\dots<t_n$ the random variables $W_{t_1}-W_{t_0}, W_{t_2}-W_{t_1},\dots,W_{t_n}-W_{t_{n-1}}$ are independent.
ii) For all $t>s\geq 0$ we have $W_t-W_s\sim\mathcal{N}(0,t-s)$.
iii) There holds $W_0=0$ almost surely.
iv) The mapping $t\mapsto W_t(\omega)$ is continuous for a.e. $\omega\in\Omega$.

Definition 8.1.5. A filtration $(\mathcal{F}_t)_{t\geq 0}$ is called *right-continuous*, if

$$\mathcal{F}_t=\bigcap_{\varepsilon>0}\mathcal{F}_{t+\varepsilon}\quad\forall t\geq 0$$

and *left-continuous*, if

$$\mathcal{F}_t=\bigcup_{s<t}\mathcal{F}_s\quad\forall t>0.$$

Definition 8.1.6. A filtration $(\mathcal{F}_t)_{t\geq 0}$ satisfies the *usual conditions*, if it is right-continuous and $\mathcal{F}_0$ contains all the $\mathbb{P}$-nullsets of $\mathcal{F}$.

Definition 8.1.7. Let X be a stochastic process and $\mathcal{T}$ a $\mathcal{F}$-measurable random variable with values in $[0,\infty]$. We define $X_{\mathcal{T}}$ in $\{\mathcal{T}<\infty\}$ by

$$X_{\mathcal{T}}(\omega)=X_{\mathcal{T}(\omega)}(\omega).$$

If $X_\infty(\omega)=\lim_{t\to\infty}X_t(\omega)$ exists for a.e. $\omega\in\Omega$ we set $X_{\mathcal{T}}(\omega)=X_\infty(\omega)$ in $\{\mathcal{T}=\infty\}$.

Definition 8.1.8. Let X be a stochastic process on a probability space $(\Omega,\mathcal{F},\mathbb{P})$ with filtration $(\mathcal{F}_t)_{t\geq 0}$. A random variable $\mathcal{T}$ is called a *stopping time* if the set $\{\mathcal{T}\leq t\}$ belongs to $\mathcal{F}_t$ for all $t\geq 0$.

Then by T induced σ-algebra $\mathcal{F}_{\mathcal{T}}$ is given by

$$\mathcal{F}_{\mathcal{T}}:=\{A\in\mathcal{F}:A\cap\{\mathcal{T}\leq t\}\in\mathcal{F}_t\ \forall t\geq 0\}.$$

Lemma 8.1.1. *Let X be a progressively $(\mathcal{F}_t)_{t\geq 0}$-measurable stochastic process. Let $f : [0,\infty) \times \mathbb{R} \to \mathbb{R}$ be $\mathfrak{B}([0,\infty)) \otimes \mathfrak{B}(\mathbb{R})$-measurable. The process*

$$Y_t := \int_0^t f(s, X_s)\,\mathrm{d}s, \qquad t \geq 0,$$

is progressively $(\mathcal{F}_t)_{t\geq 0}$-measurable. If $\mathcal{T}$ is a stopping time then $Y_{\mathcal{T}}$ is a $\mathcal{F}_{\mathcal{T}}$-measurable random variable.

Definition 8.1.9. Let $M = (M_t)_{t\geq 0}$ be a $(\mathcal{F}_t)_{t\geq 0}$-adapted stochastic process with $\mathbb{E}[|M_t|] < \infty$ for all $t \geq 0$. X is called a sub-martingale (super-martingale, respectively) if we have for all $0 \leq s \leq t < \infty$ that $\mathbb{P}$-a.s. $\mathbb{E}[M_t|\mathcal{F}_s] \geq M_s$ ($\mathbb{E}[M_t|\mathcal{F}_s] \leq M_s$, respectively). M is called a martingale if it is a sub-martingale and a super-martingale.

Definition 8.1.10. Let A be an adapted stochastic process. A is called *increasing*, if we have for $\mathbb{P}$-a.e. $\omega \in \Omega$

i) $A_0 = 0$;
ii) $t \mapsto A_t(\omega)$ is increasing and right-continuous;
iii) $\mathbb{E}\left[A_t\right] < \infty$ for all $t \in [0,\infty)$.

An increasing process is called integrable if $\mathbb{E}\left[A_\infty\right] < \infty$, where $A_\infty(\omega) = \lim_{t\to\infty} A_t(\omega)$ for $\omega \in \Omega$.

Theorem 8.1.30 (Doob–Meyer-decomposition)**.** *Let X be a positive sub-martingale with a.s. continuous trajectories. There is a continuous martingale M and an a.s. increasing continuous and adapted process A, such that*

$$X_t^2 = M_t + A_t.$$

The decomposition is unique.

Definition 8.1.11. Let X be a right-continuous martingale. We call X *quadratically integrable*, if $\mathbb{E}\left[X_t^2\right] < \infty$, for all $t \geq 0$. If we have in addition that $X_0 = 0$ $\mathbb{P}$-a.s., we write $X \in \mathcal{M}_2$ or $X \in \mathcal{M}_2^c$, if X is continuous.

Definition 8.1.12. For $X \in \mathcal{M}_2$ and $0 \leq t < \infty$ we define

$$\|X\|_t := \sqrt{\mathbb{E}\left[X_t^2\right]}.$$

We set

$$\|X\| := \sum_{n=1}^{\infty} \frac{\|X\|_n \wedge 1}{2^n}$$

which induces a metric d on $\mathcal{M}_2$ given by

$$d(X, Y) = \|X - Y\|.$$

Definition 8.1.13 (Total variation). A function $f : [0, t] \to \mathbb{R}$ is called of bounded variation if there is $M > 0$ such that $\sum_{i=1}^{n} |f(x_i) - f(x_{i-1})| \leq M$ for all finite partitions $\Pi = \{x_0, x_1, \ldots, x_n\} \subset [0, t]$ $(n \in \mathbb{N})$ with $0 = x_0 < x_1 < \cdots < x_n = t$. The quantity

$$V(f) := \sup \left\{ \sum_{i=1}^{n} |f(x_i) - f(x_{i-1})| : 0 = x_0 < x_1 < \cdots < x_n = t, n \in \mathbb{N} \right\}$$

is called total variation of f over $[0, t]$.

Definition 8.1.14 (Quadratic variation). For $X \in \mathcal{M}_2$ we define the *quadratic variation* of X, as the process $\langle\langle X \rangle\rangle_t := A_t$, where A is the increasing process from the Doob–Meyer-decomposition of X.

Definition 8.1.15 (Covariation). For $X, Y \in \mathcal{M}_2$ we define the *covariation* $\langle\langle X, Y \rangle\rangle$ by

$$\langle\langle X, Y \rangle\rangle_t := \frac{1}{4} \left[\langle\langle X + Y \rangle\rangle_t - \langle\langle X - Y \rangle\rangle_t \right],$$

for $t \geq 0$. The process $XY - \langle\langle X, Y \rangle\rangle$ is a martingale. In particular, we have $\langle\langle X, X \rangle\rangle = \langle\langle X \rangle\rangle$.

Definition 8.1.16 (p-th Variation). Let X be a stochastic process, $p \geq 1$, $t > 0$ fixed and $\Pi = \{t_0, t_1 \ldots t_n\}$, with $0 = t_0 < t_1 < \cdots < t_n = t$, $n \in \mathbb{N}$, a partition of $[0, t]$. The *p-th variation* of X in Π is defined by

$$V_t^{(p)}(\Pi) = \sum_{k=0}^{n} |X_{t_k} - X_{t_{k-1}}|^p.$$

Theorem 8.1.31. *Let $X \in \mathcal{M}_2^c$, Π be a partition of $[0, t]$ and $\|\Pi\| := \max_{1 \leq k \leq n} |t_k - t_{k-1}|$ the size of Π. Then we have $\lim_{\|\Pi\| \to 0} V_t^{(2)}(\Pi) = \langle\langle X \rangle\rangle_t$ in probability, i.e. for all $\varepsilon > 0$, $\eta > 0$ there is $\delta > 0$, such that we have that*

$$\mathbb{P}\left(|V_t^{(2)}(\Pi) - \langle X \rangle_t| > \varepsilon \right) < \eta,$$

for $\|\Pi\| < \delta$.

8.2 STOCHASTIC INTEGRATION

The aim of this section is to define stochastic integrals of the form

$$I_T(X) = \int_0^T X_t(\omega)\, \mathrm{d}M_t(\omega). \tag{8.2.1}$$

Here M is a square integrable martingale, X a stochastic process and $T > 0$. Throughout the section we assume that $M_0 = 0$ $\mathbb{P}$-a.s. Moreover, we suppose that M is a quadratically integrable $(\mathcal{F}_t)_{t\geq 0}$-adapted martingale where $(\mathcal{F}_t)_{t\geq 0}$ is a filtration which satisfies the usual conditions (see Definition 8.1.6). A process $M \in \mathcal{M}_2^c$ could be of unbounded variation in any finite subinterval of $[0, T]$. Hence integrals of the form (8.2.1) cannot be defined pointwise in $\omega \in \Omega$. However, M has finite quadratic variation given by the continuous and increasing process $\langle\langle M\rangle\rangle$ (see Theorem 8.1.31). Due to this fact, the stochastic integral can be defined with respect to continuous integrable martingales M for an appropriate class of integrands X.

The definition of the stochastic integral goes back to Itô. He studied the case where M is a Wiener process. His students Kunita and Watanabe considered the general case $M \in \mathcal{M}_2^c$. In the following we have a look at the class of integrands which are allowed in (8.2.1). We define a measure μ_M on $([0, \infty) \times \Omega, \mathfrak{B}([0, \infty)) \otimes \mathcal{F})$ by

$$\mu_M(A) = \mathbb{E}\left[\int_0^\infty \mathbb{I}_A(t, \omega)\, \mathrm{d}\langle\langle M\rangle\rangle_t(\omega)\right] \quad \text{for} \quad A \in \mathfrak{B}([0, \infty)) \otimes \mathcal{F}. \tag{8.2.2}$$

We call two $(\mathcal{F}_t)_{t\geq 0}$-adapted stochastic processes $X = (X_t)_{t\geq 0}$ and $Y = (Y_t)_{t\geq 0}$ *equivalent with respect to* M, if $X_t(\omega) = Y_t(\omega)$ μ_M-a.e. This leads to the following equivalence relation: for a $(\mathcal{F}_t)_{t\geq 0}$-adapted process X we define

$$[X]_T^2 :- \mathbb{E}\left[\int_0^T X_t^2(\omega)\, \mathrm{d}\langle\langle M\rangle\rangle_t(\omega)\right], \tag{8.2.3}$$

provided the right-hand-side exists. So $[X]_T$ is the L^2-norm of X as a function of (t, ω) with respect to the measure μ_M. We define the equivalence relation

$$X \sim Y \Leftrightarrow [X - Y]_T = 0 \quad \forall T > 0. \tag{8.2.4}$$

Our definition of the stochastic integral will imply that $I(X)$ and $I(Y)$ coincide provided X and Y are equivalent.

Definition 8.2.1. We define $\mathcal{L}^*$ as the space of equivalence classes of progressively $(\mathcal{F}_t)_{t\geq 0}$-measurable processes X with $[X]_T < \infty$ for all $T > 0$.

Remark 8.2.16. By setting $[X - Y] := \sum_{n=0}^{\infty} 2^{-n}(1 \wedge [X - Y]_n)$ we can define a metric on $\mathcal{L}^*$.

Remark 8.2.17. In the following we do not distinguish between X and the equivalence class X^* of X.

For $0 < T < \infty$ we define $\mathcal{L}_T^*$ as the space of processes $X \in \mathcal{L}^*$ with $X_t(\omega) = 0$ for all $t \geq T$ and a.e. $\omega \in \Omega$ and set

$$\mathcal{L}_\infty := \left\{ X \in \mathcal{L}_T^* : \mathbb{E}\left[\int_0^\infty X_t^2 \, \mathrm{d}\langle\langle M \rangle\rangle_t \right] < \infty \right\}.$$

A process $X \in \mathcal{L}_T^*$ can be identified with a process only defined on $[0, T] \times \Omega$. In particular we have that $\mathcal{L}_T^*$ is a closed subspace of the Hilbert space

$$\mathcal{H}_T := L^2(\Omega \times (0, T), \mathcal{F}_T \otimes \mathfrak{B}([0, T]), \mu_M). \tag{8.2.5}$$

Definition 8.2.2. A process X is called *step process* if there is a strictly increasing sequence $(t_n)_{n \in \mathbb{N}} \subseteq \mathbb{R}$ with $t_0 = 0$ and $\lim_{n \to \infty} t_n = \infty$, a sequence of random variables $(\xi_n)_{n \in \mathbb{N}_0}$ and $C < \infty$ with $\sup_{n \in \mathbb{N}_0} |\xi_n(\omega)| \leq C$ such that the following holds: for every $\omega \in \Omega$ we have that ξ_n is $\mathcal{F}_{t_n}$-measurable for every $n \in \mathbb{N}_0$ and we have the representation

$$X_t(\omega) = \xi_0(\omega)\mathbb{I}_{\{0\}}(t) + \sum_{i=0}^{\infty} \xi_i(\omega)\mathbb{I}_{(t_i, t_i+1]}(t), \tag{8.2.6}$$

for all $0 \leq t < \infty$. The space of step processes is denoted by $\mathcal{L}_0$.

Remark 8.2.18. ***(i)*** Step processes are progressively measurable and bounded. ***(ii)*** There holds $\mathcal{L}_0 \subseteq \mathcal{L}^*$.

Definition 8.2.3. Let $X \in \mathcal{L}_0$ and $M \in \mathcal{M}_2^c$. The stochastic integral of X with respect to M is the *martingale transformation*

$$I_t(X) := \sum_{i=0}^{n-1} \xi_i(M_{t_{i+1}} - M_{t_i}) + \xi_n(M_t - M_{t_n}) = \sum_{i=0}^{\infty} \xi_i(M_{t \wedge t_{i+1}} - M_{t \wedge t_i}), \tag{8.2.7}$$

for $0 \leq t < \infty$. Here $n \in \mathbb{N}$ is the unique natural number such that $t_n \leq t < t_{n+1}$.

In order to define the stochastic integral for $X \in \mathcal{L}^*$ we have to approximate the elements of $\mathcal{L}^*$ in an appropriate way by step processes, i.e. by processes in $\mathcal{L}_0$. This can be done thanks to the following theorem.

Theorem 8.2.32. *The space of step processes $\mathcal{L}_0$ is dense in $\mathcal{L}^*$ with respect to the metric defined in Remark 8.2.16.*

Definition 8.2.4. Let $X \in \mathcal{L}^*$ and $M \in \mathcal{M}_2^c$. The stochastic integral of X with respect to M is the unique quadratically integrable martingale

$$I(X) = \{I_t(X), \mathcal{F}_t, 0 \leq t < \infty\},$$

which satisfies $\lim_{n\to\infty} \|I(X^{(n)}) - I(X)\| = 0$, for every sequence $(X^{(n)})_{n\in\mathbb{N}} \subseteq \mathcal{L}_0$ with $\lim_{n\to\infty} [X^{(n)} - X] = 0$. We write

$$I_t(X) = \int_0^t X_s \,\mathrm{d}M_s; \;\; 0 \le t < \infty.$$

Theorem 8.2.33. *Let $X, Y \in \mathcal{L}^*$ and $0 \le s < t < \infty$. For the stochastic integrals $I(X), I(Y)$ we have*

a) $I_0(X) = 0$ *$\mathbb{P}$-a.s.,*

b) $\mathbb{E}[I_t(X)|\mathcal{F}_s] = I_s(X)$, *$\mathbb{P}$-a.s. (martingale property),*

c) $\mathbb{E}\left[(I_t(X))^2\right] = \mathbb{E}\left[\int_0^t X_u^2 \,\mathrm{d}\langle\langle M\rangle\rangle_u\right]$ *(Itô-isometry),*

d) $||I(X)|| = [X]$,

e) $\mathbb{E}\left[(I_t(X) - I_s(X))^2|\mathcal{F}_s\right] = \mathbb{E}\left[\int_s^t X_u^2 \;\mathrm{d}\langle\langle M\rangle\rangle_u|\mathcal{F}_s\right]$ *$\mathbb{P}$-a.s.,*

f) $I(\alpha X + \beta Y) = \alpha I(X) + \beta I(Y)$, *for $\alpha, \beta \in \mathbb{R}$.*

8.3 ITÔ'S LEMMA

One of the most important tools in stochastic analysis is Itô's Lemma. It is a chain-rule for paths of stochastic processes. In contrast to the deterministic case it can only be interpreted as an integral equation because the stochastic processes we are interested in (for instance the Wiener process) are in general not differentiable.

Definition 8.3.1 (Continuous local martingale). Let $(X_t)_{t\ge 0}$ be a continuous process adapted to $(\mathcal{F}_t)_{t\ge 0}$. Assume there is a sequence of stopping times $(T_n)_{n\in\mathbb{N}}$ of the filtration $(\mathcal{F}_t)_{t\ge 0}$, such that $(X_t^{(n)} := X_{t\wedge T_n})_{t\ge 0}$ is a $(\mathcal{F}_t)_{t\ge 0}$-martingale for all $n \in \mathbb{N}$ and $\mathbb{P}(\lim_{n\to\infty} T_n = \infty) = 1$. In this case we call X a *continuous local martingale*. If, in addition, $X_0 = 0$ $\mathbb{P}$-a.s., we write $X \in \mathcal{M}_2^{c,loc}$.

Definition 8.3.2 (Continuous semi-martingale). A *continuous semi-martingale* $(X_t)_{t\ge 0}$ is a $(\mathcal{F}_t)_{t\ge 0}$-adapted process such that the following (unique) decomposition holds:

$$X_t = X_0 + M_t + B_t, \quad 0 \le t < \infty. \tag{8.3.8}$$

In the above $M = (M_t)_{t\ge 0} \in \mathcal{M}_2^{c,loc}$ and $B = (B_t)_{t\ge 0}$ is the difference of two continuous increasing and $(\mathcal{F}_t)_{t\ge 0}$-adapted processes $A^{\pm} = (A_t^{\pm})_{t\ge 0}$, i.e. there holds

$$B_t = A_t^+ - A_t^-, \quad 0 \le t < \infty, \tag{8.3.9}$$

with $A_0^{\pm} = 0$ $\mathbb{P}$-a.s.

Theorem 8.3.34 (Itô's Lemma). *Let $f : \mathbb{R} \to \mathbb{R}$ be a C^2-function and $(X_t)_{t\geq 0}$ be a continuous $(\mathcal{F}_t)_{t\geq 0}$ semi-martingale with the decomposition* (8.3.8). *The following holds $\mathbb{P}$-a.s. for $0 \leq t < \infty$*

$$\begin{aligned} f(X_t) &= f(X_0) + \int_0^t f'(X_s)\,\mathrm{d}X_s + \frac{1}{2}\int_0^t f''(X_s)\,\mathrm{d}\langle\langle M\rangle\rangle_s \\ &= f(X_0) + \int_0^t f'(X_s)\,\mathrm{d}M_s + \int_0^t f'(X_s)\,\mathrm{d}B_s + \frac{1}{2}\int_0^t f''(X_s)\,\mathrm{d}\langle\langle M\rangle\rangle_s. \end{aligned} \tag{8.3.10}$$

Remark 8.3.19. The stochastic integral $\int_0^t f'(X_s)\,\mathrm{d}M_s$ in (8.3.10) is a continuous local martingale. The other two integrals in (8.3.10) are Lebesgue–Stieltjes integrals. They are of bounded variation as a function of t. Due to this $(f(X_t))_{t\geq 0}$ is a continuous $(\mathcal{F}_t)_{t\geq 0}$ semi-martingale.

Remark 8.3.20. Equation (8.3.10) is often written in differential form

$$\begin{aligned} \mathrm{d}f(X_t) &= f'(X_t)\,\mathrm{d}X_t + \frac{1}{2}f''(X_t)\,\mathrm{d}\langle\langle M\rangle\rangle_t \\ &= f'(X_t)\,\mathrm{d}M_t + f'(X_t)\,\mathrm{d}B_t + \frac{1}{2}f''(X_t)\,\mathrm{d}\langle\langle M\rangle\rangle_t. \end{aligned}$$

Note that this does not have a rigorous meaning. It only serves as an abbreviation of (8.3.10).

8.4 STOCHASTIC ODES

In this section we are concerned with stochastic differential equations. We seek a real-valued process $(X_t)_{t\in[0,T]}$ on a probability space $(\Omega, \mathcal{F}, \mathbb{P})$ with filtration $(\mathcal{F}_t)_{t\geq 0}$ such that

$$\begin{cases} \mathrm{d}X_t = \mu(t, X)\,\mathrm{d}t + \Sigma(t, X)\,\mathrm{d}W_t, \\ X(0) = X_0, \end{cases} \tag{8.4.11}$$

which holds true $\mathbb{P}$-a.s. and for all $t \in [0, T]$. Here W is a Wiener process with respect to $(\mathcal{F}_t)_{t\geq 0}$. The functions $\mu, \Sigma : [0, T] \times \mathbb{R} \to \mathbb{R}$ are assumed to be continuous. As in Remark 8.3.20, equation $(8.4.11)_1$ is only an abbreviation for the integral equation

$$X(t) = X(0) + \int_0^t \mu(s, X(s))\,\mathrm{d}s + \int_0^t \Sigma(s, X(s))\,\mathrm{d}W_s. \tag{8.4.12}$$

There are two different concepts of solutions to (8.4.12).

i) We talk about strong solution (in the probabilistic sense) if the solution exists on a given probability space $(\Omega, \mathcal{F}, \mathbb{P})$ with a given Wiener process W. A strong solution exists for a given initial datum $X_0 \in L^2(\Omega, \mathcal{F}_0, \mathbb{P})$ and there holds $X(0) = X_0$ a.s.

ii) We talk about weak solution (in the probabilistic sense) or martingale solution if there is a probability space and a Wiener process such that (8.4.12) holds true. The solution is usually written as

$$((\Omega, \mathcal{F}, (\mathcal{F}_t)_{t\geq 0}, \mathbb{P}), W, X).$$

This means that when seeking a weak solution, constructing the probability space (and the Wiener process on it) is part of the problem. A solution typically exists for a given initial law Λ_0 and we have $\mathbb{P} \circ X^{-1}(0) = \Lambda_0$. Even if an initial datum X_0 is given it might live on a different probability space. Hence $X(0)$ and X_0 can only coincide in law.

Theorem 8.4.35. *Let $(\Omega, \mathcal{F}, \mathbb{P})$ be a probability space with filtration $(\mathcal{F}_t)_{t\geq 0}$ and $X_0 \in L^2(\Omega, \mathcal{F}_0, \mathbb{P})$. Assume that μ and Σ are continuous on $[0, T] \times \mathbb{R}$ and globally Lipschitz continuous with respect to the second variable. Then there is a unique $(\mathcal{F}_t)_{t\geq 0}$-adapted process X such that (8.4.12) holds $\mathbb{P}$-a.s. for every $0 \leq t \leq T$ and we have $X(0) = X_0$ a.s. The trajectories of X are a.s. continuous and we have*

$$\mathbb{E}\Big[\sup_{t\in(0,T)} |X_t|^2\Big] < \infty.$$

The existence of a strong solution in the sense of Theorem 8.4.35 is classical, see e.g. [12] and [82,83]. If the assumptions on the coefficients are weakened, strong solutions might not exist, see [17]. In this case we can only hope for a weak solution. We refer to [95] for a nice proof and further references.

Theorem 8.4.36. *Let Λ_0 be a Borel probability measure on $\mathbb{R}$. Assume that μ and Σ are continuous on $[0, T] \times \mathbb{R}$ and have linear growth, i.e. there is $K \geq 0$ such that*

$$|\mu(t, X)| + |\Sigma(t, X)| \leq K(1 + |X|) \quad \forall (t, X) \in [0, T] \times \mathbb{R}.$$

There is a quantity $((\Omega, \mathcal{F}, (\mathcal{F}_t)_{t\geq 0}, \mathbb{P}), W, X)$ with the following properties.

i) *X is a $(\mathcal{F}_t)_{t\geq 0}$-adapted stochastic process with a.s. continuous trajectories such that*

$$\mathbb{E}\Big[\sup_{t\in(0,T)} |X_t|^2\Big] < \infty.$$

ii) *Equation* (8.4.12) *holds* $\mathbb{P}$*-a.s. for every* $0 \le t \le T$.
iii) *We have* $\mathbb{P} \circ X(0)^{-1} = \Lambda_0$.

The stochastic ODEs which appear later all have strong solutions. However, the concept of martingale solutions will be important for the SPDEs.

The stochastic ODEs we considered so far have two drawbacks. First, we need vector valued processes and, secondly, we have to weaken the assumptions on the drift μ (Lipschitz-continuity in X and linear growth is too strong). Everything in this chapter can be obviously extended to the multi-dimensional setting. Here a standard Wiener process in $\mathbb{R}^M$ is a vector valued stochastic process and each of its components is a real valued Wiener process (recall Definition 8.1.4). Moreover, the components are independent. Getting rid of the assumed Lipschitz continuity is more difficult. Now seek a $\mathbb{R}^N$-valued process $(\mathbf{X}_t)_{t\in[0,T]}$ on a probability space $(\Omega, \mathcal{F}, \mathbb{P})$ with filtration $(\mathcal{F}_t)_{t\geq 0}$ such that

$$\begin{cases} \mathrm{d}\mathbf{X}_t = \boldsymbol{\mu}(t, \mathbf{X})\,\mathrm{d}t + \boldsymbol{\Sigma}(t, \mathbf{X})\,\mathrm{d}\mathbf{W}_t, \\ \mathbf{X}(0) = \mathbf{X}_0. \end{cases} \tag{8.4.13}$$

Here $\mathbf{W}$ is a standard $\mathbb{R}^M$-valued Wiener process with respect to $(\mathcal{F}_t)_{t\geq 0}$ and $\mathbf{X}_0 \in L^2(\Omega, \mathcal{F}_0, \mathbb{P})$ is some initial datum. The functions

$$\begin{aligned} &\boldsymbol{\mu} : \Omega \times [0, T] \times \mathbb{R}^N \to \mathbb{R}^N, \\ &\boldsymbol{\Sigma} : \Omega \times [0, T] \times \mathbb{R}^N \to \mathbb{R}^{N\times M}, \end{aligned}$$

are continuous in $\mathbf{X} \in \mathbb{R}^N$ for each fixed $t \in [0, T]$, $\omega \in \Omega$. Moreover, they are assumed to be progressively measurable. The application in Chapter 10 requires weaker assumptions as in the classical existence theorems mentioned above. In our application we only have local Lipschitz continuity of $\boldsymbol{\mu}$. Fortunately, some more recent results apply. In the following we state the assumptions which are in fact a special case of the assumptions in [124, Thm. 3.1.1.].

(A1) We assume that the following integrability condition on $\boldsymbol{\mu}$ for all $R < \infty$

$$\int_0^T \sup_{|\mathbf{X}|\leq R} |\boldsymbol{\mu}(t, \mathbf{X})|^2 \,\mathrm{d}t < \infty \quad \text{in} \quad \Omega.$$

(A2) $\boldsymbol{\mu}$ is weakly coercive, i.e. for all $(t, \mathbf{X}) \in [0, T] \times \mathbb{R}^N$ we have that

$$\boldsymbol{\mu}(t, \mathbf{X}) \cdot \mathbf{X} \leq c(1 + \tilde{\mu}_t|\mathbf{X}|),$$

where $\tilde{\mu} \in L^2(\Omega, \mathcal{F}, \mathbb{P}; L^2(0, T))$ is $(\mathcal{F}_t)_{t\geq 0}$-adapted.

(A3) μ is locally weakly monotone, i.e. for all $t \in [0, T]$ and all $\mathbf{X}, \mathbf{Y} \in \mathbb{R}^N$ the following holds

$$\big(\mu(t, \mathbf{X}) - \mu(t, \mathbf{Y})\big) : (\mathbf{X} - \mathbf{Y}) \leq 0.$$

(A4) $\boldsymbol{\Sigma}$ is Lipschitz continuous, i.e. for all $t \in [0, T]$ and all $\mathbf{X}, \mathbf{Y} \in \mathbb{R}^N$ the following holds

$$|\boldsymbol{\Sigma}(t, \mathbf{X}) - \boldsymbol{\Sigma}(t, \mathbf{Y})|^2 \leq c|\mathbf{X} - \mathbf{Y}|^2.$$

Theorem 8.4.37. *Let μ and $\boldsymbol{\Sigma}$ satisfy* (A1)–(A4). *Assume we have a given probability space $(\Omega, \mathcal{F}, \mathbb{P})$ with filtration $(\mathcal{F}_t)_{t \geq 0}$, an initial datum $\mathbf{X}_0 \in L^2(\Omega, \mathcal{F}_0, \mathbb{P})$ and a Brownian motion $\mathbf{W}$ with respect to $(\mathcal{F}_t)_{t \geq 0}$. Then there is a unique $(\mathcal{F}_t)_{t \geq 0}$-adapted process satisfying*

$$\mathbf{X}(t) = \mathbf{X}_0 + \int_0^t \mu(\sigma, \mathbf{X}(\sigma))\,\mathrm{d}\sigma + \int_0^t \boldsymbol{\Sigma}(\sigma, \mathbf{X}(\sigma))\,\mathrm{d}\mathbf{W}_\sigma, \quad \mathbb{P}\text{-a.s.},$$

for every $t \in [0, T]$. The trajectories of $\mathbf{X}$ are $\mathbb{P}$-a.s. continuous and we have

$$\mathbb{E}\Big[\sup_{t \in (0,T)} |\mathbf{X}_t|^2\Big] < \infty.$$

Theorem 8.4.38. *Let the assumptions of Theorem 8.4.37 hold. Assume that $\mathbf{X}_0 \in L^\beta(\Omega, \mathcal{F}_0, \mathbb{P})$ and $\tilde{\mu} \in L^\beta(\Omega, \mathcal{F}, \mathbb{P}; L^\beta(0, T))$ for some $\beta > 2$. Then we have*

$$\mathbb{E}\Big[\sup_{t \in (0,T)} |\mathbf{X}_t|^\beta\Big] < \infty.$$

CHAPTER 9

Stochastic PDEs

Contents

Abstract

In this chapter we revise some well-known tools for SPDEs. In the first section we collect methods from infinite dimensional stochastic analysis, in particular stochastic integration in Hilbert spaces. After this, we give an introduction to the variational approach to SPDEs by considering the stochastic heat equation. Finally, we present the theorems of Prokhorov and Skorokhod which are essential for the stochastic compactness method.

9.1 STOCHASTIC ANALYSIS IN INFINITE DIMENSIONS

In the following we extend the setup from the previous chapter to the case of Banach or Hilbert space valued stochastic processes (see [55]).

Let $(\mathscr{V}, \|\cdot\|_{\mathscr{V}})$ be a Banach space and $1 \leq p < \infty$. We denote by $L^p(\Omega, \mathcal{F}, \mathbb{P}; \mathscr{V})$ the Banach space of all measurable mappings $v : \Omega \to \mathscr{V}$ such that

$$\mathbb{E}\big[\|v\|_{\mathscr{V}}^p\big] < \infty,$$

where the expectation is taken with respect to $(\Omega, \mathcal{F}, \mathbb{P})$. The measurability has to be understood via the approximation by step functions similar to Section 5.1. Regarding the reflexivity and the dual spaces we have the same results as in the case of Bochner spaces (see Lemma 5.1.1). The definitions of adaptivity and progressive measurability (see Chapter 8) extend in a straightforward manner to Banach space valued processes. The definition of the stochastic integral can be extended to Hilbert spaces, where the process X as well as the stochastic integral take values in some Hilbert spaces $(\mathscr{H}, \|\cdot\|_{\mathscr{H}})$. Let U be a Hilbert space with orthonormal basis $(\mathbf{e}_k)_{k\in\mathbb{N}}$ and let $L_2(U, L^2(G))$ be the set of Hilbert–Schmidt operators from U to $L^2(G)$. Recall that a bounded linear operator $\Phi : U \to L^2(G)$ is called Hilbert–Schmidt operator iff

$$\sum_{k\in\mathbb{N}} \|\Phi \mathbf{e}_k\|_{L^2(G)}^2 < \infty.$$

Existence Theory for Generalized Newtonian Fluids.
DOI: http://dx.doi.org/10.1016/B978-0-12-811044-7.00012-4

We consider a cylindrical Wiener process $\mathbf{W} = (\mathbf{W}_t)_{t\in[0,T]}$ which has the form

$$\mathbf{W}(\sigma) = \sum_{k\in\mathbb{N}} \mathbf{e}_k \beta_k(\sigma) \tag{9.1.1}$$

with a sequence $(\beta_k)_{k\in\mathbb{N}}$ of independent real valued Brownian motions on $(\Omega, \mathcal{F}, \mathbb{P})$. Define further the auxiliary space $U_0 \supset U$ as

$$\begin{aligned} U_0 &:= \left\{ \mathbf{e} = \sum_k \alpha_k \mathbf{e}_k : \sum_k \frac{\alpha_k^2}{k^2} < \infty \right\}, \\ \|\mathbf{e}\|_{U_0}^2 &:= \sum_{k=1}^{\infty} \frac{\alpha_k^2}{k^2}, \quad \mathbf{e} = \sum_k \alpha_k \mathbf{e}_k, \end{aligned} \tag{9.1.2}$$

thus the embedding $U \hookrightarrow U_0$ is Hilbert–Schmidt and trajectories of $\mathbf{W}$ belong $\mathbb{P}$-a.s. to the class $C([0,T]; U_0)$ (see [55]).

For $\psi \in L^2(\Omega, \mathcal{F}, \mathbb{P}; L^2(0,T; L_2(U, L^2(G))))$ progressively $(\mathcal{F}_t)_{t\geq 0}$-measurable we see that the equality

$$\int_0^t \psi(\sigma)\,\mathrm{d}\mathbf{W}_\sigma = \sum_{k=1}^{\infty} \int_0^t \psi(\sigma)(\mathbf{e}_k)\,\mathrm{d}\beta_k(\sigma) \tag{9.1.3}$$

defines a $\mathbb{P}$-a.s. continuous $L^2(G)$-valued $(\mathcal{F}_t)_{t\geq 0}$-martingale. Moreover, we can multiply the above with test-functions since

$$\int_G \int_0^t \psi(\sigma)\,\mathrm{d}\mathbf{W}_\sigma \cdot \boldsymbol{\varphi}\,\mathrm{d}x = \sum_{k=1}^{\infty} \int_0^t \int_G \psi(\sigma)(\mathbf{e}_k) \cdot \boldsymbol{\varphi}\,\mathrm{d}x\,\mathrm{d}\beta_k(\sigma), \quad \boldsymbol{\varphi} \in L^2(G),$$

is well-defined and $\mathbb{P}$-a.s. continuous.

In the following we define the quadratic variation of a stochastic process with values in a Hilbert space.

Definition 9.1.1. Let $(X_t)_{t\in[0,T]}$ be a continuous semi-martingale on a probability space $(\Omega, \mathcal{F}, \mathbb{P})$ with values in a separable Hilbert space $(\mathcal{H}, \langle\cdot,\cdot\rangle_{\mathcal{H}})$ with basis $(e_i)_{i\in\mathbb{N}}$. Then its quadratic variation process is defined as

$$\langle\langle X, X\rangle\rangle_t^{\mathcal{H}} := \sum_{i,j\in\mathbb{N}} \Big\langle\Big\langle \langle X, e_i\rangle_{\mathcal{H}}, \langle X, e_j\rangle_{\mathcal{H}} \Big\rangle\Big\rangle_t \langle e_j, \cdot\rangle_{\mathcal{H}}\, e_i$$

and has values in $\mathcal{N}(\mathcal{H})$ (the set of nuclear operators on $\mathcal{H}$). Moreover, we define the trace of $\langle\langle X, X\rangle\rangle_t^{\mathcal{H}}$ by

$$\mathrm{tr}\langle\langle X, X\rangle\rangle_t^{\mathcal{H}} := \sum_{i\in\mathbb{N}} \Big\langle\Big\langle \langle X, e_i\rangle_{\mathcal{H}}, \langle X, e_i\rangle_{\mathcal{H}} \Big\rangle\Big\rangle_t \langle e_i, \cdot\rangle_{\mathcal{H}}\, e_i.$$

Very useful is also the Burkholder–Davis–Gundi inequality.

Lemma 9.1.1. *Let $(\mathscr{H}, \langle\cdot,\cdot\rangle_{\mathscr{H}})$ be a separable Hilbert space, $(\Omega, \mathcal{F}, \mathbb{P})$ be a probability space and $(X_t)_{t\in[0,T]}$ be a continuous martingale on $(\Omega, \mathcal{F}, \mathbb{P})$ with values in $\mathscr{H}$. Then we have for all $p > 0$*

$$c_p \mathbb{E}\Big[\sup_{t\in(0,T)} \|X_t\|_{\mathscr{H}}\Big]^p \leq \mathbb{E}\Big[\|\operatorname{tr}\langle\langle X, X\rangle\rangle_T^{\mathscr{H}}\|_{\mathcal{N}(\mathscr{H})}\Big]^{\frac{p}{2}} \leq C_p \mathbb{E}\Big[\sup_{t\in(0,T)} \|X_t\|_{\mathscr{H}}\Big]^p,$$

where c_p, C_p are positive constants.

Now we present an infinite dimensional version of Itô's formula which is appropriate at least to obtain energy estimates for linear SPDEs, see [105, Theorem 3.1] or [127, Chapter 4.2, Theorem 2].

Theorem 9.1.39. *Let $(\mathscr{V}, \|\cdot\|_{\mathscr{V}})$ be a Banach space which is continuously embedded into a separable Hilbert space $(\mathscr{H}, \langle\cdot,\cdot\rangle_{\mathscr{H}})$. Let $(\Omega, \mathcal{F}, \mathbb{P})$ be a probability space. Assume that the processes $(X_t)_{t\in[0,T]}$ and $(Y_t)_{t\in[0,T]}$, taking values in $\mathscr{V}$ and $\mathscr{V}'$, respectively, are progressively measurable and*

$$\mathbb{P}\bigg\{\int_0^T \big(\|X\|_{\mathscr{V}}^2 + \|Y\|_{\mathscr{V}'}^2\big)\,\mathrm{d}t < \infty\bigg\} = 1.$$

Assume further that there is a continuous martingale $(M_t)_{t\in[0,T]}$, taking values in $\mathscr{H}$, such that, for $\mathbb{P}\times\mathcal{L}^1$-a.e. (ω, t), the following equality holds:

$$\langle X(t), \varphi\rangle_{\mathscr{H}} = \langle X(0), \varphi\rangle_{\mathscr{H}} + \int_0^t {}_{\mathscr{V}'}\langle Y(\sigma), \varphi\rangle_{\mathscr{V}}\,\mathrm{d}\sigma + \langle M_t, \varphi\rangle_{\mathscr{H}} \quad \forall\varphi\in\mathscr{V}.$$

Then we have

$$\begin{aligned}\|X(t)\|_{\mathscr{H}}^2 &= \|X(0)\|_{\mathscr{H}}^2 + \int_0^t {}_{\mathscr{V}'}\langle Y(\sigma), X(s)\rangle_{\mathscr{V}}\,\mathrm{d}\sigma \\ &\quad + 2\int_0^t \langle X(\sigma), \mathrm{d}M_\sigma\rangle_{\mathscr{H}} + \|\operatorname{tr}\langle\langle M, M\rangle\rangle_t^{\mathscr{H}}\|_{\mathcal{N}(\mathscr{H})}, \quad \mathbb{P}\times\mathcal{L}^1\text{-a.e.}\end{aligned}$$

In Lemma C.0.1 in Appendix C we will establish a version of Itô's formula which is appropriate for nonlinear PDEs, in particular for stochastic Navier–Stokes equations and problems with polynomial nonlinearities.

In some applications we need fractional time derivatives of stochastic integrals. The following lemma is concerned with fractional derivatives of stochastic integrals in Hilbert spaces (see [73, Lemma 2.1] for a proof).

Lemma 9.1.2. *Let $\Psi \in L^p(\Omega, \mathcal{F}, \mathbb{P}; L^p(0, T; L_2(U, L^2(G))))$ $(p \geq 2)$ be progressively $(\mathcal{F}_t)_{t\geq 0}$-measurable and $\mathbf{W}$ a cylindrical $(\mathcal{F}_t)_{t\geq 0}$-Wiener process as in (9.1.1). Then the following holds for any $\alpha \in (0, 1/2)$*

$$\mathbb{E}\bigg[\bigg\|\int_0^{\cdot} \Psi\,\mathrm{d}\mathbf{W}_\sigma\bigg\|_{W^{\alpha,p}(0,T;L^2(G))}^p\bigg] \leq c(\alpha, p)\,\mathbb{E}\bigg[\int_0^T \|\Psi\|_{L_2(U,L^2(G))}^p\,\mathrm{d}t\bigg].$$

If we have higher moments it is possible to improve the fractional time differentiability to Hölder-continuity.

Lemma 9.1.3. *Let $(\Omega, \mathcal{F}, \mathbb{P})$ be a probability space endowed with the filtration $(\mathcal{F}_t)_{t\geq 0}$.*

a) Let $\mathbf{W}$ be a M-dimensional Brownian motion with respect to $(\mathcal{F}_t)_{t\geq 0}$. Let $\mathbf{X} \in L^\beta(\Omega, \mathcal{F}, \mathbb{P}; L^\beta(0, T))$, $\beta > 2$, be a progressively $(\mathcal{F}_t)_{t\geq 0}$-measurable process with values in $\mathbb{R}^{N\times M}$. Then the paths of the process $\mathbf{Z}_t := \int_0^t \mathbf{X}\,\mathrm{d}\mathbf{W}_\sigma$ are $\mathbb{P}$-a.s. Hölder continuous with exponent $\alpha \in \left(\frac{1}{\beta}, \frac{1}{2}\right)$ and we have

$$\mathbb{E}\Big[\|\mathbf{Z}\|^\beta_{C^\alpha([0,T])}\Big] \leq c_\alpha\, \mathbb{E}\Big[\int_0^T |\mathbf{X}|^\beta \,\mathrm{d}t\Big].$$

b) Let $\psi \in L^\beta(\Omega, \mathcal{F}, \mathbb{P}; L^\beta(0, T; L_2(U, L^2(G))))$, $\beta > 2$, be progressively $(\mathcal{F}_t)_{t\geq 0}$-measurable and $\mathbf{W}$ a cylindrical $(\mathcal{F}_t)_{t\geq 0}$-Wiener process as in (9.1.1). Then the paths of the process $\mathbf{Z}_t := \int_0^t \psi\,\mathrm{d}\mathbf{W}_\sigma$ are $\mathbb{P}$-a.s. Hölder continuous with exponent $\alpha \in \left(\frac{1}{\beta}, \frac{1}{2}\right)$ and the following holds

$$\mathbb{E}\Big[\|\mathbf{Z}\|^\beta_{C^\alpha([0,T];L^2(G))}\Big] \leq c_\alpha\, \mathbb{E}\Big[\int_0^T \|\psi\|^\beta_{L_2(U,L^2(G))}\,\mathrm{d}t\Big].$$

Proof. a) We follow [94], proof of Lemma 4.6, and consider the Riemann–Liouville operator: let X be a Banach space, $p \in (1, \infty]$, $\alpha \in \left(\frac{1}{p}, 1\right]$ and $f \in L^p(0, T; X)$. Then the Riemann–Liouville operator is given by

$$(R_\alpha f)(t) := \frac{1}{\Gamma(\alpha)} \int_0^t (t-\sigma)^{\alpha-1} f(\sigma)\,\mathrm{d}\sigma, \quad t \in [0, T].$$

It is well known that R_α is a bounded linear operator from $f \in L^p(0, T; X)$ to $C^{\alpha-1/p}([0, T]; X)$ (see [130], Thm. 3.6). According to the stochastic Fubini Theorem (see [55], Thm. 4.18) we have $\mathbf{Z}_t = R_\alpha(\tilde{\mathbf{Z}}_t)$, where

$$\tilde{\mathbf{Z}}_t = \frac{1}{\Gamma(1-\alpha)} \int_0^t (t-\sigma)^{-\alpha} \mathbf{X}(t)\,\mathrm{d}\mathbf{W}_\sigma, \quad t \in [0, T].$$

For $\alpha \in \left(\frac{1}{\beta}, \frac{1}{2}\right)$ we obtain by the Burkholder–Davis–Gundi inequality and Young's inequality for convolution that

$$\begin{aligned}
\mathbb{E}\Big[\|\mathbf{Z}\|^\beta_{C^{\alpha-1/\beta}([0,T])}\Big] &\leq c\,\mathbb{E}\Big[\int_0^T |\tilde{\mathbf{Z}}(t)|^\beta \,\mathrm{d}t\Big] \leq c \int_0^T \mathbb{E}\Big[\sup_{[0,t]} |\tilde{\mathbf{Z}}(\sigma)|\Big]^\beta \mathrm{d}t \\
&\leq c \int_0^T \mathbb{E}\Big[\int_0^t (t-\sigma)^{-2\alpha} |\mathbf{X}(\sigma)|^2 \,\mathrm{d}\sigma\Big]^{\frac{\beta}{2}} \mathrm{d}t \\
&\leq c\,\mathbb{E}\Big[\int_0^T |\mathbf{X}|^\beta \,\mathrm{d}t\Big].
\end{aligned}$$

b) By exactly the same arguments we end up with

$$
\begin{aligned}
&\mathbb{E}\Big[\|\mathbf{Z}\|^{\beta}_{C^{\alpha-1/\beta}([0,T];L^2(G))}\Big] \\
&\leq c\,\mathbb{E}\bigg[\int_0^T \|\tilde{\mathbf{Z}}(t)\|^{\beta}_{L^2(G)}\,\mathrm{d}t\bigg] \leq c\int_0^T \mathbb{E}\bigg[\sup_{[0,t]}\|\tilde{\mathbf{Z}}(\sigma)\|^{\beta}_{L^2(G)}\bigg]^{\beta}\mathrm{d}t \\
&\leq c\int_0^T \mathbb{E}\bigg[\sum_i\int_0^t (t-\sigma)^{-2\alpha}\|\psi(\mathbf{e}_i)\|^2_{L^2(G)}\,\mathrm{d}\sigma\bigg]^{\frac{\beta}{2}}\mathrm{d}t \\
&= c\int_0^T \mathbb{E}\bigg[\int_0^t (t-\sigma)^{-2\alpha}\|\psi(\sigma)\|^2_{L_2(U,L^2(G))}\,\mathrm{d}\sigma\bigg]^{\frac{\beta}{2}}\mathrm{d}t \\
&\leq c\,\mathbb{E}\bigg[\int_0^T \|\psi\|^{\beta}_{L_2(U,L^2(G))}\,\mathrm{d}t\bigg].
\end{aligned}
$$

This concludes the proof. □

The following lemma is very useful in order to pass to the limit in stochastic integrals (see [56, Lemma 2.1])

Lemma 9.1.4. *Consider a sequence of cylindrical Wiener processes* ($\mathbf{W}^n$) *over* U *(see* (9.1.1)*) with respect to the filtration* $(\mathcal{F}_t)_{t\geq 0}$. *Assume that* ($\Psi^n$) *is a sequence of progressively* $(\mathcal{F}_t)_{t\geq 0}$*-measurable processes such that* $\Psi^n \in L^2(0,T;L_2(U,L^2(G)))$ $\mathbb{P}$*-a.s. Suppose there is a cylindrical* $(\mathcal{F}_t)_{t\geq 0}$*-Wiener process* $\mathbf{W}$ *and* $\Psi \in L^2(0,T;$ $L_2(U,L^2(G)))$, *progressively* $(\mathcal{F}_t)_{t\geq 0}$*-measurable, such that*

$$
\begin{aligned}
\mathbf{W}^n \to \mathbf{W} \quad &in \quad C^0([0,T];U_0),\\
\Psi^n \to \Psi \quad &in \quad L^2(0,T;L_2(U,L^2(G))),
\end{aligned}
$$

in probability. Then we have

$$
\int_0^{\cdot} \Psi^n\,\mathrm{d}\mathbf{W}^n \to \int_0^{\cdot} \Psi\,\mathrm{d}\mathbf{W} \quad in \quad L^2(0,T;L^2(G)),
$$

in probability.

9.2 STOCHASTIC HEAT EQUATION

As a preparation for the stochastic models for power law fluids we will study the stochastic heat equation by means of a Galerkin–Ansatz. So we seek for a $(\mathcal{F}_t)_{t\geq 0}$-adapted process $u:\Omega\times(0,T)\times G\to\mathbb{R}$ satisfying

$$
\begin{cases}
\mathrm{d}u_t = \Delta u\,\mathrm{d}t + \Phi\,\mathrm{d}\mathbf{W}_t,\\
u(0) = u_0.
\end{cases}
\tag{9.2.4}
$$

Here $\mathbf{W}$ is a cylindrical $(\mathcal{F}_t)_{t\geq 0}$-Wiener process (see (9.1.1)), $u_0 \in L^2(\Omega, \mathcal{F}_0, \mathbb{P}; L^2(G))$ is some initial datum and Φ is a progressively $(\mathcal{F}_t)_{t\geq 0}$-measurable process taking values in the space of Hilbert–Schmidt operators. More precisely, we assume

$$\Phi \in L^2(\Omega, \mathcal{F}, \mathbb{P}; L^2(0, T; L_2(U; L^2(G)))). \tag{9.2.5}$$

Typically one supposes some (nonlinear) dependence of Φ on u. But we neglect this here for simplicity. The Hilbert space U on which $\mathbf{W}$ is defined will most naturally be $L^2(G)$. As in the case of stochastic ODEs (9.2.4) is an abbreviation for the integral equation

$$u(t) = u_0 + \int_0^t \Delta u \,\mathrm{d}\sigma + \int_0^t \Phi \,\mathrm{d}\mathbf{W}_\sigma.$$

The stochastic integral has to be understood as in (9.1.3). The integral $\int_0^t \Delta u \,\mathrm{d}\sigma$ makes sense in case of a strong solution, i.e. if $\Delta u \in L^1((0,T)\times G)$ a.s. In order to understand the weak formulation (weak in the PDE-sense) we multiply the above with test-functions $\varphi \in C_0^\infty(G)$, thereby obtaining the following equality, which holds $\mathbb{P}$-a.s.:

$$\begin{aligned}\int_G u(t)\,\varphi \,\mathrm{d}x &= \int_G u_0\,\varphi \,\mathrm{d}x - \int_0^t\int_G \nabla u\cdot\nabla\varphi \,\mathrm{d}x\,\mathrm{d}\sigma\\ &\quad + \int_G \varphi\left(\int_0^t \Phi\,\mathrm{d}\mathbf{W}_\sigma\right)\mathrm{d}x\end{aligned} \tag{9.2.6}$$

for a.e. $t \in (0,T)$. Due to (9.2.5) this can be equivalently formulated using test-functions for $W_0^{1,2}(G)$. The natural solution space for this is obviously

$$L^2(\Omega, \mathcal{F}, \mathbb{P}; L^2(0, T; W_0^{1,2}(G))).$$

Theorem 9.2.40. *Let $(\Omega, \mathcal{F}, \mathbb{P})$ be a probability space with right-continuous filtration $(\mathcal{F}_t)_{t\geq 0}$, let $u_0 \in L^2(\Omega, \mathcal{F}_0, \mathbb{P}; L^2(G))$ and assume that Φ satisfies (9.2.5) and is progressively $(\mathcal{F}_t)_{t\geq 0}$-measurable. Then there is a progressively $(\mathcal{F}_t)_{t\geq 0}$-measurable process*

$$u \in L^2(\Omega, \mathcal{F}, \mathbb{P}; L^2(0, T; W_0^{1,2}(G)))$$

such that (9.2.6) holds. We also have $u \in C([0,T]; L^2(G))$ a.s. with

$$\mathbb{E}\left[\sup_{t\in(0,T)}\int_G |u(t)|^2\,\mathrm{d}x\right] < \infty.$$

The solution u is unique in the above class.

Proof. We will split the proof in several steps. First we approximate (9.2.6) by a Galerkin–Ansatz. Then we show a priori estimates and finally we pass to the limit. Note that the uniqueness of solutions is obvious: the difference of two potential solutions solves a deterministic heat equation with zero initial datum.

Step 1: approximation

We will solve (9.2.6) by a finite dimensional approximation. So we need an appropriate basis of $W_0^{1,2}(G)$. A good choice is the set of eigenfunctions of the Laplace operator. There is a smooth orthonormal system $(w_k)_{k\in\mathbb{N}} \subset L^2(G)$ and $(\lambda_k) \in (0, \infty)$ such that

$$\int_G \nabla w_k \cdot \nabla \varphi \,\mathrm{d}x = \lambda_k \int_G w_k \varphi \,\mathrm{d}x \quad \forall \varphi \in W_0^{1,2}(G). \tag{9.2.7}$$

We seek the approximate solution u_N such that

$$u_N = \sum_{k=1}^N c_N^i w_k = \mathbf{C}_N \cdot \mathbf{w}_N, \qquad \mathbf{w}_N = (w_1, \dots, w_N),$$

where $\mathbf{C}_N = (c_N^i) : \Omega \times (0, T) \to \mathbb{R}^N$. Therefore, we would like to solve the system

$$\int_G \mathrm{d}u_N \, w_k \,\mathrm{d}x + \int_G \nabla u_k \cdot \nabla w_k \,\mathrm{d}x \,\mathrm{d}t = \int_G \Phi \,\mathrm{d}\mathbf{W}_\sigma^N \cdot w_k \,\mathrm{d}x, \quad k = 1, \dots, N,$$
$$u_N(0) = \mathcal{P}_N u_0. \tag{9.2.8}$$

Here $\mathcal{P}^N : L^2_{\mathrm{div}}(G) \to \mathcal{X}_N := \operatorname{span}\{w_1, \dots, w_N\}$ is the orthogonal projection, i.e.

$$\mathcal{P}_N u = \sum_{k=1}^N \langle u, w_k \rangle_{L^2} w_k.$$

Equation (9.2.8) is to be understood $\mathbb{P}$-a.s. and for a.e. t and we assume

$$\mathbf{W}^N(\sigma) = \sum_{k=1}^N \mathbf{e}_k \beta_k(\sigma) = \mathbf{e}^N \cdot \boldsymbol{\beta}^N(\sigma).$$

It is equivalent to solving

$$\begin{cases} \mathrm{d}\mathbf{C}_N = \mathbf{\Lambda} \mathbf{C}_N \,\mathrm{d}t + \mathbf{\Sigma} \,\mathrm{d}\boldsymbol{\beta}_t^N, \\ \mathbf{C}_N(0) = \mathbf{C}_0, \end{cases} \tag{9.2.9}$$

where $\mathbf{\Lambda}, \mathbf{\Sigma} \in \mathbb{R}^{N\times N}$ with

$$\mathbf{\Lambda}_{ij} = \delta_{ij}\lambda_j, \quad \mathbf{\Sigma}_{ij} = \int_G \Phi \mathbf{e}_i w_j \,\mathrm{d}x.$$

As (9.2.9) is a linear system of ODEs we obtain a unique solution (with a.s. continuous trajectories) by the results of Section 8.4.

Step 2: a priori estimates

We apply Itô's formula to the function $f(\mathbf{C}) = \frac{1}{2}|\mathbf{C}|^2$, thereby obtaining

$$\begin{aligned}\frac{1}{2}\|u_N(t)\|^2_{L^2(G)} &= \frac{1}{2}\|\mathbf{C}_N(0)\|^2_{L^2(G)} + \int_0^t \mathbf{C}_N \cdot \mathrm{d}(\mathbf{C}_N)_\sigma + \frac{1}{2}\int_0^t I : \mathrm{d}\langle\langle \mathbf{C}^N \rangle\rangle_\sigma \\ &= \frac{1}{2}\|\mathcal{P}_N u_0\|^2_{L^2(G)} - \int_0^t \int_G \nabla u_N \cdot \nabla u_N \,\mathrm{d}x\,\mathrm{d}\sigma \qquad (9.2.10)\\ &+ \int_G \int_0^t u_N \Phi \,\mathrm{d}\mathbf{W}^N_\sigma \,\mathrm{d}x + \frac{1}{2}\sum_k \int_0^t \Sigma^2_{kk}\,\mathrm{d}x\,\mathrm{d}\sigma.\end{aligned}$$

In the above we used (9.2.7) and

$$\begin{aligned}\mathrm{d}c^k_N &= -\int_G \nabla u_N \cdot \nabla w_k \,\mathrm{d}x\,\mathrm{d}t + \int_G \Phi\,\mathrm{d}\mathbf{W}^N_t\, w_k\,\mathrm{d}x,\\ \langle\langle c^i_N, c^j_N \rangle\rangle &= \left\langle\left\langle \sum_{k=1}^N \int_0^\cdot \int_G \Phi\mathbf{e}_j w_k \,\mathrm{d}x\,\mathrm{d}\beta_k, \sum_{l=1}^N \int_0^\cdot \int_G \Phi\mathbf{e}_i w_l\,\mathrm{d}x\,\mathrm{d}\beta_l \right\rangle\right\rangle\\ &= \sum_{k=1}^N \sum_{l=1}^N \left\langle\left\langle \int_0^\cdot \int_G \Phi\mathbf{e}_j w_k\,\mathrm{d}x\,\mathrm{d}\beta_k, \int_0^\cdot \int_G \Phi\mathbf{e}_i w_l\,\mathrm{d}x\,\mathrm{d}\beta_l \right\rangle\right\rangle\\ &= \sum_{k=1}^N \left(\int_G \Phi\mathbf{e}_j w_k\,\mathrm{d}x\right)\left(\int_G \Phi\mathbf{e}_i w_k\,\mathrm{d}x\right) = (\mathbf{\Sigma}^2)_{ij}.\end{aligned}$$

Now we obtain, taking the supremum in time and taking expectations, that

$$\begin{aligned}&\mathbb{E}\left[\sup_{t\in(0,T)} \int_G |u_N(t)|^2\,\mathrm{d}x + \int_0^T \int_G |\nabla u_N|^2\,\mathrm{d}x\,\mathrm{d}\sigma\right]\\ &\quad \le c\,\mathbb{E}\left[\int_G |u_0|^2\,\mathrm{d}x + \sum_{k=1}^N \int_0^T \Sigma^2_{kk}\,\mathrm{d}t + \sup_{t\in(0,T)} |J(t)|\right],\end{aligned}$$

where we set

$$J(t) = \int_G \int_0^t u_N \Phi\,\mathrm{d}\mathbf{W}^N_\sigma\,\mathrm{d}x.$$

By Hölder's inequality and an account of $\|w_k\|_2 = 1$, straightforward calculations show

$$\begin{aligned}\mathbb{E}\left[\sum_{k=1}^N \int_0^T \Sigma^2_{kk}\,\mathrm{d}x\,\mathrm{d}t\right] &= \mathbb{E}\left[\sum_{k=1}^N \int_0^T \int_G |\Phi\mathbf{e}_k|^2\,\mathrm{d}x\,\mathrm{d}t\right]\\ &\le \mathbb{E}\left[\sum_{k=1}^\infty \int_0^T \int_G |\Phi\mathbf{e}_k|^2\,\mathrm{d}x\,\mathrm{d}t\right]\\ &= \mathbb{E}\left[\int_0^T \|\Phi\|^2_{L_2(U,L^2(G))}\,\mathrm{d}t\right].\end{aligned}$$

On account of Burkholder–Davis–Gundi inequality (Lemma 9.1.1) and Young's inequality we obtain

$$
\begin{aligned}
\mathbb{E}\Big[\sup_{t\in(0,T)}|J(t)|\Big] &= \mathbb{E}\Big[\sup_{t\in(0,T)}\Big|\int_0^t\int_G u_N\,\Phi\,\mathrm{d}x\,\mathrm{d}\mathbf{W}_\sigma^N\Big|\Big]\\
&= \mathbb{E}\Big[\sup_{t\in(0,T)}\Big|\int_0^t\sum_{k=1}^N\int_G u_N\,\Phi\mathbf{e}_k\,\mathrm{d}x\,\mathrm{d}\beta_k(\sigma)\Big|\Big]\\
&\leq c\,\mathbb{E}\Big[\int_0^T\sum_{k=1}^N\Big(\int_G u_N\Phi\mathbf{e}_k\,\mathrm{d}x\Big)^2\,\mathrm{d}t\Big]^{\frac{1}{2}}\\
&\leq c\,\mathbb{E}\Big[\Big(\int_0^T\Big(\sum_{k=1}^N\int_G|u_N|^2\,\mathrm{d}x\int_G|\Phi\mathbf{e}_k|^2\,\mathrm{d}x\Big)\,\mathrm{d}t\Big]^{\frac{1}{2}}\\
&\leq c\,\mathbb{E}\Big[\sup_{t\in(0,T)}\int_G|u_N|^2\,\mathrm{d}x\Big(\int_0^T\sum_{k=1}^\infty\int_G|\Phi\mathbf{e}_k|^2\,\mathrm{d}x\,\mathrm{d}t\Big)\Big]^{\frac{1}{2}}\\
&\leq \delta\,\mathbb{E}\Big[\sup_{t\in(0,T)}\int_G|u_N|^2\,\mathrm{d}x\Big]+c(\delta)\,\mathbb{E}\Big[\int_0^T\|\Phi\|^2_{L_2(U,L^2(G))}\,\mathrm{d}t\Big],
\end{aligned}
$$

for any arbitrary $\delta>0$. If δ is sufficiently small this finally proves

$$
\begin{aligned}
&\mathbb{E}\Big[\sup_{t\in(0,T)}\int_G|u_N(t)|^2\,\mathrm{d}x+\int_0^T\int_G|\nabla u_N|^2\,\mathrm{d}x\,\mathrm{d}\sigma\Big]\\
&\quad\leq c\,\mathbb{E}\Big[\int_G|u_0|^2\,\mathrm{d}x+\int_0^T\|\Phi\|^2_{L_2(U,L^2(G))}\,\mathrm{d}t\Big].
\end{aligned}
\tag{9.2.11}
$$

By our assumptions the right-hand-side is finite.

Step 3: passage to the limit

Due to (9.2.11) and passing to a subsequence we obtain a limit function u:

$$u_N \rightharpoonup u \quad \text{in} \quad L^2(\Omega,\mathcal{F},\mathbb{P};L^2(0,T;W_0^{1,2}(G))). \tag{9.2.12}$$

We also have that $u\in L^\infty(0,T;L^2(G))$ a.s. and

$$\mathbb{E}\Big[\sup_{t\in(0,T)}\int_G|u(t)|^2\,\mathrm{d}x\Big]<\infty. \tag{9.2.13}$$

We compute for $\psi\in L^2(\Omega\times(0,T))$ and $\varphi\in W_0^{1,2}(G)$

$$
\begin{aligned}
&\mathbb{E}\Big[\int_0^T\int_G u(t)\,\psi(t)\varphi\,\mathrm{d}x\,\mathrm{d}t\Big]=\lim_{N\to\infty}\mathbb{E}\Big[\int_0^T\int_G u_N(t)\,\psi(t)\varphi\,\mathrm{d}x\,\mathrm{d}t\Big]\\
&=\lim_{N\to\infty}\mathbb{E}\Big[\int_0^T\int_G u_N(t)\,\psi(t)\mathcal{P}_N\varphi\,\mathrm{d}x\,\mathrm{d}t\Big]
\end{aligned}
$$

$$= \lim_{N\to\infty} \mathbb{E}\bigg[\int_0^T \bigg(\int_G \mathcal{P}_N u_0\, \psi(t)\varphi \,\mathrm{d}x\,\mathrm{d}t + \int_0^t\int_G \nabla u_N\, \psi(t)\cdot\nabla\varphi\,\mathrm{d}x\,\mathrm{d}\sigma$$
$$+ \int_G\int_0^t \psi\mathcal{P}_N\varphi\,\Phi\,\mathrm{d}\mathbf{W}_\sigma^N\,\mathrm{d}x\bigg)\bigg].$$

We have to pass to the limit in all terms. The first integral converges as $\mathcal{P}_N w \to w$ in $L^2(G)$ for any $w \in L^2(G)$. The second term converges because of (9.2.12). Finally, we use

$$\mathbf{W}^N \to \mathbf{W} \quad \text{in} \quad L^2(\Omega, \mathcal{F}, \mathbb{P}; C([0,T]; U_0)),$$

for the stochastic integral (recall Lemma 9.1.4). All together we obtain

$$\mathbb{E}\bigg[\int_0^T\int_G u(t)\,\psi(t)\varphi\,\mathrm{d}x\,\mathrm{d}t\bigg]$$
$$= \mathbb{E}\bigg[\int_0^T\int_G\bigg(u_0\,\psi(t)\varphi + \int_0^t \nabla\,\psi(t)\cdot\nabla\varphi\,\mathrm{d}\sigma + \int_0^t \psi\varphi\,\Phi\,\mathrm{d}\mathbf{W}_\sigma\bigg)\,\mathrm{d}x\bigg],$$

which implies that $\mathbb{P}$-a.s.

$$\int_G u(t)\,\varphi\,\mathrm{d}x = \int_G u_0\,\varphi\,\mathrm{d}x - \int_0^t\int_G \nabla u\cdot\nabla\varphi\,\mathrm{d}x\,\mathrm{d}\sigma + \int_G \varphi\bigg(\int_0^t \Phi\,\mathrm{d}\mathbf{W}_\sigma\bigg)\,\mathrm{d}x$$

as ψ was arbitrary.

Step 4: continuity of u

Interpreted as an element of $W^{-1,2}(G)$ we can write $\mathbb{P}$-a.s.

$$u(t) = \int_0^t \Delta u\,\mathrm{d}\sigma + \int_0^t \Phi\,\mathrm{d}\mathbf{W}_\sigma$$

for a.e. t. For any $\alpha < \frac{1}{2}$ the deterministic integral belongs $\mathbb{P}$-a.s. to the class

$$W^{1,2}(0,T; W^{-1,2}(G)) \subset C^\alpha([0,T]; W^{-1,2}(G)).$$

For the stochastic integral we have

$$\int_0^{\cdot} \Phi\,\mathrm{d}\mathbf{W}_\sigma \in C([0,T]; (L^2(G))') \subset C([0,T]; W^{-1,2}(G)) \quad \mathbb{P}\text{-a.s.}$$

This follows from the construction and our assumption (9.2.5). Combining both facts shows that $u \in C([0,T]; W^{-1,2}(G))$. This and (9.2.11) yields

$$u \in C_w([0,T]; L^2(G)) \quad \mathbb{P}\text{-a.s.} \tag{9.2.14}$$

We want to strengthen (9.2.14) and obtain continuity with respect to the norm topology. We apply Itô's formula in infinite dimensions, Theorem 9.1.39, with $\mathcal{H} = L^2(G)$, $\mathcal{V} = W_0^{1,2}(G)$, $X = u$, $Y = {}_{\mathcal{V}'}\langle \nabla u, \nabla\cdot\rangle_{\mathcal{V}}$

and $M = \int_0^\cdot \Phi \, \mathrm{d}\mathbf{W}_\sigma$. We have

$$\int_G |u(t)|^2 \,\mathrm{d}x = \int_G |u(0)|^2 \,\mathrm{d}x - \int_0^t \int_G |\nabla u|^2 \,\mathrm{d}x \,\mathrm{d}\sigma + 2\int_G \int_0^t u\Phi \,\mathrm{d}\mathbf{W}_\sigma \,\mathrm{d}x + \int_0^t \|\Phi\|^2_{L_2(U,L^2(G))} \,\mathrm{d}t.$$

As the right-hand-side is continuous so is the left-hand-side, i.e.

$$[0,T] \ni t \mapsto \int_G |u(t)|^2 \,\mathrm{d}x$$

is $\mathbb{P}$-a.s. continuous. This and (9.2.14) implies $u \in C([0,T]; L^2(G))$ a.s. □

9.3 TOOLS FOR COMPACTNESS

In this section we present some (mainly basic) tools from probability theory which are quite crucial to obtain compactness for SPDE. Let $(\mathscr{V}, \tau)$ be a topological space. The smallest σ-field $\mathfrak{B}(\mathscr{V})$ on $(\mathscr{V}, \tau)$ which contains all open sets is called topological σ-field. A random variable with values in the topological space $(\mathscr{V}, \tau)$ is a measurable map $X : (\Omega, \mathcal{F}) \to (\mathscr{V}, \mathfrak{B}(\mathscr{V}))$. The probability law μ of X on $(\mathscr{V}, \tau)$ will be given by $\mu = \mathbb{P} \circ X^{-1}$. An important concept for applications is the pre-compactness of families of random variables. We will need the following definition.

Definition 9.3.1 (Tightness). A family $(\mu_\alpha)_{\alpha \in \mathcal{I}}$ of probability laws on a topological space $(\mathscr{V}, \mathfrak{B}(\mathscr{V}))$ is called tight if for every $\varepsilon > 0$ there is a compact subset $K \subset \mathscr{V}$ such that $\mu_\alpha(K) \geq 1 - \varepsilon$ for every $\alpha \in \mathcal{I}$.

Lemma 9.3.1 (Prokhorov; [96], Thm. 2.6). *Let $(\mu_\alpha)_{\alpha \in \mathcal{I}}$ be a family of probability laws on a metric space $(\mathscr{V}, \rho)$. If $(\mu_\alpha)_{\alpha \in \mathcal{I}}$ is tight then it is also relatively compact.*

Lemma 9.3.2 (Skorokhod; [96], Thm. 2.7). *Let $(\mu_n)_{n \in \mathbb{N}}$ be a sequence of probability laws on a complete separable metric space $(\mathscr{V}, \rho)$ such that $\mu_n \to \mu$ weakly in the sense of measures as $n \to \infty$. Then there is a probability space $(\underline{\Omega}, \underline{\mathcal{F}}, \underline{\mathbb{P}})$ and random variables $(\underline{X}_n)_{n \in \mathbb{N}}, \underline{X} : (\underline{\Omega}, \underline{\mathcal{F}}, \underline{\mathbb{P}}) \to (\mathscr{V}, \mathfrak{B}(\mathscr{V}))$ such that:*

- *The laws of $\underline{X}_n$ and $\underline{X}$ under $\underline{\mathbb{P}}$ coincide with μ_n and μ respectively, $n \in \mathbb{N}$.*
- *We have $\underline{\mathbb{P}}$ a.s. that $\underline{X}_n \to^\rho \underline{X}$ for $n \to \infty$.*

The proof of Lemma 9.3.2 in the general case is not very long but quite technical and it is hard to grasp the main ideas. We will therefore briefly outline the case of real-valued random variables, i.e. $\mathscr{V} = \mathbb{R}$ and

$\rho(x, y) = |x - y|$. Let μ_n be a probability law on $\mathbb{R}$ such that $\mu_n \to \mu$ weakly in the sense of measures as $n \to \infty$. We denote by F_n and F the distribution functions of μ_n and μ respectively. Let us assume for simplicity that they are injective (otherwise one can argue via their generalized inverse functions). In this case we have $F_n \to F$ pointwise. Now we set $(\underline{\Omega}, \underline{\mathcal{F}}, \underline{\mathbb{P}}) = ((0, 1), \mathfrak{B}((0, 1)), \mathcal{L}^1|_{(0,1)})$. Let us assume for simplicity that the distribution functions F_n ($n \in \mathbb{N}$) and F are continuous. We define random variables (for $\omega \in (0, 1)$)

$$\underline{X}_n(\omega) = F_n^{-1}(\omega), \quad \underline{X}(\omega) = F^{-1}(\omega).$$

Now one can easily see that for $n \in \mathbb{N}$

$$\mu_{\underline{X}_n} = \underline{\mathbb{P}} \circ \underline{X}_n^{-1} = \mathcal{L}^1 \circ F_n = \mu_n$$

and similarly $\mu_{\underline{X}} = \mu$. Moreover, we have

$$\underline{X}_n(\omega) = F_n^{-1}(\omega) \to F^{-1}(\omega) = \underline{X}(\omega)$$

for every $\omega \in (0, 1)$. In the general case this convergence only holds true in points where $\underline{X}$ is continuous (which is in $\mathcal{L}^1$-a.e. ω).

Lemma 9.3.2 only applies to metric spaces. Unfortunately, this does not cover Banach spaces with the weak topology. Therefore we need the following generalization.

Definition 9.3.2 (Quasi-Polish space). Let $(\mathscr{V}, \tau)$ be a topological space such that there exists a countable family

$$\{f_n : \mathscr{V} \to [-1, 1];\ n \in \mathbb{N}\}$$

of continuous functions that separates points of $\mathscr{V}$. Then $(\mathscr{V}, \tau, (f_n)_{n\in\mathbb{N}})$ is called a quasi-Polish space.

Lemma 9.3.3 (Jakubowski–Skorokhod, [99]). *Let $(\mu_n)_{n\in\mathbb{N}}$ be a family of probability laws on a quasi-Polish space $(\mathscr{V}, \tau, (f_n)_{n\in\mathbb{N}})$ and let $\mathscr{S}$ be the σ-algebra generated by the maps $(f_n)_{n\in\mathbb{N}}$. Let $(\mu_n)_{n\in\mathbb{N}}$ be a tight sequence of probability laws on $(\mathscr{V}, \mathscr{S})$. Then there is a subsequence $(\mu_{n_k})_{k\in\mathbb{N}}$ such that the following holds. There is a probability space $(\underline{\Omega}, \underline{\mathcal{F}}, \underline{\mathbb{P}})$ and random variables $(\underline{X}_k)_{k\in\mathbb{N}}, \underline{X} : (\underline{\Omega}, \underline{\mathcal{F}}, \underline{\mathbb{P}}) \to (\mathscr{V}, \mathscr{S})$ such that:*

- *The laws of $\underline{X}_k$ under $\underline{\mathbb{P}}$ coincide with μ_{n_k}, $k \in \mathbb{N}$.*
- *We have $\underline{\mathbb{P}}$-a.s. that $\underline{X}_k \to^{\tau} \underline{X}$ for $k \to \infty$.*
- *The law of $\underline{X}$ under $\underline{\mathbb{P}}$ is a Radon measure.*

CHAPTER 10

Stochastic power law fluids

Contents

Abstract

We consider the equations of motion for an incompressible non-Newtonian fluid in a bounded Lipschitz domain $G \subset \mathbb{R}^d$ during the time interval $(0, T)$ with a stochastic perturbation driven by a Brownian motion $\mathbf{W}$. The balance of momentum reads as

$$\mathrm{d}\mathbf{v} = -(\nabla\mathbf{v})\mathbf{v}\,\mathrm{d}t + \operatorname{div}\mathbf{S}\,\mathrm{d}t - \nabla\pi\,\mathrm{d}t + \mathbf{f}\mathrm{d}t + \Phi(\mathbf{v})\,\mathrm{d}\mathbf{W}_t,$$

where $\mathbf{v}$ is the velocity, π the pressure and $\mathbf{f}$ an external volume force. We assume the common power law model $\mathbf{S}(\boldsymbol{\varepsilon}(\mathbf{v})) = \big(1 + |\boldsymbol{\varepsilon}(\mathbf{v})|\big)^{p-2}\boldsymbol{\varepsilon}(\mathbf{v})$ and show the existence of weak martingale solutions provided $p > \frac{2d+2}{d+2}$. Our approach is based on the L^∞-truncation and a harmonic pressure decomposition which are adapted to the stochastic setting.

From several points of view it is reasonable to add a stochastic part to the equations of motion of a fluid.

- It can be understood as turbulence in the fluid motion.
- It can be interpreted as a perturbation from the physical model.
- Apart from the force $\mathbf{f}$ we are observing there may be further quantities with a (usually small) influence on the motion.

As in Chapter 7 we assume the power law model, i.e., the stress-strain relation

$$\mathbf{S}(\boldsymbol{\varepsilon}(\mathbf{v})) = \nu_0\big(1 + |\boldsymbol{\varepsilon}(\mathbf{v})|\big)^{p-2}\boldsymbol{\varepsilon}(\mathbf{v}), \tag{10.0.1}$$

where $\nu_0 > 0$ and $p \in (1, \infty)$. We want to establish an existence theory for the following set of equations (we neglect physical constants for simplicity):

$$\left\{\begin{array}{lll} \mathrm{d}\mathbf{v} = [-(\nabla\mathbf{v})\mathbf{v} + \operatorname{div}\mathbf{S} - \nabla\pi + \mathbf{f}]\,\mathrm{d}t + \Phi(\mathbf{u})\mathrm{d}\mathbf{W}_t & \text{in} & Q, \\ \operatorname{div}\mathbf{v} = 0 & \text{in} & Q, \\ \mathbf{v} = 0 & \text{on} & \partial G, \\ \mathbf{v}(0) = \mathbf{v}_0 & \text{in} & G. \end{array}\right. \tag{10.0.2}$$

Existence Theory for Generalized Newtonian Fluids.
DOI: http://dx.doi.org/10.1016/B978-0-12-811044-7.00013-6

We assume that $\mathbf{W}$ is a Brownian motion with values in a Hilbert (see (10.0.3) below for details). We suppose that Φ grows linearly – roughly speaking $|\Phi(\mathbf{v})| \leq c(1+|\mathbf{v}|)$ and $|D_\mathbf{v}\Phi(\mathbf{v})| \leq c$ (for a precise formulation see (10.0.4)). The idea behind this is an interaction between the solution and the random perturbation caused by the Brownian motion. For large values of $|\mathbf{v}|$ we expect a larger perturbation than for small values.

There is a huge body of literature concerning existence of weak solutions to the stochastic Navier–Stokes equations (where $\operatorname{div}\mathbf{S} = \Delta\mathbf{v}$), starting with the paper [20] by Bensoussan and Temam. A further important contribution is the existence of martingale solutions proved by Flandoli and Gątarek [73]. For a recent overview we refer to [72]. However, there seems to be very limited knowledge about the non-Newtonian fluid problem. In [45] a bipolar shear thinning fluid is observed. The authors of [45] assume the constitutive relation

$$\mathbf{S} = \nu_0\big(1 + |\boldsymbol{\varepsilon}(\mathbf{v})|\big)^{p-2}\boldsymbol{\varepsilon}(\mathbf{v}) - \nu_1\Delta\boldsymbol{\varepsilon}(\mathbf{v}),$$

where $\nu_0, \nu_1 > 0$ and $1 < p \leq 2$. Compared with our model this results in an additional bi-Laplacian $\Delta^2\mathbf{v}$ in the equations of motion. This gives enough initial regularity to argue directly with monotone operators without using any form of truncation. Moreover, the main part of the equation is linear. Hence there is no problem with passing to the limit in the approximated equation.

A further observation of stochastic power law fluids was carried out in [142] and [139]. Following the approach in [111] for the deterministic problem they consider periodic boundary conditions and obtain existence of a martingale solution for $p \geq \frac{9}{5}$ (in three dimensions). The restriction to a periodic boundary allows them to test the equation by the Laplacian of the solution (without using cut-off functions), which is not possible in general. As a consequence, the (gradient of the) constructed solution enjoys some fractional differentiability in space. A further drawback in [142] and [139] is that the noise is additive. In particular, there is no interaction between the solution and the Wiener process. This is modelled via the function Φ in (10.0.2) and quite reasonable from a physical point of view.

We will investigate an existence theory which removes all these drawbacks. Before we present the final result of this chapter we precise the assumptions on $\mathbf{W}$ and Φ. Let U be a Hilbert space with orthonormal basis $(\mathbf{e}_k)_{k\in\mathbb{N}}$ and let $L_2(U, L^2(G))$ be the set of Hilbert–Schmidt operators from U to $L^2(\Omega)$. The most natural choice is $U = L^2(G)$. We consider a cylindrical Wiener process $\mathbf{W} = (\mathbf{W}_t)_{t\in[0,T]}$ which has the form

$$\mathbf{W}_\sigma = \sum_{k\in\mathbb{N}} \mathbf{e}_k \beta_k(\sigma) \tag{10.0.3}$$

with a sequence (β_k) of independent real valued Brownian motions on a filtered probability space $(\Omega, \mathcal{F}, (\mathcal{F}_t)_{t\geq 0}, \mathbb{P})$. The filtration $(\mathcal{F}_t)_{t\geq 0}$ satisfies the *usual conditions*, see Definition 8.1.6. We suppose the following linear growth assumptions on Φ (following [94]): For each $\mathbf{z} \in L^2(G)$ there is a mapping $\Phi(\mathbf{z}) : U \to L^2(G)$ defined by $\Phi(\mathbf{z})\mathbf{e}_k = g_k(\mathbf{z}(\cdot))$. In particular, we suppose that $g_k \in C^1(\mathbb{R}^d)$ and the following conditions for some $L \geq 0$

$$\sum_{k\in\mathbb{N}} |g_k(\boldsymbol{\xi})| \leq L(1 + |\boldsymbol{\xi}|), \quad \sum_{k\in\mathbb{N}} |\nabla g_k(\boldsymbol{\xi})|^2 \leq L, \quad \boldsymbol{\xi} \in \mathbb{R}^d. \tag{10.0.4}$$

Note that the first assumption in (10.0.4) is slightly stronger than

$$\sum_{k\in\mathbb{N}} |g_k(\boldsymbol{\xi})|^2 \leq L(1 + |\boldsymbol{\xi}|^2), \quad \boldsymbol{\xi} \in \mathbb{R}^d,$$

supposed in [94] and additionally implies

$$\sup_{k\in\mathbb{N}} k^2 |g_k(\boldsymbol{\xi})|^2 \leq c(1 + |\boldsymbol{\xi}|^2). \tag{10.0.5}$$

Now we are ready to give a precise formulation of the concept of solution.

Definition 10.0.1 (Solution). Let $\Lambda_0, \Lambda_{\mathbf{f}}$ be Borel probability measures on $L^2_{\mathrm{div}}(G)$ and $L^2(Q)$ respectively. Then

$$\big((\Omega, \mathcal{F}, (\mathcal{F}_t)_{t\geq 0}, \mathbb{P}), \mathbf{v}, \mathbf{v}_0, \mathbf{f}, \mathbf{W}\big)$$

is called a weak martingale solution to (10.0.2) with $\mathbf{S}$ given by (10.0.1) with the initial datum Λ_0 and right-hand-side $\Lambda_{\mathbf{f}}$ provided the following holds.

(a) $(\Omega, \mathcal{F}, (\mathcal{F}_t)_{t\geq 0}, \mathbb{P})$ is a stochastic basis with a complete right-continuous filtration,

(b) $\mathbf{W}$ is a $(\mathcal{F}_t)_{t\geq 0}$-cylindrical Wiener process,

(c) $\mathbf{v} \in L^p(\Omega, \mathcal{F}, \mathbb{P}; L^p(0, T; W^{1,p}_{0,\mathrm{div}}(G)))$ is progressively $(\mathcal{F}_t)_{t\geq 0}$-measurable with $\mathbf{v} \in L^\infty(0, T; L^2(G))$ a.s. and

$$\mathbb{E}\bigg[\sup_{t\in(0,T)} \int_G |\mathbf{v}|^2 \, \mathrm{d}x\bigg] < \infty,$$

(d) $\mathbf{v}_0 \in L^2(\Omega, \mathcal{F}_0, \mathbb{P}; L^2(G))$ with $\Lambda_0 = \mathbb{P} \circ \mathbf{v}_0^{-1}$,

(e) $\mathbf{f} \in L^2(\Omega, \mathcal{F}, \mathbb{P}; L^2(Q))$ is adapted to $(\mathcal{F}_t)_{t\geq 0}$ and $\Lambda_{\mathbf{f}} = \mathbb{P} \circ \mathbf{f}^{-1}$,

(f) for all $\varphi \in C^{\infty}_{0,\mathrm{div}}(G)$ and all $t \in [0, T]$ we have

$$\int_G \mathbf{v}(t)\cdot\varphi\,\mathrm{d}x + \int_0^t\int_G \mathbf{S}(\boldsymbol{\varepsilon}(\mathbf{v})) : \boldsymbol{\varepsilon}(\varphi)\,\mathrm{d}x\,\mathrm{d}\sigma - \int_0^t\int_G \mathbf{v}\otimes\mathbf{v} : \boldsymbol{\varepsilon}(\varphi)\,\mathrm{d}x\,\mathrm{d}\sigma$$
$$= \int_G \mathbf{v}_0\cdot\varphi\,\mathrm{d}x + \int_G\int_0^t \mathbf{f}\cdot\varphi\,\mathrm{d}x\,\mathrm{d}\sigma + \int_G\int_0^t \Phi(\mathbf{v})\,\mathrm{d}\mathbf{W}_\sigma\cdot\varphi\,\mathrm{d}x$$

$\mathbb{P}$-a.s.

Remark 10.0.21. • As we are looking for martingale solutions (weak solutions in the probabilistic sense) we can only assume the laws of $\mathbf{v}_0$ and $\mathbf{f}$.

Theorem 10.0.41 (Existence). *Assume* (10.0.1) *with* $p > \frac{2d+2}{d+2}$ *as well as* (10.0.4) *and* (10.0.5). *Suppose further that*

$$\int_{L^2_{\mathrm{div}}(G)} \|\mathbf{u}\|^{\beta}_{L^2(G)}\,\mathrm{d}\Lambda_0(\mathbf{u}) < \infty, \qquad \int_{L^2(Q)} \|\mathbf{g}\|^{\beta}_{L^2(Q)}\,\mathrm{d}\Lambda_{\mathbf{f}}(\mathbf{g}) < \infty, \tag{10.0.6}$$

with $\beta := \max\left\{\frac{2(d+2)}{d}, \frac{p(d+2)}{d}\right\}$. *Then there is a weak martingale solution to* (10.0.2) *in the sense of Definition 10.0.1.*

Remark 10.0.22. • By Theorem 10.0.41 we extend the results from [140] to the stochastic setting. In contrast to earlier results from [139] we consider arbitrary bounded Lipschitz domains and allow a nonlinear dependence between the solution and the stochastic perturbation. The bound $p > \frac{8}{5}$ (if $d = 3$) includes a wide range of non-Newtonian fluids.
- It is not clear if it is possible to improve the result from Theorem 10.0.41 to $p > \frac{2d}{d+2}$ as in the deterministic case, see Chapter 7. The method in the deterministic case is based on the Lipschitz truncation method. Different from the L^∞-truncation the Lipschitz truncation is not only nonlinear but also nonlocal (in space-time in the parabolic case). So it seems to be impossible to perform the testing procedure by Itô's formula.
- Condition (10.0.6) ensures the existence of higher moments for initial datum and right-hand-side. This transfers to the solution, see Corollary 10.2.1.

Our procedure is as follows: in Section 10.1 we investigate the pressure. As usual the pressure disappears in the weak formulation (see Definition 10.0.1) but can be recovered. Following the ideas from [140] we relate to each term in the equation a pressure part. So also a stochastic part of the pressure is included. In Section 10.2 we study auxiliary problems which are stabilized by adding a large power of $\mathbf{v}$. This approach is based on the Galerkin method.

In Section 10.3 we prove Theorem 10.0.41. Here we follow the approach in [140] adapted to the stochastic fashion. The problems are as usual the convergences in the nonlinear parts of the approximated system. We have to combine the techniques from nonlinear PDEs with stochastic calculus for martingales. Note that it is not possible to work directly with test functions. Instead of this we apply Itô's formula to certain functions of $\mathbf{v}$. Finally we use monotone operator theory combined with L^∞-truncation to justify the limit procedure in the nonlinear tensor $\mathbf{S}$.

10.1 PRESSURE DECOMPOSITION

In this section we introduce the pressure. We decompose it in accordance to the terms in the equation. The following theorem generalizes [140, Thm. 2.6] to the stochastic case.

Theorem 10.1.42. *Let $(\Omega, \mathcal{F}, (\mathcal{F}_t)_{t\geq 0}, \mathbb{P})$ be a stochastic basis, $\mathbf{u} \in L^2(\Omega, \mathcal{F}, \mathbb{P}; L^2(Q))$ and $\mathbf{H} \in L^s(\Omega, \mathcal{F}, \mathbb{P}; L^s(Q))$ (with $s > 1$) $(\mathcal{F}_t)_{t\geq 0}$-adapted processes. Moreover, let $\mathbf{u}_0 \in L^2(\Omega, \mathcal{F}_0, \mathbb{P}; L^2_{\mathrm{div}}(G))$ and $\Phi \in L^2(\Omega, \mathcal{F}, \mathbb{P}; L^2(0, T; L_2(U, L^2(G))))$ progressively $(\mathcal{F}_t)_{t\geq 0}$-measurable such that*

$$\int_G \mathbf{u}(t)\cdot\boldsymbol{\varphi}\,\mathrm{d}x + \int_0^t\int_G \mathbf{H} : \nabla\boldsymbol{\varphi}\,\mathrm{d}x\,\mathrm{d}\sigma = \int_G \mathbf{u}_0\cdot\boldsymbol{\varphi}\,\mathrm{d}x + \int_G\int_0^t \Phi\,\mathrm{d}\mathbf{W}_\sigma\cdot\boldsymbol{\varphi}\,\mathrm{d}x$$

holds for all $\boldsymbol{\varphi} \in C^\infty_{0,\mathrm{div}}(G)$. Then there are $(\mathcal{F}_t)_{t\geq 0}$-adapted processes $\pi^{\mathbf{H}}$, π^h and a progressively $(\mathcal{F}_t)_{t\geq 0}$-measurable process Φ^π with the following properties.

a) *We have a.s. $\pi_h(0) = 0$, $\Delta\pi_h = 0$ $\mathbb{P}$-a.s. Moreover, the following holds for $\chi := \min\{2, s\}$*

$$\begin{aligned}&\mathbb{E}\bigg[\sup_{t\in(0,T)}\int_G |\pi^h|^\chi\,\mathrm{d}x\bigg]\\&\quad\leq c\,\mathbb{E}\bigg[1 + \sup_{t\in(0,T)}\int_G |\mathbf{u}|^2\,\mathrm{d}x + \int_0^T \|\Phi\|^2_{L_2(U,L^2(G))}\,\mathrm{d}t\bigg]\\&\quad\quad + c\,\mathbb{E}\bigg[\int_G |\mathbf{u}_0|^2\,\mathrm{d}x + \int_Q |\mathbf{H}|^s\,\mathrm{d}x\,\mathrm{d}t\bigg].\end{aligned}$$

b) *We have $\pi^{\mathbf{H}} \in L^s(\Omega, \mathcal{F}, \mathbb{P}; L^s(Q))$ and the following holds*

$$\mathbb{E}\bigg[\int_Q |\pi^{\mathbf{H}}|^s\,\mathrm{d}x\,\mathrm{d}t\bigg] \leq c\,\mathbb{E}\bigg[\int_Q |\mathbf{H}|^s\,\mathrm{d}x\,\mathrm{d}t\bigg].$$

c) *We have* $\Phi^\pi \in L^2(\Omega, \mathcal{F}, \mathbb{P}; L^2(0,T; L_2(U, L^2_{loc}(G))))$ *and the following holds for every* $G' \Subset G$

$$\mathbb{E}\bigg[\int_0^T \|\Phi_\pi\|^2_{L_2(U,L^2(G'))}\,\mathrm{d}t\bigg] \leq c(G')\,\mathbb{E}\bigg[\int_0^T \|\Phi\|^2_{L_2(U,L^2(G))}\,\mathrm{d}t\bigg].$$

If $\Phi \in L^\infty(0,T; L_2(U, L^2_{loc}(G)))$ *a.s. the same is true for* Φ^π *and we have*

$$\mathbb{E}\bigg[\sup_{t\in(0,T)} \|\Phi^\pi\|^2_{L_2(U,L^2(G'))}\bigg] \leq c(G')\,\mathbb{E}\bigg[\sup_{t\in(0,T)} \|\Phi\|^2_{L_2(U,L^2(G))}\bigg].$$

d) *We have for all* $t \in [0,T]$ *and all* $\boldsymbol{\varphi} \in C_0^\infty(G)$

$$\int_G \big(\mathbf{u}(t) - \nabla\pi^h(t)\big)\cdot\boldsymbol{\varphi}\,\mathrm{d}x + \int_0^t\int_G \mathbf{H} : \nabla\boldsymbol{\varphi}\,\mathrm{d}x\,\mathrm{d}\sigma - \int_0^t\int_G \pi^{\mathbf{H}}\,\mathrm{div}\,\boldsymbol{\varphi}\,\mathrm{d}x\,\mathrm{d}\sigma$$
$$= \int_G \mathbf{u}_0\cdot\boldsymbol{\varphi}\,\mathrm{d}x + \int_G\int_0^t \Phi\,\mathrm{d}\mathbf{W}_\sigma\cdot\boldsymbol{\varphi}\,\mathrm{d}x + \int_G\int_0^t \Phi^\pi\,\mathrm{d}\mathbf{W}_\sigma\cdot\boldsymbol{\varphi}\,\mathrm{d}x$$

$\mathbb{P}$*-a.s.*

Remark 10.1.23. If we put the pressure terms together by

$$\pi(t) = \pi^h(t) + \int_0^t \pi^{\mathbf{H}}\,\mathrm{d}\sigma + \int_0^t \Phi^\pi\,\mathrm{d}\mathbf{W}_\sigma$$

then we have $\pi \in L^\infty(0,T; L^\chi(G))$ a.s. as well as

$$\mathbb{E}\bigg[\sup_{t\in(0,T)}\int_G |\pi|^\chi\,\mathrm{d}x\bigg] < \infty.$$

Proof. Let $\mathbf{u}$ be a weak solution to

$$\int_G \mathbf{u}(t)\cdot\boldsymbol{\varphi}\,\mathrm{d}x + \int_0^t\int_G \mathbf{H} : \nabla\boldsymbol{\varphi}\,\mathrm{d}x\,\mathrm{d}\sigma = \int_G \mathbf{u}_0\cdot\boldsymbol{\varphi}\,\mathrm{d}x + \int_0^t\int_G \Phi\,\mathrm{d}\mathbf{W}_\sigma\cdot\boldsymbol{\varphi}\,\mathrm{d}x$$

for all $\boldsymbol{\varphi} \in W^{1,\chi'}_{0,\mathrm{div}}(G)$. Then there is a unique function $\pi(t) \in L^\chi_\perp(G)$ with $\pi(0) = 0$ such that

$$\int_G \mathbf{u}(t)\cdot\boldsymbol{\varphi}\,\mathrm{d}x + \int_0^t\int_G \mathbf{H} : \nabla\boldsymbol{\varphi}\,\mathrm{d}x\,\mathrm{d}\sigma$$
$$= \int_G \pi(t)\,\mathrm{div}\,\boldsymbol{\varphi}\,\mathrm{d}x + \int_G \mathbf{u}_0\cdot\boldsymbol{\varphi}\,\mathrm{d}x + \int_G\int_0^t \Phi\,\mathrm{d}\mathbf{W}_\sigma\cdot\boldsymbol{\varphi}\,\mathrm{d}x$$

for all $\boldsymbol{\varphi} \in W^{1,\chi'}_0(G)$. This is a consequence of the well-known Theorem by De Rahm, see also Theorem 2.2.10 with $A(t) = B(t) = t^\chi$. We will show that

$$\mathbb{E}\bigg[\sup_{t\in(0,T)}\int_G |\pi|^\chi\,\mathrm{d}x\bigg] < \infty. \tag{10.1.7}$$

The measurability of π follows from the equation. For the boundedness we write the equation as

$$\int_G \pi(t)\,\varphi\,\mathrm{d}x = \int_G \big(\mathbf{u}(t)-\mathbf{u}_0\big)\cdot \mathcal{B}(\varphi)\,\mathrm{d}x - \int_0^t\int_G \mathbf{H} : \nabla\mathscr{B}(\varphi)\,\mathrm{d}x\,\mathrm{d}\sigma$$
$$+ \int_G\int_0^t \Phi\,\mathrm{d}\mathbf{W}_\sigma \cdot \mathscr{B}(\varphi)\,\mathrm{d}x, \quad \mathscr{B}(\varphi) := \mathrm{Bog}_G\big(\varphi - (\varphi)_G\big),$$

for all $\varphi \in C_0^\infty(G)$ with the Bogovskiĭ operator Bog_G (see Section 2.1). Here $(\varphi)_G$ denotes the mean value of the function φ over G. The above implies

$$\pi(t) = \mathscr{B}^*\big(\mathbf{u}(t)-\mathbf{u}_0\big) - \int_0^t \big(\nabla\mathscr{B}\big)^*\mathbf{H}\,\mathrm{d}\sigma + \int_0^t \mathscr{B}^*\Phi\,\mathrm{d}\mathbf{W}_\sigma,$$

where $\mathscr{B}^*$ denotes the adjoint of $\mathscr{B}$ with respect to the $L^2(G)$ inner product. Using continuity of $\mathscr{B}^*$ from $L^2(G)$ to $L^2(G)$ and $\big(\nabla\mathscr{B}\big)^*$ from $L^s(G)$ to $L^s(G)$ (which follows from the properties of Bog_G) we have

$$\begin{aligned}
&\mathbb{E}\bigg[\sup_{(0,T)}\int_G |\pi|^\chi\,\mathrm{d}x\bigg] \\
&\quad\le c\,\mathbb{E}\bigg[\sup_{(0,T)}\int_G |\mathbf{u}|^2\,\mathrm{d}x + \int_G |\mathbf{u}_0|^2\,\mathrm{d}x + \int_0^T \|\Phi\|^2_{L_2(U,L^2(G))}\,\mathrm{d}t\bigg] \\
&\quad + c\,\mathbb{E}\bigg[1 + \int_Q |\mathbf{H}|^s\,\mathrm{d}x\,\mathrm{d}t\bigg],
\end{aligned} \tag{10.1.8}$$

and so (10.1.7) holds. Note that the estimate of the stochastic integral is a consequence of the infinite-dimensional Burkholder–Davis–Gundi inequality (see Lemma 9.1.1) and the continuity of $\mathscr{B}^*$ on $L^2(G)$.

We decompose pointwise on $\Omega \times (0,T)$

$$\pi = \pi^0 + \pi^h,$$
$$\pi^0 := \Delta\Delta_G^{-2}\Delta\pi, \quad \pi^h := \pi - \pi^0.$$

Here Δ_G^{-2} denotes the solution operator to the bi-Laplace equation with respect to zero boundary values for function and gradient. Since the operator $\Delta\Delta_G^{-2}\Delta$ is continuous from $L^\chi(G)$ to $L^\chi(G)$ (see Corollary 6.1.1) inequality (10.1.8) continues to hold if π is replaced by π^0 or π^h. We obtain for all $\varphi \in C_0^\infty(G)$

$$\begin{aligned}
\int_G \pi^0(t)\,\Delta\varphi\,\mathrm{d}x = &-\int_G\int_0^t \mathbf{H} : \nabla^2\varphi\,\mathrm{d}x\,\mathrm{d}\sigma \\
&+ \int_G\int_0^t \Phi\,\mathrm{d}\mathbf{W}_\sigma \cdot \nabla\varphi\,\mathrm{d}x.
\end{aligned} \tag{10.1.9}$$

Note that $\pi^0(t) \in \Delta W_0^{2,\chi}(G)$ is uniquely determined as the solution to the equation above.

There is a function $\pi^{\mathbf{H}} \in \Delta W_0^{2,q}(G)$ such that

$$\int_G \pi^{\mathbf{H}}(t)\,\Delta\varphi\,\mathrm{d}x = -\int_G \mathbf{H} : \nabla^2\varphi\,\mathrm{d}x$$

for all $\varphi \in C_0^\infty(G)$. The measurability of $\pi^{\mathbf{H}}$ follows from the measurability of the right-hand-side. Moreover, we have on account of the solvability of the bi-Laplace equation (see Corollary 6.1.1)

$$\int_G |\pi^{\mathbf{H}}|^s\,\mathrm{d}x \le c\int_G |\mathbf{H}|^s\,\mathrm{d}x \quad \mathbb{P}\otimes\mathcal{L}^1\text{-a.e.}$$

which implies

$$\int_{\Omega\times Q} |\pi^{\mathbf{H}}|^s\,\mathrm{d}x\,\mathrm{d}t\,\mathrm{d}\mathbb{P} \le c\int_{\Omega\times Q} |\mathbf{H}|^s\,\mathrm{d}x\,\mathrm{d}t\,\mathrm{d}\mathbb{P}.$$

It remains to study the stochastic part of the pressure.

We set $\pi^\Phi(t) := \pi^0(t) - \int_0^t \pi^{\mathbf{H}}\,\mathrm{d}\sigma \in \Delta W_0^{2,\chi}(G)$. This is the unique solution to

$$\int_G \pi^\Phi(t)\,\Delta\varphi\,\mathrm{d}x = \int_G\int_0^t \Phi\,\mathrm{d}\mathbf{W}_\sigma \cdot \nabla\varphi\,\mathrm{d}x, \quad \varphi \in C_0^\infty(G). \tag{10.1.10}$$

So the following holds $\mathbb{P}\otimes\mathcal{L}^1$-a.e.

$$\begin{aligned}\int_G \pi^\Phi(t)\,\operatorname{div}\boldsymbol{\varphi}\,\mathrm{d}x &= \int_G\int_0^t \Phi\,\mathrm{d}\mathbf{W}_\sigma \cdot \nabla\big(\Delta^{-2}\Delta\operatorname{div}\boldsymbol{\varphi}\big)\,\mathrm{d}x\\ &= \sum_k\int_G\int_0^t \Phi\mathbf{e}_k\,\mathrm{d}\beta_k \cdot \nabla\big(\Delta^{-2}\Delta\operatorname{div}\boldsymbol{\varphi}\big)\,\mathrm{d}x\\ &= \sum_k\int_G\int_0^t \nabla\Delta\Delta^{-2}\operatorname{div}\Phi\mathbf{e}_k\,\mathrm{d}\beta_k\cdot\boldsymbol{\varphi}\,\mathrm{d}x\\ &= \int_G\int_0^t \nabla\Delta\Delta^{-2}\operatorname{div}\Phi\,\mathrm{d}\mathbf{W}_\sigma\cdot\boldsymbol{\varphi}\,\mathrm{d}x.\end{aligned}$$

We define $\Phi_\pi = \nabla\Delta\Delta^{-2}\operatorname{div}\Phi$ and the claim in c) follows from

$$\begin{aligned}\|\Phi^\pi\|^2_{L_2(U;L^2(G'))} &= \sum_k\int_{G'} |\nabla\Delta\Delta^{-2}\operatorname{div}\Phi\mathbf{e}_k|^2\,\mathrm{d}x\\ &\le c\sum_k\int_G |\Phi\mathbf{e}_k|^2\,\mathrm{d}x = c\,\|\Phi\|^2_{L_2(U;L^2(G))}.\end{aligned}$$

The above is a consequence of local regularity theory for the bi-Laplace equation. □

Corollary 10.1.1. *Let the assumptions of Theorem 10.1.42 be satisfied. If we assume that Φ satisfies* (10.0.4) *then we have*

$$\|\Phi^{\pi}(\mathbf{u}_1)-\Phi^{\pi}(\mathbf{u}_2)\|_{L_2(U,L^2(G'))}\leq c(G')\,\|\mathbf{u}_1-\mathbf{u}_2\|_{L^2(G)}$$

for all $\mathbf{u}_1,\mathbf{u}_2\in L^2(G)$.

Proof. As in the proof of Theorem 10.1.42 c) the claim follows from local regularity theory for the bi-Laplace equation and the Lipschitz continuity of $\Phi(\mathbf{u})$ in $\mathbf{u}$ from (10.0.4). □

Remark 10.1.24. If the boundary of G is smooth then the statements of Theorem 10.1.42 c) and Corollary 10.1.1 holds globally (i.e., we can replace G' by G). In this case the operator $\nabla\Delta\Delta^{-2}\operatorname{div}$ is continuous on $L^2(G)$ (see [49], section 2.2, or [50]).

Corollary 10.1.2. *Let the assumptions of Theorem 10.1.42 be satisfied. Then we have for all $\beta\in[1,\infty)$*

$$\begin{aligned}\mathbb{E}\bigg[\sup_{(0,T)}\int_G|\pi^h|^{\chi}\,\mathrm{d}x\bigg]^{\beta}&\leq c\,\mathbb{E}\bigg[\sup_{(0,T)}\int_G|\mathbf{u}|^2\,\mathrm{d}x+\int_0^T\|\Phi\|^2_{L_2(U,L^2(G))}\,\mathrm{d}t\bigg]^{\beta}\\&\quad+c\,\mathbb{E}\bigg[1+\int_G|\mathbf{u}_0|^2\,\mathrm{d}x+\int_Q|\mathbf{H}|^{s}\,\mathrm{d}x\,\mathrm{d}t\bigg]^{\beta}\end{aligned}$$

provided the right-hand-side is finite.

Corollary 10.1.3. *Let the assumptions of Theorem 10.1.42 be satisfied. Assume further that we have the decomposition*

$$\mathbf{H}=\mathbf{H}^1+\mathbf{H}^2,$$

where $\mathbf{H}^1\in L^{s_1}(\Omega\times Q,\mathbb{P}\otimes\mathcal{L}^{d+1})$ and $\mathbf{H}^2,\nabla\mathbf{H}^2\in L^{s_2}(\Omega\times Q,\mathbb{P}\otimes\mathcal{L}^{d+1})$ with $s_1,s_2\in(1,\infty)$. Then we have

$$\pi^{\mathbf{H}}=\pi^1+\pi^2$$

and for all $\beta<\infty$ and all $G'\Subset G$ the following holds

$$\begin{aligned}\mathbb{E}\bigg[\int_Q|\pi^1|^{s_1}\,\mathrm{d}x\,\mathrm{d}t\bigg]^{\beta}&\leq c\,\mathbb{E}\bigg[\int_Q|\mathbf{H}^1|^{s_1}\,\mathrm{d}x\,\mathrm{d}t\bigg]^{\beta},\\\mathbb{E}\bigg[\int_Q|\pi^2|^{s_2}\,\mathrm{d}x\,\mathrm{d}t\bigg]^{\beta}&\leq c\,\mathbb{E}\bigg[\int_Q|\mathbf{H}_2|^{s_2}\,\mathrm{d}x\,\mathrm{d}t\bigg]^{\beta},\\\mathbb{E}\bigg[\int_0^T\int_{G'}|\nabla\pi^2|^{s_2}\,\mathrm{d}x\,\mathrm{d}t\bigg]^{\beta}&\leq c\,\mathbb{E}\bigg[\int_Q|\mathbf{H}_2|^{s_2}+|\nabla\mathbf{H}^2|^{s_2}\,\mathrm{d}x\,\mathrm{d}t\bigg]^{\beta}.\end{aligned}$$

Proof. π_1 and π_2 are the unique solutions (defined $\mathbb{P}\otimes\mathcal{L}^1$-a.e.) to

$$\int_G \pi^1(t)\,\Delta\varphi\,\mathrm{d}x = -\int_G \mathbf{H}^1 : \nabla^2\varphi\,\mathrm{d}x,$$
$$\int_G \pi^2(t)\,\Delta\varphi\,\mathrm{d}x = -\int_G \mathbf{H}^1 : \nabla^2\varphi\,\mathrm{d}x,$$

in the spaces $\Delta W_0^{2,s_1}(G)$ and $\Delta W_0^{2,s_2}(G)$. This implies immediately the claimed estimates (see [140], Lemma 2.3, for more details). □

10.2 THE APPROXIMATED SYSTEM

We stabilize the equation by adding a large power of the velocity. For $\alpha > 0$ we study the system

$$\begin{cases} \mathrm{d}\mathbf{v} = \operatorname{div}\mathbf{S}(\boldsymbol{\varepsilon}(\mathbf{v}))\,\mathrm{d}t - \alpha\,|\mathbf{v}|^{q-2}\mathbf{v}\,\mathrm{d}t - \nabla\pi\,\mathrm{d}t \\ \qquad - \operatorname{div}\big(\mathbf{v}\otimes\mathbf{v}\big)\,\mathrm{d}t + \mathbf{f}\mathrm{d}t + \Phi(\mathbf{v})\,\mathrm{d}\mathbf{W}_t \\ \mathbf{v}(0) = \mathbf{v}_0 \end{cases}, \tag{10.2.11}$$

depending on the law $\Lambda_{\mathbf{f}}$ on $L^2(Q)$ and Λ_0 on $L^2_{\mathrm{div}}(G)$. In fact, we fix some $\mathbf{f}\in L^2(\Omega,\mathcal{F},\mathbb{P};L^2(Q))$ with $\Lambda_{\mathbf{f}} = \mathbb{P}\circ\mathbf{f}^{-1}$ and some $\mathbf{v}_0 \in L^2(\Omega,\mathcal{F}_0,\mathbb{P};L^2_{\mathrm{div}}(G))$ with $\Lambda_0 = \mathbb{P}\circ\mathbf{v}_0^{-1}$. By enlarging the filtration $(\mathcal{F}_t)_{t\geq 0}$ we can assume that $\mathbf{f}$ is adapted to it. We choose $q \geq \max\{2p', 3\}$ thus the solution is also an admissible test function. We expect a solution $\mathbf{v}$ in the space

$$\begin{aligned}\mathcal{V}_{p,q} &:= L^q(\Omega\times Q;\mathbb{P}\otimes\mathcal{L}^{d+1}) \cap L^p(\Omega,\mathcal{F},\mathbb{P};L^p(0,T;W^{1,p}_{0,\mathrm{div}}(G)))\\ &\quad\cap\left\{L^1(\Omega\times Q;\mathbb{P}\otimes\mathcal{L}^{d+1}) : \mathbb{E}\Big[\sup_{t\in(0,T)}\int_G |\mathbf{w}|^2\,\mathrm{d}x\Big] < \infty\right\}.\end{aligned}$$

We will try to find a solution by separating space and time via a Galerkin-Ansatz similar to Chapter 7.1. Then we seek for an approximated solution by solving an ordinary stochastic differential equation.

There is a sequence $(\lambda_k)\subset\mathbb{R}$ and a sequence of functions $(\mathbf{w}_k)\subset W^{l,2}_{0,\mathrm{div}}(G)$, $l\in\mathbb{N}$, such that (see [111], appendix)

i) $\mathbf{w}_k$ is an eigenvector to the eigenvalue λ_k of the Stokes-operator in the sense that:

$$\langle \mathbf{w}_k, \boldsymbol{\varphi}\rangle_{W_0^{l,2}} = \lambda_k \int_G \mathbf{w}_k\cdot\boldsymbol{\varphi}\,dx \quad \text{for all} \quad \boldsymbol{\varphi}\in W^{l,2}_{0,\mathrm{div}}(G),$$

ii) $\int_G \mathbf{w}_k\mathbf{w}_m\,dx = \delta_{km}$ for all $k, m\in\mathbb{N}$,

iii) $1\leq\lambda_1\leq\lambda_2\leq\ldots$ and $\lambda_k\to\infty$,

iv) $\langle \frac{\mathbf{w}_k}{\sqrt{\lambda_k}}, \frac{\mathbf{w}_m}{\sqrt{\lambda_m}} \rangle_{W_0^{l,2}} = \delta_{km}$ for all $k, m \in \mathbb{N}$,

v) $(\mathbf{w}_k)$ is a basis of $W_{0,\mathrm{div}}^{l,2}(G)$.

We choose $l > 1 + \frac{d}{2}$ such that $W_0^{l,2}(G) \hookrightarrow W^{1,\infty}(G)$. We are looking for an approximated solution $\mathbf{v}_N$ of the form

$$\mathbf{v}_N = \sum_{k=1}^{N} c_i^N \mathbf{w}_k = \mathbf{C}_N \cdot \boldsymbol{\omega}_N, \quad \boldsymbol{\omega}_N = (\mathbf{w}_1, \ldots, \mathbf{w}_N),$$

where $\mathbf{C}_N = (c_N^i) : \Omega \times (0, T) \to \mathbb{R}^N$. Therefore, we would like to solve the system $(k = 1, \ldots, N)$

$$\begin{aligned}
&\int_G \mathrm{d}\mathbf{v}_N \cdot \mathbf{w}_k \, \mathrm{d}x + \int_G \mathbf{S}(\boldsymbol{\varepsilon}(\mathbf{v}_N)) : \boldsymbol{\varepsilon}(\mathbf{w}_k) \, \mathrm{d}x \, \mathrm{d}t + \alpha \int_G |\mathbf{v}_N|^{q-2} \mathbf{v}_N \cdot \mathbf{w}_k \, \mathrm{d}x \, \mathrm{d}t \\
&\quad = \int_G \mathbf{v}_N \otimes \mathbf{v}_N : \nabla \mathbf{w}_k \, \mathrm{d}x \, \mathrm{d}t + \int_G \mathbf{f} \cdot \mathbf{w}_k \, \mathrm{d}x \, \mathrm{d}t + \int_G \Phi(\mathbf{v}_N) \, \mathrm{d}\mathbf{W}_\sigma^N \cdot \mathbf{w}_k \, \mathrm{d}x, \\
&\quad \mathbf{v}_N(0) = \mathcal{P}_N \mathbf{v}_0.
\end{aligned} \tag{10.2.12}$$

Here $\mathcal{P}_N : L^2_{\mathrm{div}}(G) \to \mathcal{X}_N := \mathrm{span}\{\mathbf{w}_1, \ldots, \mathbf{w}_N\}$ is the orthogonal projection, i.e.,

$$\mathcal{P}_N \mathbf{u} = \sum_{k=1}^{N} \langle \mathbf{u}, \mathbf{w}_k \rangle_{L^2} \mathbf{w}_k.$$

The equation above is to be understood $\mathbb{P}$-a.s. and for a.e. t. Moreover, we have set

$$\mathbf{W}^N(\sigma) = \sum_{k=1}^{N} \mathbf{e}_k \beta_k(\sigma) = \mathbf{E}_N \cdot \boldsymbol{\beta}^N(\sigma), \quad \mathbf{E}_N = (\mathbf{e}_1, \ldots, \mathbf{e}_N).$$

The system (10.2.12) is equivalent to solving

$$\begin{cases} \mathrm{d}\mathbf{C}_N = \left[\boldsymbol{\mu}(t, \mathbf{C}_N)\right] \mathrm{d}t + \boldsymbol{\Sigma}(\mathbf{C}_N) \, \mathrm{d}\boldsymbol{\beta}_t^N, \\ \mathbf{C}_N(0) = \mathbf{C}_0, \end{cases} \tag{10.2.13}$$

with the abbreviations

$$\begin{aligned}
\boldsymbol{\mu}(\mathbf{C}^N) &= \Bigg(- \int_G \mathbf{S}(\mathbf{C}_N \cdot \boldsymbol{\varepsilon}(\boldsymbol{\omega}_N)) : \boldsymbol{\varepsilon}(\mathbf{w}_k) \, \mathrm{d}x \\
&\qquad + \int_G (\mathbf{C}_N \cdot \boldsymbol{\omega}_N) \otimes (\mathbf{C}_N \cdot \boldsymbol{\omega}_N) : \nabla \mathbf{w}_k \, \mathrm{d}x \Bigg)_{k=1}^{N} \\
&\qquad - \left(\alpha \int_G |\mathbf{C}_N \cdot \boldsymbol{\omega}_N|^{q-2} (\mathbf{C}_N \cdot \boldsymbol{\omega}_N) \cdot \mathbf{w}_k \, \mathrm{d}x \right)_{k=1}^{N} + \left(\int_G \mathbf{f}(t) \cdot \mathbf{w}_k \, \mathrm{d}x \right)_{k=1}^{N}, \\
\boldsymbol{\Sigma}(\mathbf{C}_N) &= \left(\int_G \Phi(\mathbf{C}_N \cdot \boldsymbol{\omega}_N) \mathbf{e}_l \cdot \mathbf{w}_k \, \mathrm{d}x \right)_{k,l=1}^{N},
\end{aligned}$$

$$\mathbf{C}_0 = \Big(\langle \mathbf{v}_0, \mathbf{w}_k \rangle_{L^2(G)} \Big)_{k=1}^{N}.$$

In the following we will check the assumptions of Theorem 8.4.37. We have by the monotonicity of $\mathbf{S}$ (recall (10.0.1)) that

$$\begin{aligned}
&\big(\boldsymbol{\mu}(t, \mathbf{C}_N) - \boldsymbol{\mu}(t, \tilde{\mathbf{C}}_N)\big) \cdot \big(\mathbf{C}_N - \tilde{\mathbf{C}}_N\big) \\
&\quad = - \int_G \big(\mathbf{S}(\boldsymbol{\varepsilon}(\mathbf{v}_N)) - \mathbf{S}(\boldsymbol{\varepsilon}(\tilde{\mathbf{v}}_N))\big) : \big(\boldsymbol{\varepsilon}(\mathbf{v}_N) - \boldsymbol{\varepsilon}(\tilde{\mathbf{v}}_N)\big)\, \mathrm{d}x \\
&\quad + \int_G \big(\mathbf{v}_N \otimes \mathbf{v}_N - \tilde{\mathbf{v}}_N \otimes \tilde{\mathbf{v}}_N\big) : \big(\boldsymbol{\varepsilon}(\mathbf{v}_N) - \boldsymbol{\varepsilon}(\tilde{\mathbf{v}}_N)\big)\, \mathrm{d}x \\
&\quad \leq \int_G \big(\mathbf{v}_N \otimes \mathbf{v}_N - \tilde{\mathbf{v}}_N \otimes \tilde{\mathbf{v}}_N\big) : \big(\boldsymbol{\varepsilon}(\mathbf{v}_N) - \boldsymbol{\varepsilon}(\tilde{\mathbf{v}}_N)\big)\, \mathrm{d}x.
\end{aligned}$$

If $|\mathbf{C}_N| \leq R$ and $|\tilde{\mathbf{C}}_N| \leq R$ the following holds

$$\big(\boldsymbol{\mu}(t, \mathbf{C}_N) - \boldsymbol{\mu}(t, \tilde{\mathbf{C}}_N)\big) \cdot \big(\mathbf{C}_N - \tilde{\mathbf{C}}_N\big) \leq c(R, N) |\mathbf{C}_N - \tilde{\mathbf{C}}_N|^2.$$

Here, we took into account boundedness of $\mathbf{w}_k$ and $\nabla \mathbf{w}_k$. The above implies weak monotonicity in the sense of (A3) by the Lipschitz continuity $\boldsymbol{\Sigma}$ in $\mathbf{C}_N$, cf. (10.0.4). On account of $\int_G \mathbf{v}_N \otimes \mathbf{v}_N : \boldsymbol{\varepsilon}(\mathbf{v}_N)\, \mathrm{d}x = 0$ the following holds

$$\begin{aligned}
&\boldsymbol{\mu}(t, \mathbf{C}_N) \cdot \mathbf{C}_N \\
&\quad = - \int_G \mathbf{S}(\boldsymbol{\varepsilon}(\mathbf{v}_N)) : (\boldsymbol{\varepsilon}(\mathbf{v}_N)\, \mathrm{d}x + \int_G \mathbf{f}(t) \cdot \mathbf{v}_N\, \mathrm{d}x \leq c(1 + \|\mathbf{f}(t)\|_2 \|\mathbf{v}_N\|_2) \\
&\quad \leq (1 + \|\mathbf{f}(t)\|_2)(1 + \|\mathbf{v}_N\|^2) \leq c(1 + \|\mathbf{f}(t)\|_2)(1 + |\mathbf{C}_N|^2).
\end{aligned}$$

So we have, using the linear growth of $\boldsymbol{\Sigma}$ which follows from (10.0.4), that

$$\boldsymbol{\mu}(\mathbf{C}_N) \cdot \mathbf{C}_N + |\boldsymbol{\Sigma}(\mathbf{C}_N)|^2 \leq c(+\|\mathbf{v}_N\|_2^2)\big(1 + |\mathbf{C}_N|^2\big).$$

As the integral $\int_0^T (1 + \|\mathbf{f}(t)\|_2)\, \mathrm{d}t$ is finite $\mathbb{P}$-a.s. this implies weak coercivity in the sense of (A2). We obtain a unique strong solution $\mathbf{C}_N$ to the SDE (10.2.13) with $\mathbb{P}$-a.s. continuous trajectories.

We obtain the following a priori estimate.

Theorem 10.2.43. *Assume* (10.0.1) *with* $p \in (1, \infty)$, (10.0.4), $q \geq \{2p', 3\}$ *and*

$$\int_{L^2_{\mathrm{div}}(G)} \|\mathbf{u}\|^2_{L^2(G)}\, \mathrm{d}\Lambda_0(\mathbf{u}) < \infty, \qquad \int_{L^2(Q)} \|\mathbf{g}\|^2_{L^2(Q)}\, \mathrm{d}\Lambda_{\mathbf{f}}(\mathbf{g}) < \infty. \qquad (10.2.14)$$

Then the following holds uniformly in N

$$\mathbb{E}\Big[\sup_{t \in (0,T)} \int_G |\mathbf{v}_N(t)|^2\, \mathrm{d}x + \int_Q |\nabla \mathbf{v}_N|^p\, \mathrm{d}x\, \mathrm{d}t + \alpha \int_Q |\mathbf{v}_N|^q\, \mathrm{d}x\, \mathrm{d}t \Big]$$

$$\leq c\bigg(1+\int_{L^2_{\mathrm{div}}(G)}\|\mathbf{u}\|^2_{L^2(G)}\,\mathrm{d}\Lambda_0(\mathbf{u})+\int_{L^2(Q)}\|\mathbf{g}\|^2_{L^2(Q)}\,\mathrm{d}\Lambda_{\mathbf{f}}(\mathbf{g})\bigg),$$

where c is independent of α.

Proof. We apply Itô's formula to the function $f(\mathbf{C})=\frac{1}{2}|\mathbf{C}|^2$ which shows

$$\begin{aligned}
\frac{1}{2}\|\mathbf{v}_N(t)\|^2_{L^2(G)}&=\frac{1}{2}\|\mathbf{C}_N(0)\|^2_{L^2(G)}+\sum_{k=1}^N\int_0^t c_N^k\,\mathrm{d}(c_N^k)_\sigma+\frac{1}{2}\sum_{k=1}^N\int_0^t\mathrm{d}\langle\langle c_N^k\rangle\rangle_\sigma\\
&=\frac{1}{2}\|\mathcal{P}_N\mathbf{v}_0\|^2_{L^2(G)}-\int_0^t\int_G\mathbf{S}(\boldsymbol{\varepsilon}(\mathbf{v}_N)):\boldsymbol{\varepsilon}(\mathbf{v}_N)\,\mathrm{d}x\,\mathrm{d}\sigma\\
&\quad-\alpha\int_0^t\int_G|\mathbf{v}_N|^q\,\mathrm{d}x\,\mathrm{d}\sigma+\int_0^t\int_G\mathbf{f}\cdot\mathbf{v}_N\,\mathrm{d}x\,\mathrm{d}\sigma \qquad (10.2.15)\\
&\quad+\int_G\int_0^t\mathbf{v}_N\cdot\Phi(\mathbf{v}_N)\,\mathrm{d}\mathbf{W}^N_\sigma\,\mathrm{d}x\\
&\quad+\frac{1}{2}\int_G\int_0^t\mathrm{d}\bigg\langle\bigg\langle\int_0^\cdot\Phi(\mathbf{v}_N)\,\mathrm{d}\mathbf{W}^N\bigg\rangle\bigg\rangle_\sigma\,\mathrm{d}x.
\end{aligned}$$

Here, we used $\mathrm{d}\mathbf{v}_N=\sum_{k=1}^N\mathrm{d}c_N^k\mathbf{w}_k$, $\int_G\mathbf{v}_N\otimes\mathbf{v}_N:\nabla\mathbf{v}_N\,\mathrm{d}x=0$, property (ii) of the base $(\mathbf{w}_k)$ as well as

$$\begin{aligned}
\mathrm{d}c_N^k=&-\int_G\mathbf{S}(\boldsymbol{\varepsilon}(\mathbf{v}_N)):\boldsymbol{\varepsilon}(\mathbf{w}_k)\,\mathrm{d}x\,\mathrm{d}t-\alpha\int_G|\mathbf{v}_N|^{q-2}\mathbf{v}_N\cdot\mathbf{w}_k\,\mathrm{d}x\,\mathrm{d}t\\
&+\int_G\mathbf{v}_N\otimes\mathbf{v}_N:\nabla\mathbf{w}_k\,\mathrm{d}x\,\mathrm{d}t+\int_G\mathbf{f}\cdot\mathbf{w}_k\,\mathrm{d}x\,\mathrm{d}t+\int_G\Phi(\mathbf{v}_N)\,\mathrm{d}\mathbf{W}^N_t\cdot\mathbf{w}_k\,\mathrm{d}x.
\end{aligned}$$

Now we can follow, building expectations and using (10.0.1) together with Korn's inequality, that

$$\begin{aligned}
&\mathbb{E}\bigg[\int_G|\mathbf{v}_N(t)|^2\,\mathrm{d}x+\int_0^t\int_G|\nabla\mathbf{v}_N|^p\,\mathrm{d}x\,\mathrm{d}\sigma+\alpha\int_0^t\int_G|\mathbf{v}_N|^q\,\mathrm{d}x\,\mathrm{d}\sigma\bigg]\\
&\quad\leq c\Big(1+\mathbb{E}\big[\|\mathbf{v}_0\|^2_{L^2(G)}\big]+\mathbb{E}[J_1(t)]+\mathbb{E}[J_2(t)]+\mathbb{E}[J_3(t)]\Big).
\end{aligned}$$

Here, we abbreviated

$$\begin{aligned}
J_1(t)&=\int_0^t\int_G\mathbf{f}\cdot\mathbf{v}_N\,\mathrm{d}x\,\mathrm{d}\sigma,\\
J_2(t)&=\int_G\int_0^t\mathbf{v}_N\cdot\Phi(\mathbf{v}_N)\,\mathrm{d}\mathbf{W}^N_\sigma\,\mathrm{d}x,\\
J_3(t)&=\int_G\int_0^t\mathrm{d}\bigg\langle\bigg\langle\int_0^\cdot\Phi(\mathbf{v}_N)\,\mathrm{d}\mathbf{W}^N\bigg\rangle\bigg\rangle_\sigma\,\mathrm{d}x.
\end{aligned}$$

Straightforward calculations show on account of (10.0.3) and (10.0.4)

$$\mathbb{E}[J_3]=\mathbb{E}\bigg[\sum_{i=1}^N\int_0^t\int_G|\Phi(\mathbf{v}_N)\mathbf{e}_i|^2\,\mathrm{d}x\,\mathrm{d}\sigma\bigg]$$

$$\leq \mathbb{E}\bigg[\sum_{i=1}^{\infty}\int_0^t\int_G |g_i(\mathbf{v}_N)|^2\,\mathrm{d}x\,\mathrm{d}\sigma\bigg]$$
$$\leq \mathbb{E}\bigg[1+\int_0^t\int_G |\mathbf{v}_N|^2\,\mathrm{d}x\,\mathrm{d}\sigma\bigg].$$

Using Young's inequality we obtain for arbitrary $\delta > 0$

$$\mathbb{E}[J_1] \leq \delta\mathbb{E}\bigg[\int_0^t\int_G |\mathbf{v}_N|^2\,\mathrm{d}x\,\mathrm{d}\sigma\bigg] + c(\delta)\mathbb{E}\bigg[\int_0^t\int_G |\mathbf{f}|^2\,\mathrm{d}x\,\mathrm{d}\sigma\bigg].$$

Clearly, we have $\mathbb{E}[J_2] = 0$. So interchanging the time-integral and the expectation value and applying Gronwall's Lemma leads to

$$\begin{aligned}\sup_{t\in(0,T)}\mathbb{E}\bigg[\int_G |\mathbf{v}_N(t)|^2\,\mathrm{d}x\bigg] + \mathbb{E}\bigg[\int_Q |\nabla\mathbf{v}_N|^p\,\mathrm{d}x\,\mathrm{d}t\bigg]&\\ \leq c\,\mathbb{E}\bigg[1+\int_G |\mathbf{v}_0|^2\,\mathrm{d}x + \int_Q |\mathbf{f}|^2\,\mathrm{d}x\bigg].&\end{aligned} \tag{10.2.16}$$

We want to interchange supremum and expectation value. Similar arguments as before show by (10.2.16)

$$\begin{aligned}&\mathbb{E}\bigg[\sup_{t\in(0,T)}\int_G |\mathbf{v}_N(t)|^2\,\mathrm{d}x\bigg] \\ &\leq c\,\mathbb{E}\bigg[1+\int_G |\mathbf{v}_0|^2\,\mathrm{d}x + \int_Q |\mathbf{f}|^2\,\mathrm{d}x + \int_0^T\int_G |\mathbf{v}_N|^2\,\mathrm{d}x\,\mathrm{d}t\bigg] + \mathbb{E}\bigg[\sup_{t\in(0,T)}|J_2(t)|\bigg].\end{aligned} \tag{10.2.17}$$

On account of Burkholder–Davis–Gundi inequality, Young's inequality and (10.0.4) we obtain (note that the paths of $\mathbf{v}_N$ in $L^2(G)$ are $\mathbb{P}$-a.s. continuous in time)

$$\begin{aligned}\mathbb{E}\bigg[\sup_{t\in(0,T)}|J_2(t)|\bigg] &= \mathbb{E}\bigg[\sup_{t\in(0,T)}\bigg|\int_0^t\int_G \mathbf{v}_N\cdot\Phi(\mathbf{v}_N)\,\mathrm{d}x\,\mathrm{d}\mathbf{W}_\sigma^N\bigg|\bigg]\\
&= \mathbb{E}\bigg[\sup_{t\in(0,T)}\bigg|\int_0^t\sum_i\int_G \mathbf{v}_N\cdot\Phi(\mathbf{v}_N)\mathbf{e}_i\,\mathrm{d}x\,\mathrm{d}\beta_i(\sigma)\bigg|\bigg]\\
&= \mathbb{E}\bigg[\sup_{t\in(0,T)}\bigg|\int_0^t\sum_i\int_G \mathbf{v}_N\cdot g_i(\mathbf{v}_N)\,\mathrm{d}x\,\mathrm{d}\beta_i(\sigma)\bigg|\bigg]\\
&\leq c\,\mathbb{E}\bigg[\int_0^T\sum_i\bigg(\int_G \mathbf{v}_N\cdot g_i(\mathbf{v}_N)\,\mathrm{d}x\bigg)^2\mathrm{d}t\bigg]^{\frac{1}{2}}\\
&\leq c\,\mathbb{E}\bigg[\bigg(\int_0^T\bigg(\sum_{i=1}^N\int_G |\mathbf{v}_N|^2\,\mathrm{d}x\int_G |g_i(\mathbf{v}_N)|^2\,\mathrm{d}x\bigg)\,\mathrm{d}t\bigg]^{\frac{1}{2}}\end{aligned}$$

$$\leq c\,\mathbb{E}\bigg[1+\int_0^T\bigg(\int_G|\mathbf{v}_N|^2\,\mathrm{d}x\bigg)^2\,\mathrm{d}t\bigg]^{\frac{1}{2}}$$

$$\leq \delta\,\mathbb{E}\bigg[\sup_{t\in(0,T)}\int_G|\mathbf{v}_N|^2\,\mathrm{d}x\bigg]+c(\delta)\,\mathbb{E}\bigg[1+\int_0^T\int_G|\mathbf{v}_N|^2\,\mathrm{d}x\,\mathrm{d}t\bigg].$$

This finally proves the claim for δ sufficiently small using (10.2.16) as well as $\Lambda_0=\mathbb{P}\circ\mathbf{v}_0^{-1}$ and $\Lambda_{\mathbf{f}}=\mathbb{P}\circ\mathbf{f}^{-1}$. □

Theorem 10.2.44. *Assume* (10.0.1) *with* $p\in(1,\infty)$, (10.0.4), $q\geq\{2p',3\}$ *and* (10.2.14).

a) *There is a weak martingale solution*

$$\big((\underline{\Omega},\underline{\mathcal{F}},(\underline{\mathcal{F}}_t)_{t\geq0},\underline{\mathbb{P}}),\underline{\mathbf{v}},\underline{\mathbf{v}}_0,\underline{\mathbf{f}},\underline{\mathbf{W}}\big)$$

to (10.2.11) *in the sense that*

i) $(\underline{\Omega},\underline{\mathcal{F}},(\underline{\mathcal{F}}_t)_{t\geq0},\underline{\mathbb{P}})$ *is a stochastic basis with a complete right-continuous filtration,*

ii) $\underline{\mathbf{W}}$ *is an* $(\underline{\mathcal{F}}_t)_{t\geq0}$*-cylindrical Wiener process,*

iii) $\underline{\mathbf{v}}\in\underline{\mathcal{V}}_{p,q}$ *is progressively measurable, where*

$$\underline{\mathcal{V}}_{p,q}:=L^q(\underline{\Omega}\times Q;\underline{\mathbb{P}}\otimes\mathcal{L}^{d+1})\cap L^p(\underline{\Omega},\underline{\mathcal{F}},\underline{\mathbb{P}};L^p(0,T;W^{1,p}_{0,\mathrm{div}}(G)))$$
$$\cap\bigg\{L^1(\underline{\Omega}\times Q;\underline{\mathbb{P}}\otimes\mathcal{L}^{d+1}):\underline{\mathbb{E}}\bigg[\sup_{t\in(0,T)}\int_G|\mathbf{w}|^2\,\mathrm{d}x\bigg]<\infty\bigg\}.$$

iv) $\underline{\mathbf{v}}_0\in L^2(\Omega,\mathcal{F}_0,\mathbb{P};L^2(G))$ *with* $\Lambda_0=\underline{\mathbb{P}}\circ\underline{\mathbf{v}}_0^{-1}$,

v) $\underline{\mathbf{f}}\in L^2(\underline{\Omega},\underline{\mathcal{F}},\underline{\mathbb{P}};L^2(Q))$ *is adapted to* $(\underline{\mathcal{F}}_t)_{t\geq0}$ *with* $\Lambda_{\mathbf{f}}=\underline{\mathbb{P}}\circ\underline{\mathbf{f}}^{-1}$,

vi) *for all* $\boldsymbol{\varphi}\in C^\infty_{0,\mathrm{div}}(G)$ *and all* $t\in[0,T]$ *we have*

$$\int_G\underline{\mathbf{v}}(t)\cdot\boldsymbol{\varphi}\,\mathrm{d}x+\int_0^t\int_G\mathbf{S}(\boldsymbol{\varepsilon}(\underline{\mathbf{v}})):\boldsymbol{\varepsilon}(\boldsymbol{\varphi})\,\mathrm{d}x\,\mathrm{d}\sigma+\alpha\int_0^t\int_G|\underline{\mathbf{v}}|^{q-2}\underline{\mathbf{v}}\cdot\boldsymbol{\varphi}\,\mathrm{d}x\,\mathrm{d}\sigma$$
$$=\int_0^t\int_G\underline{\mathbf{v}}\otimes\underline{\mathbf{v}}:\boldsymbol{\varepsilon}(\boldsymbol{\varphi})\,\mathrm{d}x\,\mathrm{d}\sigma+\int_G\underline{\mathbf{v}}_0\cdot\boldsymbol{\varphi}\,\mathrm{d}x+\int_G\int_0^t\underline{\mathbf{f}}\cdot\boldsymbol{\varphi}\,\mathrm{d}x\,\mathrm{d}\sigma$$
$$+\int_G\int_0^t\Phi(\underline{\mathbf{v}})\,\mathrm{d}\underline{\mathbf{W}}_\sigma\cdot\boldsymbol{\varphi}\,\mathrm{d}x.$$

$\underline{\mathbb{P}}$*-a.s.*

b) *The following holds*

$$\underline{\mathbb{E}}\bigg[\sup_{t\in(0,T)}\int_G|\underline{\mathbf{v}}(t)|^2\,\mathrm{d}x+\int_Q|\nabla\underline{\mathbf{v}}|^p\,\mathrm{d}x\,\mathrm{d}t+\alpha\int_Q|\underline{\mathbf{v}}|^q\,\mathrm{d}x\,\mathrm{d}t\bigg]$$
$$\leq c\bigg(1+\int_{L^2_{\mathrm{div}}(G)}\|\mathbf{u}\|^2_{L^2(G)}\,\mathrm{d}\Lambda_0(\mathbf{u})+\int_{L^2(Q)}\|\mathbf{g}\|^2_{L^2(Q)}\,\mathrm{d}\Lambda_{\mathbf{f}}(\mathbf{g})\bigg),$$

where c *is independent of* α.

Proof. From the a priori estimate in Theorem 10.2.43 we can follow the existence of a function $\mathbf{v} \in \mathcal{V}_{p,q}$ and functions $\tilde{\mathbf{s}}$ and $\tilde{\mathbf{S}}$ such that (after passing to a not relabelled subsequence)

$$\begin{aligned}
\mathbf{v}_N &\rightharpoonup \mathbf{v} \quad \text{in} \quad L^p(\Omega, \mathcal{F}, \mathbb{P}; L^p(0, T; W_0^{1,p}(G))), \\
\mathbf{v}_N &\rightharpoonup \mathbf{v} \quad \text{in} \quad L^q(\Omega, \mathcal{F}, \mathbb{P}; L^q(Q)), \\
\mathbf{s}(\mathbf{v}_N) &\rightharpoonup \tilde{\mathbf{s}} \quad \text{in} \quad L^{q'}(\Omega, \mathcal{F}, \mathbb{P}; L^{q'}(Q)), \\
\mathbf{S}(\boldsymbol{\varepsilon}(\mathbf{v}_N)) &\rightharpoonup \tilde{\mathbf{S}} \quad \text{in} \quad L^{p'}(\Omega, \mathcal{F}, \mathbb{P}; L^{p'}(Q)), \\
\mathbf{S}(\boldsymbol{\varepsilon}(\mathbf{v}_N)) &\rightharpoonup \tilde{\mathbf{S}} \quad \text{in} \quad L^{p'}(\Omega, \mathcal{F}, \mathbb{P}; L^{p'}(0, T; W_0^{-1,p'}(G))).
\end{aligned} \tag{10.2.18}$$

Moreover, there are $\tilde{\mathbf{V}}$ and $\tilde{\Phi}$ (recall (10.0.4) and Theorem 10.2.43) such that

$$\begin{aligned}
\mathbf{v}_N \otimes \mathbf{v}_N &\rightharpoonup \tilde{\mathbf{V}} \quad \text{in} \quad L^{\frac{q}{2}}(\Omega, \mathcal{F}, \mathbb{P}; L^{\frac{q}{2}}(Q)), \\
\Phi(\mathbf{v}_N) &\rightharpoonup \tilde{\Phi} \quad \text{in} \quad L^2(\Omega, \mathcal{F}, \mathbb{P}; L^2(0, T; L_2(U, L^2(G)))).
\end{aligned} \tag{10.2.19}$$

Step 1: compactness

We want to establish

$$\tilde{\mathbf{V}} = \mathbf{v} \otimes \mathbf{v}, \quad \tilde{\Phi} = \Phi(\mathbf{v}). \tag{10.2.20}$$

This will be a consequence of some compactness arguments. We will follow ideas from [94], section 4. We consider $\boldsymbol{\varphi} \in C^\infty_{0,\mathrm{div}}(G)$ and obtain by (10.2.12)

$$\begin{aligned}
\int_G \mathbf{v}_N(t) \cdot \boldsymbol{\varphi} \, \mathrm{d}x &= \int_G \mathbf{v}_N(t) \cdot \mathcal{P}_N^l \boldsymbol{\varphi} \, \mathrm{d}x \\
&= \int_G \mathbf{v}_0 \cdot \mathcal{P}_N^l \boldsymbol{\varphi} \, \mathrm{d}x + \int_0^t \int_G \mathbf{H}_N : \nabla \mathcal{P}_N^l \boldsymbol{\varphi} \, \mathrm{d}x \, \mathrm{d}\sigma \\
&\quad + \int_G \int_0^t \Phi(\mathbf{v}_N) \mathrm{d}\mathbf{W}_\sigma^N \cdot \mathcal{P}_N^l \boldsymbol{\varphi} \, \mathrm{d}x, \\
\mathbf{H}_N &:= -\mathbf{S}(\boldsymbol{\varepsilon}(\mathbf{v}_N)) + \nabla \Delta^{-1} \mathbf{s}(\mathbf{v}_N) + \mathbf{v}_N \otimes \mathbf{v}_N - \nabla \Delta^{-2} \mathbf{f}.
\end{aligned}$$

Here $\mathcal{P}_N^l$ denotes the projection into $\mathcal{X}_N$ with respect to the $W^{l,2}_{0,\mathrm{div}}(G)$ inner product. From the a priori estimates in Theorem 10.2.43 and the growth conditions for $\mathbf{S}$ (which follow from (10.0.1)) and $\mathbf{s}$ we obtain

$$\mathbf{H}_N \in L^{p_0}(\Omega \times Q; \mathbb{P} \otimes \mathcal{L}^{d+1}), \quad p_0 := \min\left\{p', q', \frac{q}{2}\right\} > 1, \tag{10.2.21}$$

uniformly in N. Let us consider the functional

$$\mathscr{H}_N(t, \boldsymbol{\varphi}) := \int_0^t \int_G \mathbf{H}_N : \nabla \mathcal{P}_N^l \boldsymbol{\varphi} \, \mathrm{d}x \, \mathrm{d}\sigma, \quad \boldsymbol{\varphi} \in C^\infty_{0,\mathrm{div}}(G).$$

Then we deduce from (10.2.21) and the embedding $W^{\tilde{l},p_0}(G) \hookrightarrow W_0^{l,2}(G)$ for $\tilde{l} \geq l + d\left(1 + \frac{2}{p_0}\right)$ the estimate

$$\mathbb{E}\left[\left\|\mathscr{H}_N\right\|_{W^{1,p_0}([0,T];W_{\mathrm{div}}^{-\tilde{l},p_0}(G))}\right] \leq c.$$

For the stochastic term we use Lemma 9.1.3 and (10.0.4) to estimate for all $\alpha < 1/2$

$$\begin{aligned}\mathbb{E}\left[\left\|\int_0^{\cdot} \Phi(\mathbf{v}_N)\,\mathrm{d}\mathbf{W}_\sigma^N\right\|_{W^{\alpha,2}(0,T;L^2(G))}\right] &\leq c\,\mathbb{E}\left[\int_0^T \|\Phi(\mathbf{v}_N)\|^2_{L_2(U,L^2(G))}\,\mathrm{d}t\right]\\ &\leq c\,\mathbb{E}\left[\sum_k \int_0^T \int_G |g_k(\cdot,\mathbf{v}_N)|^2\,\mathrm{d}x\,\mathrm{d}t\right] \leq c\,\mathbb{E}\left[1 + \int_Q |\mathbf{v}_N|^2\,\mathrm{d}x\,\mathrm{d}t\right].\end{aligned}$$

So we have due to Theorem 10.2.43 and $p_0 \leq 2$ that

$$\mathbb{E}\left[\left\|\int_0^{\cdot} \Phi(\mathbf{v}_N)\,\mathrm{d}\mathbf{W}_\sigma^N\right\|_{W^{\alpha,p_0}((0,T);L^2(G))}\right] \leq c.$$

Combining the both informations above shows

$$\mathbb{E}\left[\|\mathbf{v}_N\|_{W^{\alpha,p_0}(0,T;W_{0,\mathrm{div}}^{-\tilde{l},p_0}(G))}\right] \leq c. \tag{10.2.22}$$

An interpolation with $L^{p_0}(0,T;W_{0,\mathrm{div}}^{1,p_0}(G))$ implies by Theorem 10.2.43

$$\mathbb{E}\left[\|\mathbf{v}_N\|_{W^{\kappa,p_0}(0,T;L^{p_0}_{\mathrm{div}}(G))}\right] \leq c \tag{10.2.23}$$

for some $\kappa > 0$ (see Lemma 5.1.4). So, we have

$$W^{\kappa,p_0}(0,T;L^{p_0}_{\mathrm{div}}(G)) \cap \mathcal{V}_{p,q} \hookrightarrow\hookrightarrow L^r(0,T;L^r_{\mathrm{div}}(G))$$

compactly for all $r < q$. We will use this embedding in order to show compactness of $\mathbf{v}^N$. We consider the path space

$$\mathscr{V} := L^r(0,T;L^r(G)) \otimes C([0,T],U_0) \otimes L^2_{\mathrm{div}}(G) \otimes L^2(Q).$$

In the following we introduce some notations.

- $\nu_{\mathbf{v}_N}$ is the law of $\mathbf{v}_N$ on $L^r(0,T;L^r(G))$;
- $\nu_{\mathbf{W}}$ is the law of $\mathbf{W}$ on $C([0,T],U_0)$, where U_0 is defined in (9.1.2);
- $\boldsymbol{\nu}_N$ is the joint law of $\mathbf{v}_N$, $\mathbf{W}$, $\mathbf{v}_0$ and $\mathbf{f}$ on $\mathscr{V}$.

We consider the ball $\mathcal{B}_R$ in the space $W^{\kappa,p_0}(0,T;L^{p_0}_{\mathrm{div}}(G)) \cap \mathcal{V}_{p,q}$ and obtain for its complement $\mathcal{B}_R^C$ by Theorem 10.2.43 and (10.2.23)

$$\begin{aligned}\nu_{\mathbf{v}_N}(\mathcal{B}_R^C) &= \mathbb{P}\left(\|\mathbf{v}_N\|_{W^{\kappa,p_0}(0,T;L^{p_0}_{\mathrm{div}}(G))} + \|\mathbf{v}_N\|_{\mathcal{V}_{p,q}} \geq R\right)\\ &\leq \frac{1}{R}\,\mathbb{E}\left[\|\mathbf{v}_N\|_{W^{\kappa,p_0}(0,T;L^{p_0}_{\mathrm{div}}(G))} + \|\mathbf{v}_N\|_{\mathcal{V}_{p,q}}\right] \leq \frac{c}{R}.\end{aligned}$$

So, for a fixed $\eta > 0$, we find $R(\eta)$ with

$$\nu_{\mathbf{v}_N}(\mathcal{B}_{R(\eta)}) \geq 1 - \frac{\eta}{4}.$$

Since also the law $\nu_{\mathbf{W}}$ is tight, as being a Radon measure on the Polish space $C([0,T],U_0)$, there exists a compact set $C_\eta \subset C([0,T],U_0)$ such that $\nu_{\mathbf{W}}(C_\eta) \geq 1 - \frac{\eta}{4}$. For the same reason we find compact subsets of $L^2_{\mathrm{div}}(G)$ and $L^2(Q)$ such that their measures (Λ_0 and $\Lambda_{\mathbf{f}}$ respectively) are smaller than $1 - \frac{\eta}{4}$. Hence, we can find a compact subset $\mathscr{V}_\eta \subset \mathscr{V}$ such that $\boldsymbol{\nu}_N(\mathscr{V}_\eta) \geq 1 - \eta$. Thus, $\{\boldsymbol{\nu}_N,\ N \in \mathbb{N}\}$ is tight in the same space. Prokhorov's Theorem (see Lemma 9.3.1) therefore implies that $\boldsymbol{\nu}_N$ is also relatively weakly compact. This means we have a weakly convergent subsequence with limit $\boldsymbol{\nu}$. Now we use Skorokhod's representation theorem (see Lemma 9.3.2) to infer the existence of a probability space $(\underline{\Omega}, \underline{\mathcal{F}}, \underline{\mathbb{P}})$, a sequence $(\underline{\mathbf{v}}_N, \underline{\mathbf{W}}^N, \underline{\mathbf{v}}_{0,N}, \underline{\mathbf{f}}_N)$ and $(\underline{\mathbf{v}}, \underline{\mathbf{W}}, \underline{\mathbf{v}}_0, \underline{\mathbf{f}})$ on $(\underline{\Omega}, \underline{\mathcal{F}}, \underline{\mathbb{P}})$, both with values in $\mathscr{V}$, such that the following holds.

- The laws of $(\underline{\mathbf{v}}_N, \underline{\mathbf{W}}^N, \underline{\mathbf{v}}_{0,N}, \underline{\mathbf{f}}_N)$ and $(\underline{\mathbf{v}}, \underline{\mathbf{W}}, \underline{\mathbf{v}}_0, \underline{\mathbf{f}})$ under $\underline{\mathbb{P}}$ coincide with $\boldsymbol{\nu}_N$ and $\boldsymbol{\nu}$.
- We have the convergences

$$\begin{aligned} \underline{\mathbf{v}}_N &\longrightarrow \underline{\mathbf{v}} \quad \text{in} \quad L^r(0,T;L^r(G)), \\ \underline{\mathbf{W}}^N &\longrightarrow \underline{\mathbf{W}} \quad \text{in} \quad C([0,T],U_0), \\ \underline{\mathbf{v}}_{0,N} &\longrightarrow \underline{\mathbf{v}}_0 \quad \text{in} \quad L^2(G), \\ \underline{\mathbf{f}}_N &\longrightarrow \underline{\mathbf{f}} \quad \text{in} \quad L^2(0,T;L^2(G)), \end{aligned}$$

  $\underline{\mathbb{P}}$-a.s.
- The convergences in (10.2.18) and (10.2.19) remain valid for the corresponding functions defined on $(\underline{\Omega}, \underline{\mathcal{F}}, \underline{\mathbb{P}})$. Moreover, we have for all $\alpha < \infty$

$$\int_{\underline{\Omega}} \left(\sup_{[0,T]} \|\underline{\mathbf{W}}^N(t)\|_{U_0}^\alpha \right) d\underline{\mathbb{P}} = \int_{\Omega} \left(\sup_{[0,T]} \|\mathbf{W}(t)\|_{U_0}^\alpha \right) d\mathbb{P}.$$

After choosing a subsequence we obtain by Vitali's convergence Theorem

$$\underline{\mathbf{W}}^N \longrightarrow \underline{\mathbf{W}} \quad \text{in} \quad L^2(\underline{\Omega}, \underline{\mathcal{F}}, \underline{\mathbb{P}}; C([0,T],U_0)), \tag{10.2.24}$$

$$\underline{\mathbf{v}}_N \longrightarrow \underline{\mathbf{v}} \quad \text{in} \quad L^r(\underline{\Omega} \times Q; \underline{\mathbb{P}} \otimes \mathcal{L}^{d+1}), \tag{10.2.25}$$

$$\underline{\mathbf{v}}_{0,N} \longrightarrow \underline{\mathbf{v}}_0 \quad \text{in} \quad L^2(\underline{\Omega} \times G, \underline{\mathbb{P}} \otimes \mathcal{L}^{d+1}), \tag{10.2.26}$$

$$\underline{\mathbf{f}}_N \longrightarrow \underline{\mathbf{f}} \quad \text{in} \quad L^2(\underline{\Omega} \times Q, \underline{\mathbb{P}} \otimes \mathcal{L}^{d+1}), \tag{10.2.27}$$

for all $r < q$. Now we introduce the filtration on the new probability space which ensures the correct measurabilities of the new variables. We denote by $\mathbf{r}_t$ the operator of restriction to the interval $[0,t]$ acting on various path spaces. In particular, if X stands for one of the path spaces $L^r(0,T;L^r(G))$, $L^2(Q)$ or $C([0,T],U_0)$ and $t\in[0,T]$, we define

$$\mathbf{r}_t : X \to X|_{[0,t]}, \quad f \mapsto f|_{[0,t]}. \tag{10.2.28}$$

Clearly, $\mathbf{r}_t$ is a continuous mapping. Let $(\underline{\mathcal{F}}_t)_{t\geq 0}$ be the $\underline{\mathbb{P}}$-augmented canonical filtration of the process $(\underline{\mathbf{v}}, \underline{\mathbf{f}}, \underline{\mathbf{W}})$, respectively, that is

$$\underline{\mathcal{F}}_t = \sigma\Big(\sigma\big(\mathbf{r}_t\underline{\mathbf{v}}, \mathbf{r}_t\underline{\mathbf{f}}, \mathbf{r}_t\underline{\mathbf{W}}\big) \cup \big\{N\in\underline{\mathcal{F}};\ \underline{\mathbb{P}}(N)=0\big\}\Big), \quad t\in[0,T].$$

Step 2: the system on the new probability space

Now we are going to show that the approximated equations also hold on the new probability space. We use a general and elementary method that was recently introduced in [38] and already generalized to different settings (see for instance [36,94]). The keystone is to identify not only the quadratic variation of the corresponding martingale but also its cross variation with the limit Wiener process obtained through compactness. First we notice that $\underline{\mathbf{W}}^N$ has the same law as $\mathbf{W}$. As a consequence, there exists a collection of mutually independent real-valued $(\underline{\mathcal{F}}_t)_{t\geq 0}$-Wiener processes $(\underline{\beta}_k^N)_k$ such that $\underline{\mathbf{W}}^N = \sum_k \underline{\beta}_k^N e_k$. In particular, there exists a collection of mutually independent real-valued $(\underline{\mathcal{F}}_t)_{t\geq 0}$-Wiener processes $(\underline{\beta}_k)_{k\geq 1}$ such that $\underline{\mathbf{W}} = \sum_k \underline{\beta}_k e_k$. We abbreviate $\underline{\mathbf{W}}^{N,N} := \sum_{k=1}^N \mathbf{e}_k \underline{\beta}_k^N$. Let us now define for all $t\in[0,T]$ and $\boldsymbol{\varphi}\in C^\infty_{0,\mathrm{div}}(G)$ the functionals

$$\begin{aligned}
\mathfrak{M}(\mathbf{v}_N,\mathbf{v}_0,\mathbf{f})_t &= \int_G \mathbf{v}_N(t)\cdot\boldsymbol{\varphi}\,\mathrm{d}x - \int_G \mathbf{v}_0\cdot\boldsymbol{\varphi}\,\mathrm{d}x \\
&\quad + \int_0^t\int_G \mathbf{v}_N\otimes\mathbf{v}_N : \nabla\mathcal{P}_N\boldsymbol{\varphi}\,\mathrm{d}x\,\mathrm{d}\sigma \\
&\quad + \int_0^t\int_G \mathbf{S}(\boldsymbol{\varepsilon}(\mathbf{v}_N)) : \boldsymbol{\varepsilon}(\mathcal{P}_N\boldsymbol{\varphi})\,\mathrm{d}x\,\mathrm{d}\sigma + \int_0^t\int_G \mathbf{f}\cdot\mathcal{P}_N\boldsymbol{\varphi}\,\mathrm{d}x\,\mathrm{d}\sigma, \\
\mathfrak{N}(\mathbf{v}_N)_t &= \sum_{k=1}^N \int_0^t\left(\int_G g_k(\mathbf{v})\cdot\mathcal{P}_N\boldsymbol{\varphi}\,\mathrm{d}x\right)^2\mathrm{d}\sigma, \\
\mathfrak{N}_k(\mathbf{v}_N)_t &= \int_0^t\int_G g_k(\mathbf{v})\cdot\mathcal{P}_N\boldsymbol{\varphi}\,\mathrm{d}x\,\mathrm{d}\sigma,
\end{aligned}$$

let $\mathfrak{M}(\mathbf{v}_N,\mathbf{v}_0,\mathbf{f})_{s,t}$ denote the increment $\mathfrak{M}(\mathbf{v}_N,\mathbf{v}_0,\mathbf{f})_t - \mathfrak{M}(\mathbf{v}_N,\mathbf{v}_0,\mathbf{f})_s$ and similarly for $\mathfrak{N}(\mathbf{v}_N)_{s,t}$ and $\mathfrak{N}_k(\mathbf{v}_N)_{s,t}$. Note that the proof will be complete once we show that the process $\mathfrak{M}(\underline{\mathbf{v}}_N)$ is an $(\underline{\mathcal{F}}_t)_{t\geq 0}$-martingale and its quadratic and cross variations satisfy, respectively,

$$\langle\langle\mathfrak{M}(\underline{\mathbf{v}}_N,\underline{\mathbf{v}}_0,\underline{\mathbf{f}})\rangle\rangle = \mathfrak{N}(\underline{\mathbf{v}}^N), \quad \langle\langle\mathfrak{M}(\underline{\mathbf{v}}_N,\underline{\mathbf{v}}_0,\underline{\mathbf{f}}), \underline{\beta}_k\rangle\rangle = \mathfrak{N}_k(\underline{\mathbf{v}}_N). \tag{10.2.29}$$

Indeed, in that case we have

$$\left\langle\!\!\left\langle \mathfrak{M}(\underline{\mathbf{v}}_N, \underline{\mathbf{v}}_0, \underline{\mathbf{f}}) - \int_0^{\cdot} \int_G \Phi(\underline{\mathbf{v}}_N)\, \mathrm{d}\underline{\mathbf{W}}^{N,N} \cdot \mathcal{P}_N \boldsymbol{\varphi}\, \mathrm{d}x \right\rangle\!\!\right\rangle = 0 \qquad (10.2.30)$$

which implies the desired equation on the new probability space. Let us verify (10.2.29). To this end, we claim that with the above uniform estimates in hand, the mappings

$$(\mathbf{v}_N, \mathbf{v}_0, \mathbf{f}) \mapsto \mathfrak{M}(\mathbf{v}_N, \mathbf{v}_0, \mathbf{f})_t, \qquad \mathbf{v}_N \mapsto \mathfrak{N}(\mathbf{v}_N)_t, \qquad \mathbf{v}_N \mapsto \mathfrak{N}_k(\mathbf{v}_N)_t$$

are well-defined and measurable on a subspace of the path space where the joint law of $(\underline{\mathbf{v}}_N, \underline{\mathbf{v}}_0, \underline{\mathbf{f}})$ is supported, i.e. the uniform estimates from Theorem 10.2.43 hold true. Indeed, in the case of $\mathfrak{N}(\rho, \mathbf{q})_t$ we have by (10.0.4) and the continuity of $\mathcal{P}_N$ in $L^2(G)$

$$\begin{aligned}\sum_{k=1}^{N} \int_0^t \left(\int_G g_k(\mathbf{v}_N) \cdot \mathcal{P}_N \varphi \,\mathrm{d}x \right)^2 \mathrm{d}\sigma &\leq c(\boldsymbol{\varphi}) \sum_{k=1}^{\infty} \int_0^t \int_G |g_k(\mathbf{v}_N)|^2 \,\mathrm{d}x \,\mathrm{d}\sigma \\ &\leq c(\boldsymbol{\varphi}) \left(1 + \int_Q |\mathbf{v}_N|^2 \,\mathrm{d}x \,\mathrm{d}t \right)\end{aligned}$$

which is finite. $\mathfrak{M}(\rho, \mathbf{v}, \mathbf{q})$ and $\mathfrak{N}_k(\rho, \mathbf{v})_t$ can be handled similarly and therefore, the following random variables have the same laws

$$\begin{aligned}\mathfrak{M}(\mathbf{v}_N, \mathbf{v}_0, \mathbf{f}) &\sim \mathfrak{M}(\underline{\mathbf{v}}_N, \underline{\mathbf{v}}_0, \underline{\mathbf{f}}), \\ \mathfrak{N}(\mathbf{v}_N) &\sim \mathfrak{N}(\underline{\mathbf{v}}_N), \\ \mathfrak{N}_k(\mathbf{v}_N) &\sim \mathfrak{N}_k(\underline{\mathbf{v}}_N).\end{aligned}$$

Let us now fix times $s, t \in [0, T]$ such that $s < t$ and let

$$h : \mathcal{V}\big|_{[0,s]} \to [0, 1]$$

be a continuous function. Since

$$\mathfrak{M}(\mathbf{v}_N, \mathbf{v}_0, \mathbf{f})_t = \int_0^t \int_G \Phi(\mathbf{v}_N)\, \mathrm{d}\mathbf{W}_\sigma^N \cdot \mathcal{P}_N \boldsymbol{\varphi}\, \mathrm{d}x = \sum_{k=1}^{N} \int_0^t \int_G g_k(\mathbf{v}_N) \cdot \mathcal{P}_N \boldsymbol{\varphi}\, \mathrm{d}x\, \mathrm{d}\beta_k$$

is a square integrable $(\mathcal{F}_t)_{t \geq 0}$-martingale, we infer that

$$\left[\mathfrak{M}(\mathbf{v}_N, \mathbf{v}_0, \mathbf{f})\right]^2 - \mathfrak{N}(\mathbf{v}^N), \qquad \mathfrak{M}(\mathbf{v}_N)\beta_k - \mathfrak{N}_k(\mathbf{v}_N),$$

are $(\mathcal{F}_t)_{t \geq 0}$-martingales. Let $\mathbf{r}_s$ be the restriction of a function to the interval $[0, s]$. Then it follows from the equality of laws that

$$\begin{aligned}&\underline{\mathbb{E}}\big[h\big(\mathbf{r}_s \underline{\mathbf{v}}_N, \mathbf{r}_s \underline{\mathbf{W}}^N, \mathbf{r}_s \underline{\mathbf{f}}, \underline{\mathbf{v}}_0\big) \mathfrak{M}(\underline{\mathbf{v}}_N, \underline{\mathbf{v}}_0, \underline{\mathbf{f}})_{s,t} \big] \\ &\quad = \mathbb{E}\big[h\big(\mathbf{r}_s \mathbf{v}_N, \mathbf{r}_s \mathbf{W}^N, \mathbf{r}_s \mathbf{f}, \mathbf{v}_0\big) \mathfrak{M}(\mathbf{v}_N, \mathbf{v}_0, \mathbf{f})_{s,t} \big] = 0,\end{aligned}$$

$$\underline{\mathbb{E}}\Big[h\big(\mathbf{r}_s\underline{\mathbf{v}}_N,\mathbf{r}_s\underline{\mathbf{W}}^N,\mathbf{r}_s\underline{\mathbf{f}},\underline{\mathbf{v}}_0\big)\Big([\mathfrak{M}(\underline{\mathbf{v}}_N,\underline{\mathbf{v}}_0,\underline{\mathbf{f}})^2]_{s,t}-\mathfrak{N}(\underline{\mathbf{v}}_N)_{s,t}\Big)\Big]$$
$$=\mathbb{E}\Big[h\big(\mathbf{r}_s\mathbf{v}^N,\mathbf{r}_s\mathbf{W}^N,\mathbf{r}_s\mathbf{f},\mathbf{v}_0\big)\Big([\mathfrak{M}(\mathbf{v}_N,\mathbf{v}_0,\mathbf{f})^2]_{s,t}-\mathfrak{N}(\mathbf{v}_N)_{s,t}\Big)\Big]=0,$$
$$\underline{\mathbb{E}}\Big[h\big(\mathbf{r}_s\underline{\mathbf{v}}_N,\mathbf{r}_s\underline{\mathbf{W}}^N,\mathbf{r}_s\underline{\mathbf{f}},\underline{\mathbf{v}}_0\big)\Big([\mathfrak{M}(\underline{\mathbf{v}}_N,\underline{\mathbf{v}}_0,\underline{\mathbf{f}})\underline{\beta}_k^N]_{s,t}-\mathfrak{N}_k(\underline{\mathbf{v}}_N)_{s,t}\Big)\Big]$$
$$=\mathbb{E}\Big[h\big(\mathbf{r}_s\mathbf{v}_N,\mathbf{r}_s\mathbf{W}^N,\mathbf{r}_s\mathbf{f},\mathbf{v}_0\big)\Big([\mathfrak{M}(\mathbf{v}_N,\mathbf{v}_0,\mathbf{f})\beta_k]_{s,t}-\mathfrak{N}_k(\mathbf{v}_N)_{s,t}\Big)\Big]=0.$$

So we have shown (10.2.29) and hence (10.2.30). This means on the new probability space $(\underline{\Omega},\underline{\mathcal{F}},\underline{\mathbb{P}})$ we have the equations ($k=1,\dots,N$)

$$\begin{aligned}&\int_G \mathrm{d}\underline{\mathbf{v}}_N\cdot\mathbf{w}_k\,\mathrm{d}x+\int_G \mathbf{S}(\boldsymbol{\varepsilon}(\underline{\mathbf{v}}_N)):\boldsymbol{\varepsilon}(\mathbf{w}_k)\,\mathrm{d}x\,\mathrm{d}t+\alpha\int_G|\underline{\mathbf{v}}_N|^{q-2}\underline{\mathbf{v}}_N\cdot\mathbf{w}_k\,\mathrm{d}x\,\mathrm{d}t\\&\quad=\int_G\underline{\mathbf{v}}_N\otimes\underline{\mathbf{v}}_N:\nabla\mathbf{w}_k\,\mathrm{d}x\,\mathrm{d}t+\int_G\underline{\mathbf{f}}\cdot\mathbf{w}_k\,\mathrm{d}x\,\mathrm{d}t+\int_G\Phi(\underline{\mathbf{v}}_N)\,\mathrm{d}\underline{\mathbf{W}}_\sigma^{N,N}\cdot\mathbf{w}_k\,\mathrm{d}x,\\&\quad\underline{\mathbf{v}}_N(0)=\mathcal{P}_N\underline{\mathbf{v}}_0,\end{aligned}\tag{10.2.31}$$

and the convergences

$$\begin{aligned}\underline{\mathbf{v}}_N&\rightharpoonup\underline{\mathbf{v}}\quad\text{in}\quad L^p(\underline{\Omega},\underline{\mathcal{F}},\underline{\mathbb{P}};L^p(0,T;W_0^{1,p}(G))),\\\underline{\mathbf{v}}_N&\rightharpoonup\underline{\mathbf{v}}\quad\text{in}\quad L^q(\underline{\Omega},\underline{\mathcal{F}},\underline{\mathbb{P}};L^q(Q)),\\\mathbf{s}(\underline{\mathbf{v}}_N)&\rightharpoonup\mathbf{s}(\underline{\mathbf{v}})\quad\text{in}\quad L^{q'}(\underline{\Omega},\underline{\mathcal{F}},\underline{\mathbb{P}};L^{q'}(Q)),\\\mathbf{S}(\boldsymbol{\varepsilon}(\underline{\mathbf{v}}_N))&\rightharpoonup\tilde{\underline{\mathbf{S}}}\quad\text{in}\quad L^{p'}(\underline{\Omega},\underline{\mathcal{F}},\underline{\mathbb{P}};L^{p'}(Q)),\\\mathbf{S}(\boldsymbol{\varepsilon}(\underline{\mathbf{v}}_N))&\rightharpoonup\tilde{\tilde{\underline{\mathbf{S}}}}\quad\text{in}\quad L^{p'}(\underline{\Omega},\underline{\mathcal{F}},\underline{\mathbb{P}};L^{p'}(0,T;W_0^{-1,p'}(G))),\\\underline{\mathbf{v}}_N\otimes\underline{\mathbf{v}}_N&\rightharpoonup\underline{\mathbf{v}}\otimes\underline{\mathbf{v}}\quad\text{in}\quad L^{\frac{q}{2}}(\underline{\Omega},\underline{\mathcal{F}},\underline{\mathbb{P}};L^{\frac{q}{2}}(Q)),\\\Phi(\underline{\mathbf{v}}_N)&\rightharpoonup\Phi(\underline{\mathbf{v}})\quad\text{in}\quad L^2(\underline{\Omega},\underline{\mathcal{F}},\underline{\mathbb{P}};L^2(0,T;L_2(U,L^2(G)))).\end{aligned}\tag{10.2.32}$$

We obtain from (10.2.24)–(10.2.32) the limit equation

$$\begin{aligned}&\int_G\underline{\mathbf{v}}(t)\cdot\boldsymbol{\varphi}\,\mathrm{d}x+\int_0^t\int_G\tilde{\underline{\mathbf{S}}}:\nabla\boldsymbol{\varphi}\,\mathrm{d}x\,\mathrm{d}\sigma+\int_0^t\int_G\mathbf{s}(\underline{\mathbf{v}})\cdot\boldsymbol{\varphi}\,\mathrm{d}x\,\mathrm{d}\sigma\\&\quad+\int_0^t\int_G\underline{\mathbf{v}}\otimes\underline{\mathbf{v}}:\nabla\boldsymbol{\varphi}\,\mathrm{d}x\,\mathrm{d}\sigma=\int_0^t\int_G\underline{\mathbf{f}}\cdot\boldsymbol{\varphi}\,\mathrm{d}x\,\mathrm{d}\sigma+\int_G\int_0^t\Phi(\underline{\mathbf{v}})\,\mathrm{d}\underline{\mathbf{W}}_\sigma\cdot\boldsymbol{\varphi}\,\mathrm{d}x\end{aligned}\tag{10.2.33}$$

for all $\boldsymbol{\varphi}\in C^\infty_{0,\mathrm{div}}(G)$. The limit in the stochastic term needs some explanations. We have the convergences

$$\begin{aligned}\underline{\mathbf{W}}^N&\longrightarrow\underline{\mathbf{W}}\quad\text{in}\quad C([0,T],U_0),\\\Phi(\underline{\mathbf{v}}_N)&\longrightarrow\Phi(\underline{\mathbf{v}})\quad\text{in}\quad L^2(0,T;L_2(U,L^2(G))),\end{aligned}$$

in probability. For the second one we use (10.0.4) and (10.2.25). These convergences imply

$$\int_0^t \Phi(\underline{\mathbf{v}}_N)\,\mathrm{d}\underline{\mathbf{W}}_\sigma^N \longrightarrow \int_0^t \Phi(\underline{\mathbf{v}})\,\mathrm{d}\underline{\mathbf{W}}_\sigma \quad \text{in} \quad L^2(0,T;L^2(G))$$

in probability by Lemma 9.1.2. So we can pass to the limit in the stochastic integral.

Step 3: monotone operator theory

Now, it remains to show

$$\tilde{\underline{\mathbf{S}}} = \mathbf{S}(\boldsymbol{\varepsilon}(\underline{\mathbf{v}})). \tag{10.2.34}$$

We will apply monotone operator theory to verify (10.2.34). On account of $\int_G \underline{\mathbf{v}} \otimes \underline{\mathbf{v}} : \nabla \underline{\mathbf{v}}\,\mathrm{d}x = 0$, equation (10.2.33) implies

$$\begin{aligned}
\frac{1}{2}\|\underline{\mathbf{v}}(t)\|_{L^2(G)}^2 &= \frac{1}{2}\|\underline{\mathbf{v}}_0\|_{L^2(G)}^2 - \int_0^t\int_G \tilde{\underline{\mathbf{S}}} : \boldsymbol{\varepsilon}(\underline{\mathbf{v}})\,\mathrm{d}x\,\mathrm{d}\sigma - \int_0^t\int_G \mathbf{s}(\underline{\mathbf{v}})\cdot\underline{\mathbf{v}}\,\mathrm{d}x\,\mathrm{d}\sigma \\
&\quad + \int_0^t\int_G \underline{\mathbf{f}}\cdot\underline{\mathbf{v}}\,\mathrm{d}x\,\mathrm{d}\sigma + \int_G\int_0^t \underline{\mathbf{v}}\cdot\Phi(\underline{\mathbf{v}})\,\mathrm{d}\underline{\mathbf{W}}_\sigma\,\mathrm{d}x \\
&\quad + \frac{1}{2}\int_G\int_0^t \mathrm{d}\Big\langle\!\Big\langle \int_0^{\cdot} \Phi(\underline{\mathbf{v}})\,\mathrm{d}\underline{\mathbf{W}}\Big\rangle\!\Big\rangle_\sigma\,\mathrm{d}x.
\end{aligned}$$

Here, we applied Itô's formula to $f(\mathbf{w}) = \frac{1}{2}\|\mathbf{w}\|_{L^2(G)}^2$, see Lemma C.0.1. Subtracting this from the formula for $\|\underline{\mathbf{v}}^N(t)\|_{L^2(G)}^2$ and applying expectation shows

$$\begin{aligned}
&\mathbb{E}\Big[\int_Q \big(\mathbf{S}(\nabla\underline{\mathbf{v}}_N) - \mathbf{S}(\nabla\underline{\mathbf{v}})\big) : \nabla\big(\underline{\mathbf{v}}_N - \underline{\mathbf{v}}\big)\,\mathrm{d}x\,\mathrm{d}\sigma\Big] \\
&\quad + \mathbb{E}\Big[\int_Q \big(\mathbf{s}(\underline{\mathbf{v}}_N) - \mathbf{s}(\underline{\mathbf{v}})\big) : \big(\underline{\mathbf{v}}_N - \underline{\mathbf{v}}\big)\,\mathrm{d}x\,\mathrm{d}\sigma\Big] \\
&= \frac{1}{2}\mathbb{E}\Big[-\int_G |\underline{\mathbf{v}}_N(T)|^2\,\mathrm{d}x + \int_G |\underline{\mathbf{v}}(T)|^2\,\mathrm{d}x + \int_G |\mathcal{P}_N\underline{\mathbf{v}}_0|^2\,\mathrm{d}x - \int_G |\underline{\mathbf{v}}_0|^2\,\mathrm{d}x\Big] \\
&\quad + \mathbb{E}\Big[\int_Q \big(\tilde{\underline{\mathbf{S}}} - \mathbf{S}(\boldsymbol{\varepsilon}(\underline{\mathbf{v}}_N))\big) : \boldsymbol{\varepsilon}(\underline{\mathbf{v}})\,\mathrm{d}x\,\mathrm{d}\sigma - \int_Q \mathbf{S}(\boldsymbol{\varepsilon}(\underline{\mathbf{v}})) : \boldsymbol{\varepsilon}\big(\underline{\mathbf{v}}_N - \underline{\mathbf{v}}\big)\,\mathrm{d}x\,\mathrm{d}\sigma\Big] \\
&\quad + \mathbb{E}\Big[\int_Q \big(\mathbf{s}(\underline{\mathbf{v}}) - \mathbf{s}(\underline{\mathbf{v}}_N)\big) : \underline{\mathbf{v}}\,\mathrm{d}x\,\mathrm{d}\sigma - \int_Q \mathbf{s}(\underline{\mathbf{v}})\cdot\big(\underline{\mathbf{v}}_N - \underline{\mathbf{v}}\big)\,\mathrm{d}x\,\mathrm{d}\sigma\Big] \\
&\quad + \mathbb{E}\Big[\int_Q \mathbf{f}\cdot\big(\underline{\mathbf{v}}_N - \underline{\mathbf{v}}\big)\,\mathrm{d}x\,\mathrm{d}\sigma + \frac{1}{2}\int_G\int_0^T \mathrm{d}\Big\langle\!\Big\langle \int_0^{\cdot} \Phi(\underline{\mathbf{v}}_N)\,\mathrm{d}\underline{\mathbf{W}}^{N,N}\Big\rangle\!\Big\rangle_\sigma\,\mathrm{d}x\Big] \\
&\quad - \frac{1}{2}\mathbb{E}\Big[\int_G\int_0^T \mathrm{d}\Big\langle\!\Big\langle \int_0^{\cdot} \Phi(\underline{\mathbf{v}})\,\mathrm{d}\underline{\mathbf{W}}\Big\rangle\!\Big\rangle_\sigma\,\mathrm{d}x\Big].
\end{aligned}$$

Letting $N \to \infty$ shows using (10.2.32) and monotonicity of $\mathbf{S}$

$$\lim_N \left(\mathbb{E}\left[\int_Q (\mathbf{S}(\boldsymbol{\varepsilon}(\underline{\mathbf{v}}_N)) - \mathbf{S}(\boldsymbol{\varepsilon}(\underline{\mathbf{v}}))) : \boldsymbol{\varepsilon}(\underline{\mathbf{v}}_N - \underline{\mathbf{v}})\,\mathrm{d}x\,\mathrm{d}t\right]\right.$$
$$\leq \frac{1}{2}\lim_N \mathbb{E}\left[\int_G\int_0^T \mathrm{d}\left(\left\langle\!\left\langle \int_0^\cdot \Phi(\underline{\mathbf{v}}_N)\,\mathrm{d}\underline{\mathbf{W}}^{N,N}\right\rangle\!\right\rangle_\sigma - \left\langle\!\left\langle \int_0^\cdot \Phi(\underline{\mathbf{v}})\,\mathrm{d}\underline{\mathbf{W}}\right\rangle\!\right\rangle_\sigma\right)\mathrm{d}x\right].$$

Here we used $\liminf_N \mathbb{E}\big[\int_G \big(|\underline{\mathbf{v}}_N(T)|^2 - |\underline{\mathbf{v}}(T)|^2\big)\,\mathrm{d}x\,\mathrm{d}\sigma\big] \geq 0$ which follows by lower semi-continuity and weak convergence of $\underline{\mathbf{v}}_N(T)$. On account of (10.2.24) and (10.2.25) together with (10.0.4) we obtain for the last integral

$$\mathbb{E}\left[\int_G\int_0^T \mathrm{d}\left\langle\!\left\langle \int_0^\cdot \Phi(\underline{\mathbf{v}}_N)\,\mathrm{d}\underline{\mathbf{W}}^{N,N}\right\rangle\!\right\rangle_\sigma \mathrm{d}x\right] \longrightarrow \mathbb{E}\left[\int_G\int_0^T \mathrm{d}\left\langle\!\left\langle \int_0^\cdot \Phi(\underline{\mathbf{v}})\,\mathrm{d}\underline{\mathbf{W}}\right\rangle\!\right\rangle_\sigma \mathrm{d}x\right]$$

as $N \to \infty$. This finally implies

$$\lim_{N\to\infty} \mathbb{E}\left[\int_Q \big(\mathbf{S}(\boldsymbol{\varepsilon}(\underline{\mathbf{v}}_N)) - \mathbf{S}(\boldsymbol{\varepsilon}(\underline{\mathbf{v}}))\big) : \boldsymbol{\varepsilon}\big(\underline{\mathbf{v}}_N - \underline{\mathbf{v}}\big)\,\mathrm{d}x\,\mathrm{d}t\right] = 0.$$

As a consequence of the monotonicity of $\mathbf{S}$ and $\mathbf{s}$ we have established

$$\boldsymbol{\varepsilon}(\underline{\mathbf{v}}_N) \longrightarrow \boldsymbol{\varepsilon}(\underline{\mathbf{v}}) \quad \underline{\mathbb{P}} \otimes \mathcal{L}^{d+1}\text{-a.e.}$$

This shows (10.2.34) and the proof of Theorem 10.2.44 is hereby complete. □

Remark 10.2.25. According to the remarks in [96] (beginning of the proof of Thm. 2.7 on p. 9) it is possible to choose the new probability space obtained in the proof of Theorem 10.2.44 as

$$(\underline{\Omega}, \underline{\mathcal{F}}, \underline{\mathbb{P}}) = ([0,1); \mathcal{B}[0,1); \mathcal{L}^1);$$

especially it does not depend on the choice of α.

Corollary 10.2.1. *Let the assumptions of Theorem 10.2.44 be satisfied and in addition*

$$\int_{L^2_{\mathrm{div}}(G)} \|\mathbf{u}\|^\beta_{L^2(G)}\,\mathrm{d}\Lambda_0(\mathbf{u}) < \infty, \qquad \int_{L^2(Q)} \|\mathbf{g}\|^\beta_{L^2(Q)}\,\mathrm{d}\Lambda_{\mathbf{f}}(\mathbf{g}) < \infty,$$

for some $\beta \geq 2$. Then there is a weak martingale solution to (10.2.11) *such that*

$$\mathbb{E}\left[\sup_{t\in(0,T)}\int_G |\underline{\mathbf{v}}(t)|^2\,\mathrm{d}x + \int_Q |\nabla\underline{\mathbf{v}}|^p\,\mathrm{d}x\,\mathrm{d}t + \alpha\int_Q |\underline{\mathbf{v}}|^q\,\mathrm{d}x\,\mathrm{d}t\right]^{\frac{\beta}{2}}$$
$$\leq c\,\mathbb{E}\left[1 + \int_{L^2_{\mathrm{div}}(G)} \|\mathbf{u}\|^2_{L^2(G)}\,\mathrm{d}\Lambda_0(\mathbf{u}) + \int_{L^2(Q)} \|\mathbf{g}\|^2_{L^2(Q)}\,\mathrm{d}\Lambda_{\mathbf{f}}(\mathbf{g})\,\mathrm{d}x\,\mathrm{d}t\right]^{\frac{\beta}{2}},$$

where c is independent of α.

Proof. Taking the supremum with respect to time and the $\frac{\beta}{2}$-th power of (10.2.15) implies

$$\begin{aligned}
&\frac{1}{2}\mathbb{E}\bigg[\sup_{t\in(0,T)}\int_G |\mathbf{v}_N(t)|^2\,\mathrm{d}x\bigg]^{\frac{\beta}{2}} + \mathbb{E}\bigg[\int_0^T\int_G |\nabla\mathbf{v}_N|^p + \alpha|\mathbf{v}_N|^q\,\mathrm{d}x\,\mathrm{d}\sigma\bigg]^{\frac{\beta}{2}}\\
&\leq c\,\mathbb{E}\bigg[1+\int_G |\mathbf{v}_0|^2\,\mathrm{d}x + \int_0^T\int_G |\mathbf{f}||\mathbf{v}_N|\,\mathrm{d}x\,\mathrm{d}\sigma\bigg]^{\frac{\beta}{2}}\\
&+c\,\mathbb{E}\bigg[\sup_{(0,T)}\bigg|\int_G\int_0^t \mathbf{v}_N\cdot\Phi(\mathbf{v}_N)\,\mathrm{d}\mathbf{W}_\sigma^N\,\mathrm{d}x\bigg|\bigg]^{\frac{\beta}{2}}\\
&+c\,\mathbb{E}\bigg[\int_G\int_0^T \mathrm{d}\Big\langle\!\Big\langle\int_0^\cdot \Phi(\mathbf{v}_N)\,\mathrm{d}\mathbf{W}^N\Big\rangle\!\Big\rangle_\sigma\,\mathrm{d}x\bigg]^{\frac{\beta}{2}}.
\end{aligned}$$

Obviously, the following holds

$$\begin{aligned}
&\mathbb{E}\bigg[\int_0^T\int_G |\mathbf{f}||\mathbf{v}_N|\,\mathrm{d}x\,\mathrm{d}\sigma\bigg]^{\frac{\beta}{2}}\\
&\quad\leq c\,\mathbb{E}\bigg[\int_0^T\Big(\int_G|\mathbf{v}_N|^2\,\mathrm{d}x\Big)^{\frac{\beta}{2}}\mathrm{d}\sigma\bigg] + c\,\mathbb{E}\bigg[\int_Q|\mathbf{f}|^2\,\mathrm{d}x\,\mathrm{d}t\bigg]^{\frac{\beta}{2}}.
\end{aligned}$$

Moreover, we have as a consequence of Burkholder–Davis–Gundi inequality, (10.0.4) and Young's inequality (similarly to the proof of Theorem 10.2.43)

$$\begin{aligned}
&\mathbb{E}\bigg[\sup_{t\in(0,T)}|\mathcal{K}(t)|\bigg]^{\frac{\beta}{2}}\\
&:=\mathbb{E}\bigg[\sup_{t\in(0,T)}\bigg|\int_0^t\int_G \mathbf{v}_N\cdot\Phi(\mathbf{v}_N)\,\mathrm{d}x\,\mathrm{d}\mathbf{W}_\sigma^N\bigg|\bigg]^{\frac{\beta}{2}}\\
&=\mathbb{E}\bigg[\sup_{t\in(0,T)}\bigg|\int_0^t\sum_i\int_G \mathbf{v}_N\cdot g_i(\mathbf{v}_N)\,\mathrm{d}x\,\mathrm{d}\beta_i(\sigma)\bigg|\bigg]^{\frac{\beta}{2}}\\
&\leq c\,\mathbb{E}\bigg[\int_0^T\sum_i\Big(\int_G \mathbf{v}_N\cdot g_i(\mathbf{v}_N)\,\mathrm{d}x\Big)^2\mathrm{d}t\bigg]^{\frac{\beta}{4}}\\
&\leq c\,\mathbb{E}\bigg[\bigg(\int_0^T\Big(\sum_{i=1}^N\int_G|\mathbf{v}_N|^2\,\mathrm{d}x\int_G|g_i(\mathbf{v}_N)|^2\,\mathrm{d}x\Big)\,\mathrm{d}t\bigg]^{\frac{\beta}{4}}\\
&\leq c\,\mathbb{E}\bigg[1+\int_0^T\Big(\int_G|\mathbf{v}_N|^2\,\mathrm{d}x\Big)^2\mathrm{d}t\bigg]^{\frac{\beta}{2}}\\
&\leq \delta\,\mathbb{E}\bigg[\sup_{t\in(0,T)}\int_G|\mathbf{v}_N|^2\,\mathrm{d}x\bigg]^{\frac{\beta}{2}} + c(\delta)\,\mathbb{E}\bigg[1+\int_0^T\int_G|\mathbf{v}_N|^2\,\mathrm{d}x\,\mathrm{d}t\bigg]^{\frac{\beta}{2}},
\end{aligned}$$

for arbitrary $\delta > 0$. So we have shown (choosing δ small enough)

$$\mathbb{E}\bigg[\sup_{t\in(0,T)}\bigg(\int_G |\mathbf{v}_N(t)|^2\,\mathrm{d}x\bigg)^{\frac{\beta}{2}}\bigg]+\mathbb{E}\bigg[\int_Q |\nabla\mathbf{v}_N|^p+\alpha|\mathbf{v}_N|^q\,\mathrm{d}x\,\mathrm{d}t\bigg]^{\frac{\beta}{2}}$$
$$\leq c\,\mathbb{E}\bigg[1+\int_G|\mathbf{v}_0|^2\,\mathrm{d}x+\int_Q|\mathbf{f}|^2\,\mathrm{d}x\,\mathrm{d}t+c\,\mathbb{E}\bigg[\int_0^T\bigg(\int_G|\mathbf{v}_N|^2\,\mathrm{d}x\bigg)^{\frac{\beta}{2}}\mathrm{d}\sigma\bigg].$$

Gronwall's Lemma implies

$$\begin{aligned}&\mathbb{E}\bigg[\sup_{t\in(0,T)}\int_G |\mathbf{v}_N(t)|^2\,\mathrm{d}x\bigg]^{\frac{\beta}{2}}+\mathbb{E}\bigg[\int_Q |\nabla\mathbf{v}_N|^p+\alpha|\mathbf{v}_N|^q\,\mathrm{d}x\,\mathrm{d}t\bigg]^{\frac{\beta}{2}}\\&\quad\leq c\,\mathbb{E}\bigg[1+\int_G|\mathbf{v}_0|^2\,\mathrm{d}x+\int_Q|\mathbf{f}|^2\,\mathrm{d}x\,\mathrm{d}t\bigg]^{\frac{\beta}{2}}\end{aligned}\tag{10.2.35}$$

which implies the claimed inequality. □

10.3 NON-STATIONARY FLOWS

In this section we prove Theorem 10.0.41. The proof is divided into several steps. First, we approximate the equation of interest by an equation satisfying the assumptions from the last section. Due to Theorem 10.2.43 we have a solution to this approximated system. Then we obtain uniform a priori estimates and hence the weak convergence of a subsequence. In the second step we prove compactness of the approximated velocity. In order to pass to the limit in the nonlinear stress deviator we use the L^∞-truncation and monotone operator theory.

Step 1: a priori estimates and weak convergence

Let us consider the equation

$$\begin{cases}\mathrm{d}\mathbf{v}=\operatorname{div}\mathbf{S}(\boldsymbol{\varepsilon}(\mathbf{v}))\,\mathrm{d}t-\frac{1}{m}|\mathbf{v}|^{q-2}\mathbf{v}\,\mathrm{d}t-\nabla\pi\,\mathrm{d}t,\\ \qquad-\operatorname{div}\big(\mathbf{v}\otimes\mathbf{v}\big)\,\mathrm{d}t+\mathbf{f}\mathrm{d}t+\Phi(\mathbf{v})\,\mathrm{d}\mathbf{W}_t\\ \mathbf{v}(0)=\mathbf{v}_0.\end{cases}\tag{10.3.36}$$

By Theorem 10.2.43 and Theorem 10.2.44 (for $\alpha=\frac{1}{m}$) we know that there is a weak martingale solution

$$\big((\Omega,\mathcal{F},(\mathcal{F}_t)_{t\geq0},\mathbb{P}),\mathbf{v}_m,\mathbf{v}_{0,m},\mathbf{f}_m,\mathbf{W}\big)$$

to (10.3.36) with $\mathbf{v}_m\in\mathcal{V}_{p,q}$, $\Lambda_0=\mathbb{P}\circ(\mathbf{v}_{0,m})^{-1}$ and $\Lambda_{\mathbf{f}}=\mathbb{P}\circ(\mathbf{f}_m)^{-1}$ (we skip the underlines for simplicity). To be precise, we have for all $\boldsymbol{\varphi}\in C^\infty_{0,\mathrm{div}}(G)$

$$\int_G\mathbf{v}_m(t)\cdot\boldsymbol{\varphi}\,\mathrm{d}x+\int_0^t\int_G\mathbf{S}(\boldsymbol{\varepsilon}(\mathbf{v}_m)):\boldsymbol{\varepsilon}(\boldsymbol{\varphi})\,\mathrm{d}x\,\mathrm{d}\sigma+\frac{1}{m}\int_0^t\int_G|\mathbf{v}_m|^{q-2}\mathbf{v}_m\cdot\boldsymbol{\varphi}\,\mathrm{d}x\,\mathrm{d}\sigma$$

$$= \int_0^t \int_G \mathbf{v}_m \otimes \mathbf{v}_m : \boldsymbol{\varepsilon}(\boldsymbol{\varphi})\,\mathrm{d}x\,\mathrm{d}\sigma + \int_G \mathbf{v}_{0,m}\cdot\boldsymbol{\varphi}\,\mathrm{d}x + \int_G\int_0^t \mathbf{f}_m\cdot\boldsymbol{\varphi}\,\mathrm{d}x\,\mathrm{d}\sigma$$
$$+ \int_G\int_0^t \Phi(\mathbf{v}_m)\,\mathrm{d}\mathbf{W}_\sigma\cdot\boldsymbol{\varphi}\,\mathrm{d}x$$

$\mathbb{P}\otimes\mathcal{L}^1$-a.e. Note that due to Remark 10.2.25 the probability space can be chosen independently of m. The same is true for the Brownian motion $\mathbf{W}$. Theorem 10.2.43 yields uniform estimates for $\mathbf{v}_m$. Using Corollary 10.2.1 and the assumptions on $\Lambda_{\mathbf{f}}$ and Λ_0 in (10.0.6) they can be improved such that

$$\mathbb{E}\bigg[\sup_{t\in(0,T)}\int_G |\mathbf{v}_m(t)|^2\,\mathrm{d}x + \int_Q |\nabla\mathbf{v}_m|^p + \frac{|\mathbf{v}_m|^q}{m}\,\mathrm{d}x\,\mathrm{d}t\bigg]^{\frac{\beta}{2}} \le c(\beta), \tag{10.3.37}$$

where $\beta := \max\left\{\frac{2(d+2)}{d}, \frac{p(d+2)}{d}\right\}$. This and a parabolic interpolation imply

$$\mathbb{E}\bigg[\int_Q |\mathbf{v}_m|^{r_0}\,\mathrm{d}x\,\mathrm{d}t\bigg] \le c \quad \text{for} \quad r_0 := p\frac{d+2}{d} \tag{10.3.38}$$

uniformly in m. By (10.3.37), (10.3.38) and the assumption $p > \frac{2d+2}{d+2}$ we find that

$$\mathbb{E}\bigg[\int_Q |\mathbf{v}_m\otimes\mathbf{v}_m|^{p_0} + \int_Q \big|\nabla\big(\mathbf{v}_m\otimes\mathbf{v}_m\big)\big|^{p_0}\,\mathrm{d}x\,\mathrm{d}t\bigg] \le c \tag{10.3.39}$$

for some $p_0 > 1$. We obtain limit functions $\mathbf{v}$, $\tilde{\mathbf{S}}$, $\mathbf{V}$, $\tilde{\Phi}$ such that, after passing to a subsequence,

$$\begin{aligned}
\mathbf{v}_m &\rightharpoonup \mathbf{v} && \text{in} \quad L^{\frac{\beta}{2}p}(\Omega,\mathcal{F},\mathbb{P}; L^p(0,T; W^{1,p}_{0,\mathrm{div}}(G))),\\
\mathbf{v}_m &\rightharpoonup \mathbf{v} && \text{in} \quad L^{\beta}(\Omega,\mathcal{F},\mathbb{P}; L^r(0,T; L^2(G))),\\
\frac{1}{m}|\mathbf{v}_m|^{q-2}\mathbf{v}_m &\to 0 && \text{in} \quad L^{\frac{\beta}{2}q'}(\Omega,\mathcal{F},\mathbb{P}; L^{q'}(Q)),\\
\mathbf{S}(\boldsymbol{\varepsilon}(\mathbf{v}_m)) &\rightharpoonup \tilde{\mathbf{S}} && \text{in} \quad L^{p'}(\Omega,\mathcal{F},\mathbb{P}; L^{p'}(Q)),\\
\mathbf{S}(\boldsymbol{\varepsilon}(\mathbf{v}_m)) &\rightharpoonup \tilde{\mathbf{S}} && \text{in} \quad L^{p'}(\Omega,\mathcal{F},\mathbb{P}; L^{p'}(0,T; W^{-1,p'}(G))),\\
\mathbf{v}^m\otimes\mathbf{v}_m &\rightharpoonup \mathbf{V} && \text{in} \quad L^{p_0}(\Omega,\mathcal{F},\mathbb{P}; L^{p_0}(0,T; W^{1,p_0}(G))),\\
\Phi(\mathbf{v}_m) &\rightharpoonup \tilde{\Phi} && \text{in} \quad L^{\beta}(\Omega,\mathcal{F},\mathbb{P}; L^r(0,T; L_2(U, L^2(G)))),
\end{aligned} \tag{10.3.40}$$

where $r < \infty$ is arbitrary. Moreover, we know

$$\mathbb{E}\bigg[\sup_{t\in(0,T)}\int_G |\mathbf{v}|^2\,\mathrm{d}x\bigg]^{\frac{\beta}{2}} < \infty, \quad \mathbb{E}\bigg[\sup_{t\in(0,T)} \|\tilde{\Phi}\|_{L_2(U,L^2(G))}\,\mathrm{d}x\bigg]^{\frac{\beta}{2}} < \infty.$$

In order to introduce the pressure we set

$$\mathbf{H}^1_m := \mathbf{S}(\boldsymbol{\varepsilon}(\mathbf{v}_m)),$$

$$\mathbf{H}_m^2 := \nabla\Delta^{-1}\mathbf{f}_m + \nabla\Delta^{-1}\Big(\frac{1}{m}|\mathbf{v}_m|^{q-2}\mathbf{v}^m\Big) + \mathbf{v}_m \otimes \mathbf{v}_m,$$
$$\Phi_m := \Phi(\mathbf{v}_m).$$

Using Theorem 10.1.42, Corollary 10.1.1 and Corollary 10.1.3 we obtain functions π_m^h, π_m^1, π_m^2 which are adapted to $(\underline{\mathcal{F}_t})_{t\geq 0}$, and Φ_m^π progressively measurable, such that

$$\begin{aligned}
&\int_G (\mathbf{v}_m - \nabla\pi_m^h)(t)\cdot\boldsymbol{\varphi}\,\mathrm{d}x\\
&= \int_G \mathbf{v}_{0,m}\cdot\boldsymbol{\varphi}\,\mathrm{d}x - \int_G\int_0^t \big(\mathbf{H}_m^1 - \pi_m^1 I\big) : \nabla\boldsymbol{\varphi}\,\mathrm{d}x\,\mathrm{d}\sigma\\
&+ \int_G\int_0^t \operatorname{div}\big(\mathbf{H}_m^2 - \pi_m^2 I\big)\cdot\boldsymbol{\varphi}\,\mathrm{d}x\,\mathrm{d}\sigma + \int_G\int_0^t \Phi_m\,\mathrm{d}\mathbf{W}_\sigma\cdot\boldsymbol{\varphi}\,\mathrm{d}x\\
&+ \int_G\int_0^t \Phi_m^\pi\,\mathrm{d}\mathbf{W}_\sigma\cdot\boldsymbol{\varphi}\,\mathrm{d}x.
\end{aligned}\tag{10.3.41}$$

The following bounds hold uniformly in m

$$\begin{aligned}
&\mathbf{H}_m^1 \in L^{\frac{\beta}{2}p'}(\Omega,\mathcal{F},\mathbb{P},L^{p'}(Q)),\\
&\mathbf{H}_m^2 \in L^{p_0}(\Omega,\mathcal{F},\mathbb{P},L^{p_0}(0,T;W^{1,p_0}(G))),\\
&\Phi_m \in L^\beta(\Omega,\mathcal{F},\mathbb{P},L^r(0,T;L_2(U,L^2(G)))),
\end{aligned}\tag{10.3.42}$$

where $r<\infty$ is arbitrary (here we use the continuity of $\nabla\Delta^{-1}$ from $L^{p_0}(G)$ to $W^{1,p_0}(G)$). We have the same uniform bounds for the pressure functions, i.e.,

$$\begin{aligned}
&\pi_m^h \in L^\beta(\Omega,\mathcal{F},\mathbb{P},L^r(0,T;L^2(G))),\\
&\pi_m^1 \in L^{\frac{\beta}{2}p'}(\Omega,\mathcal{F},\mathbb{P},L^{p'}(Q)),\\
&\pi_m^2 \in L^{p_0}(\Omega,\mathcal{F},\mathbb{P},L^{p_0}(0,T;W^{1,p_0}(G)))),\\
&\Phi_m^\pi \in L^\beta(\Omega,\mathcal{F},\mathbb{P},L^r(0,T;L_2(U,L^2(G)))).
\end{aligned}\tag{10.3.43}$$

Here we used Corollary 10.1.1 and Corollary 10.1.3. For the harmonic pressure we obtain, using regularity theory for harmonic functions as well as Corollary 10.1.2,

$$\pi_m^h \in L^\beta(\Omega,\mathcal{F},\mathbb{P};L^r(0,T;W^{k,r}_{loc}(G)))\tag{10.3.44}$$

for all $k\in\mathbb{N}$. After passing to a subsequence (not relabelled) we have the following convergences

$$\begin{aligned}
&\pi_m^h \rightharpoonup \pi^h && \text{in}\quad L^\beta(\Omega,\mathcal{F},\mathbb{P};L^r(0,T;W^{k,r}_{loc}(G))),\\
&\pi_m^1 \rightharpoonup \pi^1 && \text{in}\quad L^{\frac{\beta}{2}p'}(\Omega,\mathcal{F},\mathbb{P},L^{p'}(Q)),\\
&\pi_m^2 \rightharpoonup \pi^2 && \text{in}\quad L^{p_0}(\Omega,\mathcal{F},\mathbb{P},L^{p_0}(0,T;W^{1,p_0}(G)))),\\
&\Phi_m^\pi \rightharpoonup \Phi^\pi && \text{in}\quad L^\beta(\Omega,\mathcal{F},\mathbb{P},L^r(0,T;L_2(U,L^2(G)))).
\end{aligned}\tag{10.3.45}$$

In the following we need to show that the limit functions in (10.3.40) satisfy $\mathbf{V} = \mathbf{v} \otimes \mathbf{v}$ and $\tilde{\Phi} = \Phi(\mathbf{v})$. We will prove this by compactness arguments and a change of the probability space similar to the proof of Theorem 10.2.44. After this, in a final step, we will show $\tilde{\mathbf{S}} = \mathbf{S}(\boldsymbol{\varepsilon}(\mathbf{v}))$.

Step 2: compactness

Now we will show compactness of $\mathbf{v}_m$. In order to include the pressure in the compactness method we have to deal with weak convergences. This situation is not covered by the classical Skorokhod Theorem. However, a generalization of it – the Jakubowski–Skorokhod Theorem, see Lemma 9.3.3 – applies to quasi-Polish spaces. This includes weak topologies of separable reflexive Banach spaces.

First we deduce from (10.3.36)–(10.3.38)

$$\mathbb{E}\bigg[\Big\|\mathbf{v}_m(t) - \int_0^t \Phi(\mathbf{v}_m)\,\mathrm{d}\mathbf{W}_\sigma\Big\|_{W^{1,p_0}([0,T];\,W^{-1,p_0}_{\mathrm{div}}(G))}\bigg] \leq c.$$

For the stochastic term we use Lemma 9.1.3 and (10.0.4) to estimate

$$\begin{aligned}\mathbb{E}\bigg[\Big\|\int_0^{\cdot} \Phi(\mathbf{v}_m)\,\mathrm{d}\mathbf{W}_\sigma\Big\|_{W^{\alpha,2}(0,T;L^2(G))}\bigg] &\leq c\,\mathbb{E}\bigg[\int_0^T \|\Phi(\mathbf{v}_m)\|^2_{L_2(U,L^2(G))}\,\mathrm{d}t\bigg]\\ &\leq c\,\mathbb{E}\bigg[\sum_k \int_0^T\int_G |g_k(\cdot,\mathbf{v}_m)|^2\,\mathrm{d}x\,\mathrm{d}t\bigg] \leq c\,\mathbb{E}\bigg[1 + \int_Q |\mathbf{v}_m|^2\,\mathrm{d}x\,\mathrm{d}t\bigg]\end{aligned}$$

for all $\alpha < 1/2$. Due to (10.3.37) and $p_0 \leq 2$ this implies

$$\mathbb{E}\bigg[\Big\|\int_0^{\cdot} \Phi(\mathbf{v}_m)\,\mathrm{d}\mathbf{W}_\sigma\Big\|_{W^{\alpha,p_0}((0,T);L^2(G))}\bigg] \leq c.$$

Combining the both informations above shows

$$\mathbb{E}\Big[\|\mathbf{v}_m\|_{W^{\alpha,p_0}(0,T;\,W^{-1,p_0}_{\mathrm{div}}(G))}\Big] \leq c. \tag{10.3.46}$$

On account of (10.3.37), an interpolation with $L^{p_0}(0, T;\, W^{1,p_0}_{0,\mathrm{div}}(G))$ implies (see Lemma 5.1.4)

$$\mathbb{E}\Big[\|\mathbf{v}_m\|_{W^{\kappa,p_0}(0,T;L^{p_0}_{\mathrm{div}}(G))}\Big] \leq c, \tag{10.3.47}$$

for some $\kappa > 0$. As a consequence of $p > \frac{2d+2}{d+2}$ we have

$$W^{\kappa,p_0}(0, T;\, L^{p_0}_{\mathrm{div}}(G)) \cap L^p(0, T;\, W^{1,p}_{0,\mathrm{div}}(G)) \hookrightarrow L^r(0, T;\, L^r_{\mathrm{div}}(G))$$

compactly for all $r < p\frac{d+2}{d}$. We will use this embedding in order to show compactness of $\mathbf{v}_m$. In order to obtain compactness for the harmonic pressure π_m^h we introduce the quantity $\mathbf{q}_m = \mathbf{v}_m - \nabla\pi_m^h$. On account of (10.3.40)$_1$ and (10.3.43)$_1$ the following holds by local regularity theory for harmonic functions

$$\mathbb{E}\Big[\|\mathbf{q}_m\|_{L^p(0,T;W^{1,p}_{loc}(G))}\Big] \leq c.$$

Arguing as in (10.3.47) and using (10.3.41) together with the uniform bounds stated after we obtain

$$\mathbb{E}\Big[\|\mathbf{q}_m\|_{W^{\kappa,p_0}(0,T;L^{p_0}_{loc}(G))}\Big] \leq c$$

for some $p_0 > 1$. So we can use compactness of the embedding

$$\begin{aligned}
&L^p(0,T;W^{1,p}_{loc}(G)) \cap W^{\kappa,p_0}(0,T;L^{p_0}_{loc}(G))\\
&\quad + W^{\kappa,p_0}(0,T;L^{p_0}_{\mathrm{div}}(G)) \cap L^p(0,T;W^{1,p}_{0,\mathrm{div}}(G)) \hookrightarrow L^r(0,T;L^r(G))
\end{aligned}$$

in order to handle π^h_m. We consider the path space

$$\begin{aligned}
\mathscr{V} := &\, L^r(0,T;L^r_{\mathrm{div}}(G)) \otimes L^r(0,T;L^r_{loc}(G)) \otimes \big(L^{p'}(Q), w\big)\\
&\otimes \big(L^{p_0}(0,T;W^{1,p_0}(G))), w\big) \otimes \big(L^r(0,T;L_2(U,L^2(G))), w\big)\\
&\otimes C([0,T],U_0) \otimes L^2(G) \otimes L^2(Q).
\end{aligned}$$

We will use the following notations (w refers to the weak topology):

- $\nu_{\mathbf{v}_m}$ is the law of $\mathbf{v}_m$ on $L^r(0,T;L^r(G))$;
- $\nu_{\pi^h_m}$ is the law of π^h_m on $L^r(0,T;L^r_{loc}(G))$;
- $\nu_{\pi^1_m}$ is the law of π^1_m on $\big(L^{p'}(Q), w\big)$;
- $\nu_{\pi^2_m}$ is the law of π^2_m on $\big(L^{p_0}(0,T;W^{1,p_0}(G))), w\big)$;
- $\nu_{\Phi^\pi_m}$ is the law of Φ^π_m on $\big(L^r(0,T;L_2(U,L^2(G))), w\big)$;
- $\nu_{\mathbf{W}}$ is the law of $\mathbf{W}$ on $C([0,T],U_0)$, where U_0 is defined in (9.1.2);
- $\boldsymbol{\nu}_m$ is the joint law of $\mathbf{v}_m$, π^h_m, π^1_m, π^2_m, Φ^π_m, $\mathbf{W}$, $\mathbf{v}_0$ and $\mathbf{f}$ on $\mathscr{V}$.

We need to show tightness of the measure $\boldsymbol{\nu}_m$.

We consider the ball $\mathcal{B}_R$ in the space $W^{\kappa,p_0}(0,T;L^{p_0}_{\mathrm{div}}(G)) \cap L^p(0,T;W^{1,p}_{\mathrm{div}}(G))$ and obtain for its complement $\mathcal{B}^C_R$ by (10.3.38) and (10.3.47)

$$\begin{aligned}
\nu_{\mathbf{v}_m}(\mathcal{B}^C_R) &= \mathbb{P}\big(\|\mathbf{v}_m\|_{W^{\kappa,p_0}(0,T;L^{p_0}_{\mathrm{div}}(G))} + \|\mathbf{v}_m\|_{L^p(0,T;W^{1,p}_{\mathrm{div}}(G))} \geq R\big)\\
&\leq \frac{1}{R}\mathbb{E}\Big(\|\mathbf{v}_m\|_{W^{\kappa,p_0}(0,T;L^{p_0}_{\mathrm{div}}(G))} + \|\mathbf{v}_m\|_{L^p(0,T;W^{1,p}_{\mathrm{div}}(G))}\Big) \leq \frac{c}{R}.
\end{aligned}$$

So for a fixed $\eta > 0$ we find $R(\eta)$ with

$$\nu_{\mathbf{v}_m}(\mathcal{B}_{R(\eta)}) \geq 1 - \frac{\eta}{8}.$$

Using (10.3.44) we can show that also the law of π^h_m is tight, i.e., there exists a compact set $C_\pi \subset L^r(0,T;L^r_{loc}(G))$ such that $\nu_{\pi^h_m}(C_\pi) \geq 1 - \frac{\eta}{8}$. Due to the reflexivity of the corresponding spaces we find compact sets for π^1_m, π^2_m and Φ^π_m with measures greater or equal to $1 - \frac{\eta}{8}$. The law

$\nu_{\mathbf{W}}$ is tight as it coincides with the law of $\mathbf{W}$ which is a Radon measure on the Polish space $C([0,T],U_0)$. So, there exists a compact set $C_\eta \subset C([0,T],U_0)$ such that $\nu_{\mathbf{W}}(C_\eta) \geq 1 - \frac{\eta}{8}$. By the same argument we can find compact subsets of $L^2_{\mathrm{div}}(G)$ and $L^2(Q)$ such that their measures (Λ_0 and $\Lambda_{\mathbf{f}}$ respectively) are smaller than $1 - \frac{\eta}{8}$. Hence, we can find a compact subset $\mathscr{V}_\eta \subset \mathscr{V}$ such that $\boldsymbol{\nu}_m(\mathscr{V}_\eta) \geq 1 - \eta$. Thus, $\{\boldsymbol{\nu}_m,\ m \in \mathbb{N}\}$ is tight on the same space. On account of the Jakubowski–Skorokhod Theorem from Lemma 9.3.3 we conclude the existence of a probability space $(\underline{\Omega}, \underline{\mathcal{F}}, \underline{\mathbb{P}})$, a sequence $(\underline{\mathbf{v}}_m, \underline{\pi}^h_m, \underline{\pi}^1_m, \underline{\pi}^2_m, \underline{\Phi}^m_\pi, \underline{\mathbf{W}}^m, \underline{\mathbf{v}}_{0,m}, \underline{\mathbf{f}}^m)$ and $(\underline{\mathbf{v}}, \underline{\pi}^h, \underline{\pi}^1, \underline{\pi}^2, \underline{\Phi}^\pi, \underline{\mathbf{W}}, \underline{\mathbf{v}}_0, \underline{\mathbf{f}})$ on $(\underline{\Omega}, \underline{\mathcal{F}}, \underline{\mathbb{P}})$ both with values in $\mathscr{V}$ such that the following holds.

- The laws of $(\underline{\mathbf{v}}_m, \underline{\pi}^h_m, \underline{\pi}^1_m, \underline{\pi}^2_m, \underline{\Phi}^\pi_m, \underline{\mathbf{W}}^m, \underline{\mathbf{v}}_{0,m}, \underline{\mathbf{f}}_m)$ and $(\underline{\mathbf{v}}, \underline{\pi}^h, \underline{\pi}^1, \underline{\pi}^2, \underline{\Phi}^\pi, \underline{\mathbf{W}}, \underline{\mathbf{v}}_0, \underline{\mathbf{f}})$ under $\underline{\mathbb{P}}$ coincide with $\boldsymbol{\nu}_m$ and $\boldsymbol{\nu} := \lim_m \boldsymbol{\nu}_m$.
- We have $\underline{\mathbb{P}}$-a.s. the weak convergences

$$\begin{aligned} \underline{\pi}^1_m &\rightharpoonup \underline{\pi}^1 \quad \text{in} \quad L^{p'}(Q), \\ \underline{\pi}^2_m &\rightharpoonup \underline{\pi}^2 \quad \text{in} \quad L^{p_0}(0,T;W^{1,p_0}(G))), \\ \underline{\Phi}^\pi_m &\rightharpoonup \underline{\Phi}^\pi \quad \text{in} \quad L^r(0,T;L_2(U,L^2(G))). \end{aligned}$$

- We have $\underline{\mathbb{P}}$-a.s. the strong convergences

$$\begin{aligned} \underline{\mathbf{v}}_m &\to \underline{\mathbf{v}} \quad \text{in} \quad L^r(0,T;L^r(G)), \\ \underline{\pi}^h_m &\to \underline{\pi}_h \quad \text{in} \quad L^r(0,T;L^r_{loc}(G)), \\ \underline{\mathbf{W}}^m &\to \underline{\mathbf{W}} \quad \text{in} \quad C([0,T],U_0), \\ \underline{\mathbf{v}}_{0,m} &\to \underline{\mathbf{v}}_0 \quad \text{in} \quad L^2(G), \\ \underline{\mathbf{f}}_m &\to \underline{\mathbf{f}} \quad \text{in} \quad L^2(0,T;L^2(G)). \end{aligned}$$

- We have for all $\alpha < \infty$

$$\int_{\underline{\Omega}} \left(\sup_{[0,T]} \|\underline{\mathbf{W}}^m(t)\|^\alpha_{U_0} \right) \mathrm{d}\underline{\mathbb{P}} = \int_{\underline{\Omega}} \left(\sup_{[0,T]} \|\mathbf{W}(t)\|^\alpha_{U_0} \right) \mathrm{d}\mathbb{P}.$$

On account of the equality of laws we obtain the weak convergences (after choosing a subsequence)

$$\begin{aligned} \underline{\pi}^1_m &\rightharpoonup \underline{\pi}^1 \quad \text{in} \quad L^{p'}(\underline{\Omega}, \underline{\mathcal{F}}, \underline{\mathbb{P}}, L^{p'}(Q), \\ \underline{\pi}^2_m &\rightharpoonup \underline{\pi}^2 \quad \text{in} \quad L^{p_0}(\underline{\Omega}, \underline{\mathcal{F}}, \underline{\mathbb{P}}, L^{p_0}(0,T;W^{1,p_0}(G))), \\ \underline{\Phi}^\pi_m &\rightharpoonup \underline{\Phi}^\pi \quad \text{in} \quad L^{p_0}(\underline{\Omega}, \underline{\mathcal{F}}, \underline{\mathbb{P}}, L^r(0,T;L_2(U,L^2(G)))). \end{aligned}$$

By Vitali's convergence Theorem we obtain the strong convergences

$$\underline{\mathbf{W}}^m \longrightarrow \underline{\mathbf{W}} \quad \text{in} \quad L^2(\underline{\Omega}, \underline{\mathcal{F}}, \underline{\mathbb{P}}; C([0,T],U_0)), \tag{10.3.48}$$

$$\underline{\mathbf{v}}_m \longrightarrow \underline{\mathbf{v}} \quad \text{in} \quad L^r(\underline{\Omega} \times Q; \underline{\mathbb{P}} \otimes \mathcal{L}^{d+1}), \tag{10.3.49}$$

$$\nabla^k \underline{\pi}_m^h \longrightarrow \nabla^k \underline{\pi}^h \quad \text{in} \quad L^r(\underline{\Omega} \times (0,T) \times G'; \underline{\mathbb{P}} \otimes \mathcal{L}^{d+1}), \tag{10.3.50}$$

$$\underline{\mathbf{v}}_{0,m} \longrightarrow \underline{\mathbf{v}}_0 \quad \text{in} \quad L^2(\underline{\Omega} \times G, \underline{\mathbb{P}} \otimes \mathcal{L}^{d+1}), \tag{10.3.51}$$

$$\underline{\mathbf{f}}_m \longrightarrow \underline{\mathbf{f}} \quad \text{in} \quad L^2(\underline{\Omega} \times Q, \underline{\mathbb{P}} \otimes \mathcal{L}^{d+1}), \tag{10.3.52}$$

for all $r < p\frac{d+2}{d}$ and all $G' \Subset G$. For the harmonic pressure we used local regularity theory for harmonic maps. The convergences above imply

$$\begin{aligned}
\underline{\mathbf{v}}_m \otimes \underline{\mathbf{v}}_m &\rightharpoondown \underline{\mathbf{v}} \otimes \underline{\mathbf{v}} \quad \text{in} \quad L^{p_0}(\underline{\Omega}, \underline{\mathcal{F}}, \underline{\mathbb{P}}, L^{p_0}(0,T; W^{1,p_0}(G))), \\
\Phi(\underline{\mathbf{v}}_m) &\rightharpoondown \Phi(\underline{\mathbf{v}}) \quad \text{in} \quad L^{\beta}(\underline{\Omega}, \underline{\mathcal{F}}, \underline{\mathbb{P}}, L^{\alpha}(0,T; L_2(U, L^2(G)))), \\
\Phi^{\pi}(\underline{\mathbf{v}}_m) &\rightharpoondown \Phi^{\pi}(\underline{\mathbf{v}}) \quad \text{in} \quad L^{\beta}(\underline{\Omega}, \underline{\mathcal{F}}, \underline{\mathbb{P}}, L^{\alpha}(0,T; L_2(U, L^2(G)))),
\end{aligned} \tag{10.3.53}$$

for all $\alpha < \infty$. Again we define $(\underline{\mathcal{F}}_t)_{t \geq 0}$ to be the $\underline{\mathbb{P}}$-augmented canonical filtration of the process $(\underline{\mathbf{v}}, \underline{\pi}^h, \underline{\pi}^1, \underline{\pi}^2, \underline{\Phi}^{\pi}, \underline{\mathbf{W}}, \underline{\mathbf{f}})$, respectively, that is

$$\underline{\mathcal{F}}_t = \sigma\Big(\sigma\big(\mathbf{r}_t\underline{\mathbf{v}}, \mathbf{r}_t\underline{\pi}^h, \mathbf{r}_t\underline{\pi}^1, \mathbf{r}_t\underline{\pi}^2, \mathbf{r}_t\underline{\Phi}^{\pi}, \mathbf{r}_t\underline{\mathbf{W}}, \mathbf{r}_t\underline{\mathbf{f}}\big) \cup \big\{N \in \underline{\mathcal{F}};\ \underline{\mathbb{P}}(N) = 0\big\}\Big),$$
$$t \in [0,T].$$

As done in the proof of Theorem 10.2.44 (but using test-functions from $C_0^\infty(G)$ instead of $C_{0,\mathrm{div}}^\infty(G)$) we can show that the equation also holds on the new probability space, i.e., we have $\underline{\mathbb{P}} \otimes \mathcal{L}^1$-a.e.

$$\begin{aligned}
&\int_G (\underline{\mathbf{v}}_m - \nabla \underline{\pi}_m^h)(t) \cdot \boldsymbol{\varphi}\, \mathrm{d}x \\
&= \int_G \underline{\mathbf{v}}_0^m \cdot \boldsymbol{\varphi}\, \mathrm{d}x - \int_G \int_0^t \big(\underline{\mathbf{H}}_m^1 - \underline{\pi}_m^1 I\big) : \nabla \boldsymbol{\varphi}\, \mathrm{d}x\, \mathrm{d}\sigma \\
&+ \int_G \int_0^t \mathrm{div}\,\big(\underline{\mathbf{H}}_m^2 - \underline{\pi}_m^2 I\big) \cdot \boldsymbol{\varphi}\, \mathrm{d}x\, \mathrm{d}\sigma + \int_G \int_0^t \Phi(\underline{\mathbf{v}}_m)\, \mathrm{d}\underline{\mathbf{W}}_\sigma^m \cdot \boldsymbol{\varphi}\, \mathrm{d}x \\
&+ \int_G \int_0^t \underline{\Phi}_m^{\pi}\, \mathrm{d}\underline{\mathbf{W}}_\sigma^m \cdot \boldsymbol{\varphi}\, \mathrm{d}x
\end{aligned}$$

for all $\boldsymbol{\varphi} \in C_0^\infty(G)$. We have used the abbreviations

$$\begin{aligned}
\underline{\mathbf{H}}_m^1 &:= \mathbf{S}(\boldsymbol{\varepsilon}(\underline{\mathbf{v}}_m)), \\
\underline{\mathbf{H}}_m^2 &:= \nabla\Delta^{-1}\underline{\mathbf{f}}_m + \nabla\Delta^{-1}\Big(\frac{1}{m}|\underline{\mathbf{v}}_m|^{q-2}\underline{\mathbf{v}}_m\Big) + \mathbf{v}_m \otimes \mathbf{v}_m.
\end{aligned}$$

From the convergences above we obtain the limit equation (using again Lemma 9.1.2 for the convergence of the stochastic integral)

$$\begin{aligned}
&\int_G \big(\underline{\mathbf{v}} - \nabla \underline{\pi}^h\big)(t) \cdot \boldsymbol{\varphi} \,\mathrm{d}x \\
&\quad = \int_G \underline{\mathbf{v}}_0 \cdot \boldsymbol{\varphi} \,\mathrm{d}x - \int_G \int_0^t \big(\underline{\mathbf{H}}^1 - \underline{\pi}^1 I\big) : \nabla \boldsymbol{\varphi} \,\mathrm{d}x \,\mathrm{d}\sigma \\
&\quad + \int_G \int_0^t \operatorname{div}\big(\underline{\mathbf{H}}^2 - \underline{\pi}^2 I\big) \cdot \boldsymbol{\varphi} \,\mathrm{d}x \,\mathrm{d}\sigma \\
&\quad + \int_G \int_0^t \Phi(\underline{\mathbf{v}}) \,\mathrm{d}\underline{\mathbf{W}}_\sigma \cdot \boldsymbol{\varphi} \,\mathrm{d}x + \int_G \int_0^t \underline{\Phi}^\pi \,\mathrm{d}\underline{\mathbf{W}}_\sigma \cdot \boldsymbol{\varphi} \,\mathrm{d}x
\end{aligned} \tag{10.3.54}$$

for all $\boldsymbol{\varphi} \in C_0^\infty(G)$, where

$$\underline{\mathbf{H}}^1 := \underline{\tilde{\mathbf{S}}}, \quad \underline{\mathbf{H}}^2 := \nabla \Delta^{-1} \underline{\mathbf{f}} + \mathbf{v} \otimes \mathbf{v}.$$

It remains to show $\underline{\tilde{\mathbf{S}}} = \mathbf{S}(\boldsymbol{\varepsilon}(\underline{\mathbf{v}}))$. Now we let

$$\begin{aligned}
&\underline{\mathbf{G}}_m^1 := \mathbf{S}(\boldsymbol{\varepsilon}(\underline{\mathbf{v}}_m)) - \underline{\tilde{\mathbf{S}}}, \\
&\underline{\mathbf{G}}_m^2 := \nabla \Delta^{-1}(\underline{\mathbf{f}}_m - \underline{\mathbf{f}}) + \nabla \Delta^{-1}\Big(\frac{1}{m} |\underline{\mathbf{v}}_m|^{q-2} \underline{\mathbf{v}}_m\Big) + \underline{\mathbf{v}}_m \otimes \underline{\mathbf{v}}_m - \underline{\mathbf{v}} \otimes \underline{\mathbf{v}}, \\
&\underline{\Phi}_m := \big(\Phi(\underline{\mathbf{v}}_m), -\Phi(\underline{\mathbf{v}})\big), \quad \underline{\Phi}_m^\vartheta := \big(\Phi^\pi(\underline{\mathbf{v}}_m), -\Phi^\pi(\underline{\mathbf{v}})\big), \\
&\underline{\vartheta}_m^h = \underline{\pi}_m^h - \underline{\pi}^h, \quad \underline{\vartheta}_m^1 = \underline{\pi}_m^1 - \underline{\pi}^1, \quad \underline{\vartheta}_m^2 = \underline{\pi}_m^2 - \underline{\pi}^2.
\end{aligned}$$

We have the following convergences

$$\begin{aligned}
\underline{\mathbf{v}}_m - \underline{\mathbf{v}} \to 0 &\quad \text{in} \quad L^{\frac{\beta}{2}p}(\underline{\Omega}, \underline{\mathcal{F}}, \underline{\mathbb{P}}, L^p(0, T; W_0^{1,p}(G))), \\
\underline{\mathbf{v}}_m - \underline{\mathbf{v}} \to 0 &\quad \text{in} \quad L^\beta(\underline{\Omega}, \underline{\mathcal{F}}, \underline{\mathbb{P}}, L^r(0, T; L^2(G))), \\
\underline{\mathbf{G}}_m^1 \to 0 &\quad \text{in} \quad L^{\frac{\beta}{2}p'}(\underline{\Omega}, \underline{\mathcal{F}}, \underline{\mathbb{P}}, L^{p'}(Q)), \\
\underline{\mathbf{G}}_m^2 \to 0 &\quad \text{in} \quad L^{p_0}(\underline{\Omega}, \underline{\mathcal{F}}, \underline{\mathbb{P}}, L^{p_0}(0, T; W^{1,p_0}(G))), \\
\underline{\Phi}_m - \underline{\Phi} \to 0 &\quad \text{in} \quad L^\beta(\underline{\Omega}, \underline{\mathcal{F}}, \underline{\mathbb{P}}, L^r(0, T; L_2(U, L^2(G)))),
\end{aligned} \tag{10.3.55}$$

where $\underline{\Phi} = (\Phi(\underline{\mathbf{v}}), -\Phi(\underline{\mathbf{v}}))$ and $r < \infty$ is arbitrary. We have the same convergences for the pressure functions, i.e.,

$$\begin{aligned}
\underline{\vartheta}_m^h \to 0 &\quad \text{in} \quad L^\beta(\underline{\Omega}, \underline{\mathcal{F}}, \underline{\mathbb{P}}, L^r(0, T; W_{loc}^{k,r}(G))), \\
\underline{\vartheta}_m^1 \to 0 &\quad \text{in} \quad L^{\frac{\beta}{2}p'}(\underline{\Omega}, \underline{\mathcal{F}}, \underline{\mathbb{P}}, L^{p'}(Q)), \\
\underline{\vartheta}_m^2 \to 0 &\quad \text{in} \quad L^{p_0}(\underline{\Omega}, \underline{\mathcal{F}}, \underline{\mathbb{P}}, L^{p_0}(0, T; W^{1,p_0}(G))), \\
\underline{\Phi}_m^\vartheta - \underline{\Phi}^\vartheta \to 0 &\quad \text{in} \quad L^\beta(\underline{\Omega}, \underline{\mathcal{F}}, \underline{\mathbb{P}}, L^r(0, T; L_2(U, L^2(G)))),
\end{aligned} \tag{10.3.56}$$

where $r < \infty$ is arbitrary. Moreover, we have

$$\begin{aligned}
&\underline{\vartheta}_m^h \in L^\beta(\underline{\Omega}, \underline{\mathcal{F}}, \underline{\mathbb{P}}, L^r(0, T; L^2(G))), \\
&\underline{\Phi}_m \in L^\beta(\underline{\Omega}, \underline{\mathcal{F}}, \underline{\mathbb{P}}, L^r(0, T; L_2(U, L^2(G)))), \\
&\underline{\Phi}_m^\vartheta \in L^\beta(\underline{\Omega}, \underline{\mathcal{F}}, \underline{\mathbb{P}}, L^r(0, T; L_2(U, L^2(G)))),
\end{aligned} \tag{10.3.57}$$

uniformly in m.

The difference of approximated equation and limit equation reads as

$$\begin{aligned}
&\int_G (\underline{\mathbf{v}}_m - \underline{\mathbf{v}} + \nabla\underline{\vartheta}_m^h)(t)\cdot\boldsymbol{\varphi}\,\mathrm{d}x \\
&\quad = \int_G (\underline{\mathbf{v}}_m(0) - \underline{\mathbf{v}}_0)\cdot\boldsymbol{\varphi}\,\mathrm{d}x - \int_G\int_0^t (\underline{\mathbf{G}}_m^1 - \underline{\vartheta}_m^1 I) : \nabla\boldsymbol{\varphi}\,\mathrm{d}x\,\mathrm{d}\sigma \\
&\quad + \int_G\int_0^t \operatorname{div}(\underline{\mathbf{G}}_m^2 - \underline{\vartheta}_m^2 I)\cdot\boldsymbol{\varphi}\,\mathrm{d}x\,\mathrm{d}\sigma + \int_G\int_0^t \underline{\Phi}_m\,\mathrm{d}(\underline{\mathbf{W}}_\sigma^m, \underline{\mathbf{W}}_\sigma)\cdot\boldsymbol{\varphi}\,\mathrm{d}x \\
&\quad + \int_G\int_0^t \underline{\Phi}_m^\vartheta\,\mathrm{d}(\underline{\mathbf{W}}_\sigma^m, \underline{\mathbf{W}}_\sigma)\cdot\boldsymbol{\varphi}\,\mathrm{d}x, \qquad (10.3.58)
\end{aligned}$$

for all $\boldsymbol{\varphi} \in C_0^\infty(G)$. In the following we will show that $\tilde{\underline{\mathbf{S}}} = \mathbf{S}(\boldsymbol{\varepsilon}(\underline{\mathbf{v}}))$ which finishes the proof of Theorem 10.0.41. In order to do so we introduce the sequence $\underline{\mathbf{u}}_m := \underline{\mathbf{v}}_m - \nabla\underline{\vartheta}_m^h$ and the double sequence $\underline{\mathbf{u}}_{m,k} := \underline{\mathbf{u}}_m - \underline{\mathbf{u}}_k$, $m \geq k$, for which we have the convergences

$$\underline{\mathbf{u}}_{m,k} \rightharpoonup 0 \quad \text{in} \quad L^p(\underline{\Omega}, \underline{\mathcal{F}}, \underline{\mathbb{P}}, L^p(0,T; W_0^{1,p}(G))), \qquad (10.3.59)$$

$$\underline{\mathbf{u}}_{m,k} \to 0 \quad \text{in} \quad L^r(\underline{\Omega}\times(0,T)\times G'; \underline{\mathbb{P}}\otimes\mathcal{L}^{d+1}), \qquad (10.3.60)$$

as $m, k \to \infty$. Moreover, the following holds for all $\boldsymbol{\varphi} \in C_0^\infty(G)$

$$\begin{aligned}
&\int_G \underline{\mathbf{u}}_{m,k}(t)\cdot\boldsymbol{\varphi}\,\mathrm{d}x = \int_G (\underline{\mathbf{u}}_{m,k})_0\cdot\boldsymbol{\varphi}\,\mathrm{d}x \\
&\quad - \int_G\int_0^t (\underline{\mathbf{G}}_{m,k}^1 - \underline{\vartheta}_{m,k}^1 I) : \nabla\boldsymbol{\varphi}\,\mathrm{d}x\,\mathrm{d}\sigma + \int_G\int_0^t \operatorname{div}(\underline{\mathbf{G}}_{m,k}^2 - \underline{\vartheta}_{m,k}^2 I)\cdot\boldsymbol{\varphi}\,\mathrm{d}x\,\mathrm{d}\sigma \\
&\quad + \int_G\int_0^t \underline{\Phi}_{m,k}\,\mathrm{d}(\underline{\mathbf{W}}_\sigma^m, \underline{\mathbf{W}}_\sigma^k)\cdot\boldsymbol{\varphi}\,\mathrm{d}x + \int_G\int_0^t \underline{\Phi}_{m,k}^\vartheta\,\mathrm{d}(\underline{\mathbf{W}}_\sigma^m, \underline{\mathbf{W}}_\sigma^k)\cdot\boldsymbol{\varphi}\,\mathrm{d}x. \\
&\hfill (10.3.61)
\end{aligned}$$

All involved quantities with subscript $_{m,k}$ are defined analogously to $\underline{\mathbf{u}}_{m,k}$ by taking an appropriate difference.

Step 3: monotone operator theory and L^∞-truncation

By density arguments we are allowed to test with $\boldsymbol{\varphi} \in W_0^{1,p} \cap L^\infty(G)$. Since the function $\underline{\mathbf{v}}(\omega, t, \cdot)$ does not belong to this class, the L^∞-truncation was used for the deterministic problem (see [78] for the steady case and [140] for the unsteady problem). We will apply a variant of it adapted to the stochastic fashion.

We define h_L and H_L, $L \in \mathbb{N}_0$, by

$$h_L(s) := \int_0^s \Psi_L(\theta)\theta\,\mathrm{d}\theta, \quad H_L(\boldsymbol{\xi}) := h_L(|\boldsymbol{\xi}|),$$

$$\Psi_L := \sum_{\ell=1}^{L} \psi_{2^{-\ell}}, \quad \psi_\delta(s) := \psi(\delta s),$$

where $\psi \in C^\infty([0,\infty))$ with $0 \le \psi \le 1$, $\psi \equiv 1$ on $[0,1]$, $\psi = 0$ on $[2,\infty)$ and $0 \le -\psi' \le 2$. Now, we consider for $\eta \in C_0^\infty(G)$ the function

$$f_L(\mathbf{v}) := \int_G \eta H_L(\mathbf{v})\, \mathrm{d}x$$

and apply Itô's formula (see Lemma C.0.1). This implies

$$\begin{aligned}
&\int_G \eta H_L(\underline{\mathbf{u}}_{m,k}(t))\, \mathrm{d}x \\
&= f_L(\underline{\mathbf{u}}_{m,k}(0)) + \int_0^t f_L'(\underline{\mathbf{u}}_{m,k})\, \mathrm{d}\underline{\mathbf{u}}_{m,k} + \frac{1}{2}\int_0^t f_L''(\underline{\mathbf{u}}_{m,k})\, \mathrm{d}\langle \underline{\mathbf{u}}_{m,k}\rangle_\sigma \\
&= \int_G \eta H_L(\underline{\mathbf{u}}_{0,m} - \underline{\mathbf{u}}_{0,k})\, \mathrm{d}x \\
&\quad - \int_G\int_0^t \eta\big(\underline{\mathbf{G}}^1_{m,k} - \underline{\vartheta}^1_{m,k} I\big) : \nabla\big(\Psi_L(|\underline{\mathbf{u}}_{m,k}|)\underline{\mathbf{u}}_{m,k}\big)\, \mathrm{d}x\, \mathrm{d}\sigma \\
&\quad - \int_G\int_0^t \big(\underline{\mathbf{G}}^1_{m,k} - \underline{\vartheta}^1_{m,k} I\big) : \nabla\eta \otimes \Psi_L(|\underline{\mathbf{u}}_{m,k}|)\underline{\mathbf{u}}_{m,k}\, \mathrm{d}x\, \mathrm{d}\sigma \\
&\quad + \int_G\int_0^t \eta\Psi_L(|\underline{\mathbf{u}}_{m,k}|)\, \mathrm{div}\,\big(\underline{\mathbf{G}}^2_{m,k} - \underline{\vartheta}^2_{m,k} I\big) \cdot \underline{\mathbf{u}}_{m,k}\, \mathrm{d}x\, \mathrm{d}\sigma \\
&\quad + \int_G\int_0^t \eta\Psi_L(|\underline{\mathbf{u}}_{m,k}|)\underline{\mathbf{u}}_{m,k} \cdot \big(\Phi(\underline{\mathbf{v}}_m)\, \mathrm{d}\underline{\mathbf{W}}^m_\sigma - \Phi(\underline{\mathbf{v}}_k)\, \mathrm{d}\underline{\mathbf{W}}^k_\sigma\big)\, \mathrm{d}x \\
&\quad + \int_G\int_0^t \eta\Psi_L(|\underline{\mathbf{u}}_{m,k}|)\underline{\mathbf{u}}_{m,k} \cdot \big(\Phi^\vartheta(\underline{\mathbf{v}}_m)\, \mathrm{d}\underline{\mathbf{W}}^m_\sigma - \Phi^\vartheta(\underline{\mathbf{v}}_k)\, \mathrm{d}\underline{\mathbf{W}}^k_\sigma\big)\, \mathrm{d}x \\
&\quad + \frac{1}{2}\int_G\int_0^t \eta D^2 H_L(\underline{\mathbf{u}}_{m,k})\, \mathrm{d}\Big\langle\!\Big\langle \int_0^\cdot \Phi(\underline{\mathbf{v}}_m)\, \mathrm{d}\underline{\mathbf{W}}^m - \int_0^\cdot \Phi(\underline{\mathbf{v}}_k)\, \mathrm{d}\underline{\mathbf{W}}^k \Big\rangle\!\Big\rangle_\sigma \mathrm{d}x \\
&\quad + \frac{1}{2}\int_G\int_0^t \eta D^2 H_L(\underline{\mathbf{u}}_{m,k})\, \mathrm{d}\Big\langle\!\Big\langle \int_0^\cdot \Phi^\vartheta(\underline{\mathbf{v}}_m)\, \mathrm{d}\underline{\mathbf{W}}^m - \int_0^\cdot \Phi^\vartheta(\underline{\mathbf{v}}_k)\, \mathrm{d}\underline{\mathbf{W}}^k \Big\rangle\!\Big\rangle_\sigma \mathrm{d}x \\
&=: (O) + (I) + (II) + (III) + (IV) + (V) + (VI) + (VII).
\end{aligned}$$

Equation (10.3.51) and $\underline{\mathbf{u}}_m(0) - \underline{\mathbf{u}}_k(0) = \underline{\mathbf{v}}_m(0) - \underline{\mathbf{v}}_k(0)$ (see Theorem 10.1.42 b) imply that $\underline{\mathbb{E}}\big[(O)\big] \to 0$ if $m, k \to \infty$. The aim of the following observations is to show that the expectation values of (II)–(VI) vanish for $m, k \to \infty$ which gives the same for (I). By monotone operator theory this proves $\boldsymbol{\varepsilon}(\mathbf{v}_m) \to \boldsymbol{\varepsilon}(\mathbf{v})$ a.e. Although the rough ideas are clear their rigorous proof is quite technical.

By construction of Ψ_L we obtain, after passing to a subsequence,

$$\Psi_L(|\underline{\mathbf{u}}_{m,k}|)\underline{\mathbf{u}}_{m,k} \longrightarrow 0 \quad \text{in} \quad L^r(\underline{\Omega} \times Q, \underline{\mathbb{P}} \times \mathcal{L}^{d+1}), \quad m, k \to 0, \quad (10.3.62)$$

for all $r < \infty$ (first, we have boundedness in L^r, then the strong convergence follows in combination with (10.3.49)). This implies

$$\underline{\mathbb{E}}\big[(II)\big],\ \underline{\mathbb{E}}\big[(III)\big] \longrightarrow 0, \quad m, k \to \infty,$$

as a consequence of (10.3.55) and (10.3.56). Since $\underline{\mathbb{E}}[(IV)] = \underline{\mathbb{E}}[(V)] = 0$, only (VI) and (VII) remain. We obtain by $|D^2 H_L| \le c(L)$ that

$$\begin{aligned}
(VI) &\le c\sum_{\ell=1}^{d}\int_G\int_0^t \mathrm{d}\Big\langle\!\Big\langle \int_0^{\cdot} \big(\Phi(\underline{\mathbf{v}}_m) - \Phi(\underline{\mathbf{v}}_k)\big)\,\mathrm{d}\underline{\mathbf{W}}^m\Big\rangle\!\Big\rangle_\sigma^{\ell\ell}\,\mathrm{d}x\\
&+ c\sum_{\ell=1}^{d}\int_G\int_0^t \mathrm{d}\Big\langle\!\Big\langle \int_0^{\cdot} \Phi(\underline{\mathbf{v}}_k)\,\mathrm{d}\big(\underline{\mathbf{W}}^m - \underline{\mathbf{W}}^k\big)\Big\rangle\!\Big\rangle_\sigma^{\ell\ell}\,\mathrm{d}x\\
&+ c\sum_{\ell=1}^{d}\int_G\int_0^t \mathrm{d}\Big\langle\!\Big\langle \int_0^{\cdot} \big(\Phi(\underline{\mathbf{v}}_m) - \Phi(\underline{\mathbf{v}}_k)\big)\,\mathrm{d}\underline{\mathbf{W}}^m\big), \int_0^{\cdot} \big(\Phi(\underline{\mathbf{v}}_k)\,\mathrm{d}\big(\underline{\mathbf{W}}^m - \underline{\mathbf{W}}^k\big)\Big\rangle\!\Big\rangle_\sigma^{\ell\ell}\,\mathrm{d}x\\
&\le c\sum_{\ell=1}^{d}\int_G\int_0^t \mathrm{d}\Big\langle\!\Big\langle \int_0^{\cdot} \big(\Phi(\underline{\mathbf{v}}_m) - \Phi(\underline{\mathbf{v}}_k)\big)\,\mathrm{d}\underline{\mathbf{W}}^m\Big\rangle\!\Big\rangle_\sigma^{\ell\ell}\,\mathrm{d}x\\
&+ c\sum_{\ell=1}^{d}\int_G\int_0^t \mathrm{d}\Big\langle\!\Big\langle \int_0^{\cdot} \Phi(\underline{\mathbf{v}}_k)\,\mathrm{d}\big(\underline{\mathbf{W}}^m - \underline{\mathbf{W}}^k\big)\Big\rangle\!\Big\rangle_\sigma^{\ell\ell}\,\mathrm{d}x\\
&=: c(VI)_1 + c(VI)_2.
\end{aligned}$$

We have by (10.0.4) and (10.3.49)

$$\begin{aligned}
\underline{\mathbb{E}}\big[(VI)_1\big] &\le c\,\underline{\mathbb{E}}\bigg[\int_0^t \|\Phi(\underline{\mathbf{v}}_m) - \Phi(\underline{\mathbf{v}}_k)\|^2_{L_2(U,L^2(G))}\,\mathrm{d}\sigma\bigg]\\
&\le c\,\underline{\mathbb{E}}\bigg[\int_0^t\int_G |\underline{\mathbf{v}}_m - \underline{\mathbf{v}}_k|^2\,\mathrm{d}x\,\mathrm{d}\sigma\bigg] \longrightarrow 0, \quad m, k \to 0.
\end{aligned}$$

Moreover, since $\underline{\mathbf{v}}_k \in L^2(\underline{\Omega} \times Q, \underline{\mathbb{P}} \otimes \mathcal{L}^{d+1})$ uniformly in k we obtain by (10.0.5) and (10.3.48)

$$\begin{aligned}
\underline{\mathbb{E}}\big[(VI)_2\big] &= \underline{\mathbb{E}}\bigg[\int_0^T \sum_i \Big(\int_G |g_i(\underline{\mathbf{v}}_k)|^2\,\mathrm{Var}\big(\underline{\beta}_i^m(1) - \underline{\beta}_i^k(1)\big)\,\mathrm{d}x\Big)\,\mathrm{d}t\bigg]\\
&\le \underline{\mathbb{E}}\bigg[\int_0^T \Big(\int_G \sup_i i^2 |g_i(\underline{\mathbf{v}}_k)|^2\,\mathrm{d}x\Big)\,\mathrm{d}t\bigg]\sum_i \frac{1}{i^2}\,\mathrm{Var}\big(\underline{\beta}_i^m(1) - \underline{\beta}_i^k(1)\big)\\
&\le c\,\underline{\mathbb{E}}\bigg[\int_0^T\int_G \big(1 + |\underline{\mathbf{v}}_k|^2\big)\,\mathrm{d}x\,\mathrm{d}t\bigg]\underline{\mathbb{E}}\Big[\|\underline{\mathbf{W}}^m - \underline{\mathbf{W}}^k\|^2_{C([0,T],U_0)}\Big]\\
&\longrightarrow 0, \quad m, k \to \infty.
\end{aligned}$$

As a consequence of Corollary 10.1.1 (and the usage of the cut-off function η) we know that Φ^ϑ inherits the properties of Φ. So (VII) can be estimated following the same ideas. Plugging all together, we have shown

$$\begin{aligned}
&\limsup_{m,k}\underline{\mathbb{E}}\bigg[\int_Q \eta\big(\mathbf{S}(\boldsymbol{\varepsilon}(\underline{\mathbf{v}}_m))-\mathbf{S}(\boldsymbol{\varepsilon}(\underline{\mathbf{v}}_k))\big):\Psi_L(|\underline{\mathbf{u}}_{m,k}|)\boldsymbol{\varepsilon}(\underline{\mathbf{u}}_{m,k})\,\mathrm{d}x\,\mathrm{d}\sigma\bigg]\\
&\quad\leq \limsup_{m,k}\underline{\mathbb{E}}\bigg[\int_Q \eta\big(\mathbf{S}(\boldsymbol{\varepsilon}(\underline{\mathbf{v}}_m))-\mathbf{S}(\boldsymbol{\varepsilon}(\underline{\mathbf{v}}_k))\big):\nabla\big\{\Psi_L(|\underline{\mathbf{u}}_{m,k}|)\big\}\otimes\underline{\mathbf{u}}_{m,k}\,\mathrm{d}x\,\mathrm{d}\sigma\bigg]\\
&\quad+\limsup_{m,k}\underline{\mathbb{E}}\bigg[\int_Q \eta\,\underline{\vartheta}^1_{m,k}\operatorname{div}\big(\Psi_L(|\underline{\mathbf{u}}_{m,k}|)\underline{\mathbf{u}}_{m,k}\big)\,\mathrm{d}x\,\mathrm{d}\sigma\bigg]. \qquad (10.3.63)
\end{aligned}$$

Now we want to prove that the right-hand-side is bounded in L. Since $\operatorname{div}\underline{u}_{m,k}=0$ the following holds

$$\begin{aligned}
&\limsup_{m,k}\underline{\mathbb{E}}\bigg[\int_Q \eta\,\underline{\vartheta}^1_{m,k}\operatorname{div}\big(\Psi_L(|\underline{\mathbf{u}}_{m,k}|)\underline{\mathbf{u}}_{m,k}\big)\,\mathrm{d}x\,\mathrm{d}\sigma\bigg]\\
&\quad=\limsup_{m,k}\underline{\mathbb{E}}\bigg[\int_Q \eta\,\underline{\vartheta}^1_{m,k}\nabla\big\{\Psi_L(|\underline{\mathbf{u}}_{m,k}|)\big\}\cdot\underline{\mathbf{u}}_{m,k}\,\mathrm{d}x\,\mathrm{d}\sigma\bigg].
\end{aligned}$$

So, by (10.3.55) and (10.3.56), we only need to show

$$\nabla\Psi_L(|\underline{\mathbf{u}}_{m,k}|)\underline{\mathbf{u}}_{m,k}\in L^p(\underline{\Omega}\times Q,\underline{\mathbb{P}}\otimes\mathcal{L}^{d+1}) \qquad (10.3.64)$$

uniformly in L, m and k to conclude

$$\limsup_{m,k}\underline{\mathbb{E}}\bigg[\int_Q \eta\big(\mathbf{S}(\boldsymbol{\varepsilon}(\underline{\mathbf{v}}_m))-\mathbf{S}(\boldsymbol{\varepsilon}(\underline{\mathbf{v}}_k))\big):\Psi_L(|\underline{\mathbf{u}}_{m,k}|)\boldsymbol{\varepsilon}(\underline{\mathbf{u}}_{m,k})\,\mathrm{d}x\,\mathrm{d}\sigma\bigg]\leq K. \qquad (10.3.65)$$

We have for all $\ell\in\mathbb{N}_0$

$$\begin{aligned}
\big|\nabla\{\psi_{2^{-\ell}}(|\underline{\mathbf{u}}_{m,k}|)\}\underline{\mathbf{u}}_{m,k}\big| &\leq \big|\psi'_{2^{-\ell}}(|\underline{\mathbf{u}}_{m,k}|)\underline{\mathbf{u}}_{m,k}\otimes\nabla\underline{\mathbf{u}}_{m,k}\big|\\
&\leq 2^{-\ell}|\underline{\mathbf{u}}_{m,k}|\psi'(2^{-\ell}|\underline{\mathbf{u}}_{m,k}|)|\nabla\underline{\mathbf{u}}_{m,k}|\\
&\leq c|\nabla\underline{\mathbf{u}}_{m,k}|\chi_{A_\ell},\\
A_\ell &:=\big\{2^\ell<|\underline{\mathbf{u}}_{m,k}|\leq 2^{\ell+1}\big\}.
\end{aligned}$$

Finally, we have

$$\begin{aligned}
\big|\nabla\Psi_L(|\underline{\mathbf{u}}_{m,k}|)\underline{\mathbf{u}}_{m,k}\big| &\leq \sum_{\ell=0}^{L}\big|\nabla\{\psi_{2^{-\ell}}(|\underline{\mathbf{u}}_{m,k}|)\}\underline{\mathbf{u}}_{m,k}\big|\\
&\leq c\sum_{\ell=0}^{L}\big|\nabla\underline{\mathbf{u}}_{m,k}\big|\chi_{A_\ell}\leq c|\nabla\underline{\mathbf{u}}_{m,k}|.
\end{aligned}$$

This implies (10.3.64) and hence (10.3.65) is shown. Now we consider the quantity

$$\Sigma_{L,m,k}:=\underline{\mathbb{E}}\bigg[\int_Q \eta\big(\mathbf{S}(\boldsymbol{\varepsilon}(\underline{\mathbf{v}}_m))-\mathbf{S}(\boldsymbol{\varepsilon}(\underline{\mathbf{v}}_k))\big):\Psi_L(|\underline{\mathbf{u}}_{m,k}|)\boldsymbol{\varepsilon}(\underline{\mathbf{u}}_{m,k})\,\mathrm{d}x\,\mathrm{d}\sigma\bigg].$$

On account of (10.3.65) we have $\Sigma_{L,m} \leq K$ independent of L and m. Thus, using Cantor's diagonalizing principle we obtain a subsequence such that for all $\ell \in \mathbb{N}_0$

$$\sigma_{\ell,m_l,k_l} := \underline{\mathbb{E}}\bigg[\int_Q \eta\big(\mathbf{S}(\boldsymbol{\varepsilon}(\underline{\mathbf{v}}_{m_l})) - \mathbf{S}(\boldsymbol{\varepsilon}(\underline{\mathbf{v}}_{k_l}))\big) : \psi_{2^{-\ell}}(|\underline{\mathbf{u}}_{m_l,k_l}|)\boldsymbol{\varepsilon}(\underline{\mathbf{u}}_{m_l,k_l})\,\mathrm{d}x\,\mathrm{d}\sigma\bigg] \longrightarrow \sigma_\ell$$

as $l \to \infty$. We have, as a consequence of the monotonicity of $\mathbf{S}$, that $\sigma_\ell \geq 0$ for all $\ell \in \mathbb{N}$. Moreover, σ_ℓ is increasing in ℓ. On account of (10.3.65) this implies

$$0 \leq \sigma_0 \leq \frac{\sigma_0 + \sigma_1 + \ldots + \sigma_\ell}{\ell} \leq \frac{K}{\ell}$$

for all $\ell \in \mathbb{N}$. Hence we have $\sigma_0 = 0$ and therefore

$$\underline{\mathbb{E}}\bigg[\int_Q \big(\mathbf{S}(\boldsymbol{\varepsilon}(\underline{\mathbf{v}}_m)) - \mathbf{S}(\boldsymbol{\varepsilon}(\underline{\mathbf{v}}_k))\big) : \psi_1(|\underline{\mathbf{u}}_{m,k}|)\boldsymbol{\varepsilon}(\underline{\mathbf{u}}_{m,k})\,\mathrm{d}x\,\mathrm{d}\sigma\bigg] \longrightarrow 0$$

as $m, k \to 0$. Due to (10.3.50) we conclude

$$\underline{\mathbb{E}}\bigg[\int_Q \big(\mathbf{S}(\boldsymbol{\varepsilon}(\underline{\mathbf{v}}_m)) - \mathbf{S}(\boldsymbol{\varepsilon}(\underline{\mathbf{v}}_k))\big) : \psi_1(|\underline{\mathbf{u}}_{m,k}|)\boldsymbol{\varepsilon}(\underline{\mathbf{v}}_{m,k})\,\mathrm{d}x\,\mathrm{d}\sigma\bigg] \longrightarrow 0 \qquad (10.3.66)$$

as $m, k \to 0$. We obtain

$$\begin{aligned}
&\underline{\mathbb{E}}\bigg[\int_Q \Big(\big(\mathbf{S}(\boldsymbol{\varepsilon}(\underline{\mathbf{v}}_m)) - \mathbf{S}(\boldsymbol{\varepsilon}(\underline{\mathbf{v}}_k))\big) : \boldsymbol{\varepsilon}(\underline{\mathbf{v}}_{m,k})\Big)^\theta \,\mathrm{d}x\,\mathrm{d}\sigma\bigg] \\
&\quad = \int_{\Omega\times Q} \chi_{\{|\underline{\mathbf{u}}_{m,k}|>1\}}\Big(\big(\mathbf{S}(\boldsymbol{\varepsilon}(\underline{\mathbf{v}}_m)) - \mathbf{S}(\boldsymbol{\varepsilon}(\underline{\mathbf{v}}_k))\big) : \boldsymbol{\varepsilon}(\underline{\mathbf{v}}_{m,k})\Big)^\theta \,\mathrm{d}x\,\mathrm{d}\sigma\,\mathrm{d}\mathbb{P} \\
&\quad + \int_{\Omega\times Q} \chi_{\{|\underline{\mathbf{u}}_{m,k}|\leq 1\}}\Big(\big(\mathbf{S}(\boldsymbol{\varepsilon}(\underline{\mathbf{v}}_m)) - \mathbf{S}(\boldsymbol{\varepsilon}(\underline{\mathbf{v}}_k))\big) : \boldsymbol{\varepsilon}(\underline{\mathbf{v}}_{m,k})\Big)^\theta \,\mathrm{d}x\,\mathrm{d}\sigma\,\mathrm{d}\mathbb{P} \\
&\quad =: (A) + (B)
\end{aligned}$$

for all $\theta \in (0, 1)$. By (10.3.59) and (10.3.60) the following holds

$$\begin{aligned}
(A) &\leq \mathbb{P} \otimes \mathcal{L}^{d+1}\big([|\underline{\mathbf{u}}_{m,k}| \geq 1]\big)^{1-\theta} \\
&\quad \times \bigg(\int_{\Omega\times Q} \big(\mathbf{S}(\boldsymbol{\varepsilon}(\underline{\mathbf{v}}_m)) - \mathbf{S}(\boldsymbol{\varepsilon}(\underline{\mathbf{v}}_k))\big) : \boldsymbol{\varepsilon}(\underline{\mathbf{u}}_{m,k})\,\mathrm{d}x\,\mathrm{d}\sigma\,\mathrm{d}\mathbb{P}\bigg)^\theta \\
&\leq c\bigg(\underline{\mathbb{E}}\bigg[\int_Q |\underline{\mathbf{u}}_m - \underline{\mathbf{u}}_k|^2\,\mathrm{d}x\,\mathrm{d}\sigma\bigg]\bigg)^{1-\theta} \longrightarrow 0, \quad m, k \to 0.
\end{aligned}$$

Here we took into account Hölder's inequality. Since (B) vanishes for $m, k \to 0$ by (10.3.66) we finally have shown

$$\underline{\mathbb{E}}\bigg[\int_Q \Big(\big(\mathbf{S}(\boldsymbol{\varepsilon}(\underline{\mathbf{v}}_m)) - \mathbf{S}(\boldsymbol{\varepsilon}(\underline{\mathbf{v}}_k))\big) : \boldsymbol{\varepsilon}\big(\underline{\mathbf{v}}_m - \underline{\mathbf{v}}_k\big)\Big)^\theta \,\mathrm{d}x\,\mathrm{d}\sigma\bigg] \longrightarrow 0, \quad m, k \to 0,$$

for all $\theta < 1$. The monotonicity of $\mathbf{S}$ implies that $\boldsymbol{\varepsilon}(\underline{\mathbf{v}}_m)$ is $\underline{\mathbb{P}} \otimes \mathcal{L}^{d+1}$-a.e. a Cauchy sequence. The limit function therefore exists $\underline{\mathbb{P}} \otimes \mathcal{L}^{d+1}$. On account of $(10.3.55)_1$ it has to be equal to $\boldsymbol{\varepsilon}(\underline{\mathbf{v}})$. This justifies the limit procedure in the energy integral, e.g. $\underline{\tilde{\mathbf{S}}} = \mathbf{S}(\boldsymbol{\varepsilon}(\underline{\mathbf{v}}))$ is shown and the proof of Theorem 10.0.41 is therefore complete.

APPENDIX A

Function spaces

A.1 FUNCTION SPACES INVOLVING THE DIVERGENCE

Given a Young function A, denote by $H^A(G)$ the Banach space of those vector-valued functions $\mathbf{u} : G \to \mathbb{R}^n$ such that the norm

$$\|\mathbf{u}\|_{H^A(G)} = \|\mathbf{u}\|_{L^A(G,\mathbb{R}^n)} + \|\operatorname{div} \mathbf{u}\|_{L^A(G)} \tag{A.1.1}$$

is finite. We also denote by $H_0^A(G)$ its subspace of those functions $\mathbf{u} \in H^A(G)$ whose normal component on ∂G vanishes, in the sense that

$$\int_G \varphi \operatorname{div} \mathbf{u} \, dx = -\int_G \mathbf{u} \cdot \nabla\varphi \, dx \tag{A.1.2}$$

for every $\varphi \in C^\infty(\overline{G})$. It is easy to see that both $H^A(G)$ and $H_0^A(G)$ are Banach spaces. We are interested in some smooth approximation theorems.

Theorem A.1.45. *Let $G \subset \mathbb{R}^d$ be an open set with Lipschitz boundary. Let A be a Young function satisfying the Δ_2-condition. Then $C^\infty(\overline{G})$ is dense in $H^A(G)$.*

Proof. We follow the lines of the proof in [137, Thm. 1.1] and split the proof into three parts. Let $\mathbf{u} \in H^A(G)$ be given.

i) We show that $\mathbf{u}$ can be approximated by functions with compact support. Take a function $\varphi_n \in C_0^\infty(B_{2n}(0))$ with $0 \le \varphi \le 1$ and $\varphi = 1$ in $B_n(0)$. It is easy to check that $\varphi_n \mathbf{u}|_G$ converges to $\mathbf{u}$ in $H^A(G)$.

ii) Let $G = \mathbb{R}^d$, so $\mathbf{u} \in H^A(\mathbb{R}^d)$ and by i) we can assume that $\mathbf{u}$ has compact support. Let ϱ be a standard mollifier, i.e. $\varrho \in C_0^\infty(B_1(0))$, $\varrho \ge 0$ and $\int \varrho \, dx = 1$. We set $\varrho_\varepsilon(x) = \varepsilon^{-d}\varrho\left(\frac{x}{\varepsilon}\right)$ and $\mathbf{u}_\varepsilon = \mathbf{u} * \varrho_\varepsilon$. Of course this means that $\mathbf{u}_\varepsilon \in C_0^\infty(\mathbb{R}^d)$. Moreover, due to Δ_2-condition of A we have for $\varepsilon \to 0$

$$\mathbf{u}_\varepsilon \longrightarrow \mathbf{u} \quad \text{in} \quad L^A(\mathbb{R}^d).$$

As $\operatorname{div} \mathbf{u}_\varepsilon = \operatorname{div}(\mathbf{u} * \varrho_\varepsilon) = (\operatorname{div} \mathbf{u}) * \varrho_\varepsilon$ we have

$$\operatorname{div} \mathbf{u}_\varepsilon \longrightarrow \operatorname{div} \mathbf{u} \quad \text{in} \quad L^A(\mathbb{R}^d)$$

as well. Both together yields that $\mathbf{u}_\varepsilon$ converges to $\mathbf{u}$ in $H^A(G)$.

iii) For the general case $G \neq \mathbb{R}^d$ we can assume that G is locally star-shaped. There is an open covering G, $(\mathcal{O}_j)_{j\in J}$ of $\overline{G}$. Let φ, $(\varphi_j)_{j\in J}$ be

Existence Theory for Generalized Newtonian Fluids.
DOI: http://dx.doi.org/10.1016/B978-0-12-811044-7.00022-7

a partition of unity subordinated to this covering, where $\varphi \in C_0^\infty(G)$ and $\varphi_j \in C_0^\infty(\mathcal{O}_j)$. We have

$$\mathbf{u} = \varphi\mathbf{u} + \sum_{j\in J} \varphi_j\mathbf{u}.$$

The sum is finite as we can assume that $\mathbf{u}$ has compact support by i). The function $\varphi\mathbf{u}$ has compact support in G and by standard mollification as in ii) it can be approximated by a sequence in $C_0^\infty(G)$. So it remains to show that we can approximate the functions $\varphi_j\mathbf{u}$. The set $\mathcal{O}_j' := \mathcal{O}_j \cap G$ is star-shaped with respect to some of its points; for simplicity we assume that this point is 0. We consider the function $\mathbf{u}_j = \varphi_j\mathbf{u}$ and for $\mathbf{u}_j^\lambda = \mathbf{u}_j(\lambda\cdot)$ for $\lambda > 1$. It is easy to show that $\mathbf{u}_j^\lambda|_{\mathcal{O}_j'}$ converges to $\mathbf{u}_j$ in $H^A(\mathcal{O}_j')$ for $\lambda \to 1$. So it is enough to approximate $\mathbf{u}_j^\lambda$ instead of $\mathbf{u}_j$. The function $\mathbf{u}_j^\lambda$ as compact support and hence belongs to $H^A(\mathbb{R}^d)$. On account of ii) it can be approximated in $H^A(\mathbb{R}^d)$ by functions from $C_0^\infty(\mathbb{R}^d) \subset C^\infty(\overline{G})$. Restricting to G yields the convergence in $H^A(G)$ we are looking for. □

Theorem A.1.46. *Let $G \subset \mathbb{R}^d$ be an open set with Lipschitz boundary. Let A be a Young function satisfying the Δ_2-condition. Then there holds*

$$H_0^A(G) \subset \overline{C_0^\infty(G)}^{\|\cdot\|_{H^A(G)}}. \tag{A.1.3}$$

Proof. Let $\tilde{\mathbf{u}}$ be the extension of $\mathbf{u}$ to $\mathbb{R}^d$ by 0. By (A.1.2) we have

$$\int_{\mathbb{R}^d} \tilde{\mathbf{u}} \cdot \nabla\varphi \,\mathrm{d}x + \int_{\mathbb{R}^d} \operatorname{div}\mathbf{u}\,\varphi \,\mathrm{d}x = 0$$

for every $\varphi \in C_0^\infty(\mathbb{R}^d)$. Hence $\operatorname{div}\tilde{\mathbf{u}} = \chi_G \operatorname{div}\mathbf{u}$ and so

$$\int_{\mathbb{R}^d} \tilde{\mathbf{u}} \cdot \nabla\varphi \,\mathrm{d}x + \int_{\mathbb{R}^d} \operatorname{div}\tilde{\mathbf{u}}\,\varphi \,\mathrm{d}x = 0$$

and $\tilde{\mathbf{u}} \in H^A(G)$. We now follow the ideas from the proof of Theorem A.1.45 and can reduce the situation to the case that G is star-shaped with respect to 0. For $\lambda > 1$ we consider $\tilde{\mathbf{u}}_\lambda = \tilde{\mathbf{u}}(\lambda\cdot)$ and have $\tilde{\mathbf{u}}_\lambda \to \tilde{\mathbf{u}}$ in $H^A(G)$ for $\lambda \to 1$. But $\tilde{\mathbf{u}}_\lambda$ has compact support in G and so has the mollification $(\tilde{\mathbf{u}}_\lambda)_\varepsilon$ for ε small enough. On account of the properties of the mollification (using the Δ_2-condition) we are able to approximate $\mathbf{u}$ by a sequence of $C_0^\infty(G)$-functions. □

Remark A.1.26. It is possible to show $\overline{C_0^\infty(G)}^{\|\cdot\|_{H^A(G)}} \subset H_0^A(G)$ provided the boundary of G is smooth enough. In fact one has to introduce the trace $\gamma_{\mathcal{N}}(\mathbf{u}) = \mathcal{N} \cdot \mathbf{u}|_{\partial G}$ in a negative Sobolev space and follow the ideas in [137, Thm. 1.3]. In the case studied there the surjectivity of the trace map $W^{1,2}(G) \to W^{1/2,2}(\partial G)$ is used. An analogone to this in Orlicz spaces is not known. So a Lipschitz boundary might not be sufficient.

A.2 FUNCTION SPACES INVOLVING SYMMETRIC GRADIENTS

We start with a survey about the space BD(G) containing all functions of bounded deformation introduced by Suquet [136] and by Matthies, Strang, Christiansen [110]. The class BD(G) has been widely considered in the literature in connection with problems from plasticity, we refer to the works of Anzellotti and Giaquinta [11], Teman and Strang [135] and Teman [138]. The space BD(G) is defined as the set of L^1-functions $\mathbf{u}$ with

$$\sup_{\eta \in C_0^1(G,\mathbb{R}^{d\times d}_{sym}),\|\eta\|_\infty=1} \int_G \mathbf{u} \cdot \operatorname{div} \eta \,\mathrm{d}x < \infty.$$

By Riesz' representation Theorem it can be shown $\mathbf{u} \in L^1(G)$ belongs to BD(G) if and only if $\boldsymbol{\varepsilon}(\mathbf{u})$ generates a bounded Radon measure on G. This means every component of $\boldsymbol{\varepsilon}(\mathbf{u})$ belongs to the class

$$\begin{aligned} \mathcal{M}(G) &:= \text{set of all signed measures } \mu \text{ defined on } \mathcal{B}(G) \\ &\qquad \text{such that } |\mu|(G) < \infty, \\ \mathfrak{B}(G) &:= \sigma\text{-algebra of all sets } A = B \cap G \text{ with } B \subset \mathbb{R}^d \text{ Borel.} \end{aligned}$$

If $\mathbf{u} \in$ BD(G) then there is $\boldsymbol{\mu} = (\mu_{ij})_{i,j=1}^d$ with $\mu_{ij} \in \mathcal{M}(G)$ for $i,j = 1, ..., d$ such that the distributions

$$C_0^\infty(G) \ni \varphi \mapsto \langle \varepsilon_{ij}(\mathbf{u}), \varphi \rangle = -\frac{1}{2} \int_G \left(\mathbf{u}^i \partial_j \varphi + \mathbf{u}^j \partial_i \varphi\right) \mathrm{d}x$$

can be represented as

$$\langle \varepsilon_{ij}(\mathbf{u}), \varphi \rangle = \int_G \varphi \,\mathrm{d}\mu_{ij}$$

and we have

$$\begin{aligned} |\mu_{ij}|(A) = |\varepsilon_{ij}(\mathbf{u})|(A) &= \sup_{\varphi \in C_0^\infty(A),\|\varphi\|_\infty=1} \langle \varepsilon_{ij}(\mathbf{u}), \varphi \rangle, \\ |\boldsymbol{\mu}|(A) = |\boldsymbol{\varepsilon}(\mathbf{u})|(A) &= \left(\sum_{i,j=1}^d |\varepsilon_{ij}(\mathbf{u})|^2(A) \right)^{\frac{1}{2}}, \end{aligned}$$

for all $A \in \mathfrak{B}(G)$. The space BD(G) is a Banach space equipped with the norm

$$\|\mathbf{u}\|_{\mathrm{BD}(G)} := \|\mathbf{u}\|_{L^1(G)} + |\boldsymbol{\varepsilon}(\mathbf{u})|(G), \tag{A.2.4}$$

where $|\boldsymbol{\varepsilon}(\mathbf{u})|(G)$ is the total variation of the matrix valued measure $\boldsymbol{\varepsilon}(\mathbf{u})$. From the above references we deduce the following basic properties of $\mathrm{BD}(G)$.

Lemma A.2.1. *a) The space* $\mathrm{BD}(G)$ *is continuously embedded into the Lebesgue space* $L^{d/(d-1)}(G)$.
b) For $1 \le p < d/(d-1)$ *the embedding* $\mathrm{BD}(G) \hookrightarrow L^p(G)$ *is compact.*

Lemma A.2.2. *a) There is continuous linear operator*

$$\gamma : \mathrm{BD}(G) \to L^1(\partial G, \mathcal{H}^{d-1})$$

with $\gamma(\mathbf{u}) = \mathbf{u}|_{\partial G}$ *for all* $\mathbf{u} \in C(\overline{G})$.
b) Let $\mathbf{u}, \mathbf{u}_k \in \mathrm{BD}(G)$ *with* $\mathbf{u}_k \to \mathbf{u}$ *in* $L^1(G)$ *and* $|\boldsymbol{\varepsilon}(\mathbf{u}_k)|(G) \to |\boldsymbol{\varepsilon}(\mathbf{u})|(G)$ *for* $k \to \infty$. *Then we have*

$$\gamma(\mathbf{u}_k) \to \gamma(\mathbf{u}) \quad \text{in} \quad L^1(\partial G, \mathcal{H}^{d-1}).$$

c) If $\gamma(\mathbf{u}) = 0$ *then we have*

$$\|\mathbf{u}\|_{L^{d/(d-1)}(G)} \le c(d, G)|\boldsymbol{\varepsilon}(\mathbf{u})|(G). \tag{A.2.5}$$

Lemma A.2.3. *Let* $G \subset \mathbb{R}^d$ *be a bounded star-shaped domain with Lipschitz boundary* ∂G. *Then, for every* $\mathbf{u} \in BD(G)$ *there exists a sequence* $(\mathbf{u}_k) \subset \mathrm{BD}(G) \cap C_0^\infty(G)$ *such that*

$$\mathbf{u}_k \to \mathbf{u} \quad \text{in} \quad L^{d/(d-1)}(G), \quad k \to \infty, \tag{A.2.6}$$

$$\int_G |\boldsymbol{\varepsilon}(\mathbf{u}_k)|\, \mathrm{d}x \to |\boldsymbol{\varepsilon}(\mathbf{u})|(G) + \int_{\partial G} |\gamma(\mathbf{u}) \odot \mathcal{N}|\, \mathrm{d}\mathcal{H}^{d-1}, \quad k \to \infty. \tag{A.2.7}$$

Having in mind the results about the space $\mathrm{BD}(G)$ we introduce the space (this part presents results from [33])

$$E^h(G) := \left\{ \mathbf{u} \in L^1(G) : \int_G h(|\boldsymbol{\varepsilon}(\mathbf{u})|)\, \mathrm{d}x < \infty \right\}, \qquad h(t) = t \ln(1+t),$$

$$\|\mathbf{u}\|_{E^h(G)} := \|\mathbf{u}\|_{L^1(G)} + \|\boldsymbol{\varepsilon}(\mathbf{u})\|_{L^h(G)}.$$

Note that we can similarly introduce spaces $E^A(G)$ with an arbitrary N-function A. We will use the notation $E^p(G)$ for the space of functions with symmetric gradients in $L^p(G)$, $1 \le p \le \infty$. From (A.2.5) it follows that on the subspace $\mathrm{BD}(\Omega) \cap \{\mathbf{u} : \mathbf{u}|_{\partial\Omega} = 0\}$ the BD-norm defined in (A.2.4) can be replaced by the equivalent norm $|\boldsymbol{\varepsilon}(\cdot)|(G)$. We observe that (cf. [77], Lemma 4.1.6)

$$\begin{aligned} E_0^h(G) &:= \overline{C_0^\infty(G)}^{E^h(G)} = \{\mathbf{u} \in E^h(G) : \mathbf{u}|_{\partial G} = 0\}, \\ E_0^h(G) &:= \overline{C_{0,\mathrm{div}}^\infty(G)}^{E^h(G)} = \{\mathbf{u} \in E_0^h(G) : \operatorname{div} \mathbf{u} = 0\}, \end{aligned} \tag{A.2.8}$$

where $\mathbf{u}|_{\partial\Omega}$ has to be understood in the BD-trace sense. We therefore have inequality (A.2.5) for functions $\mathbf{u} \in E_0^h(\Omega)$, which means that

$$\|\mathbf{u}\|_{E_0^h(G)} := \|\boldsymbol{\varepsilon}(\mathbf{u})\|_{L^h(G)} \tag{A.2.9}$$

is a norm equivalent to $\|\cdot\|_{E^{1,h}(G)}$ on the class $E_0^h(G)$.

From Korn's inequality (see Theorem 2.3.11) it follows that $E_0^h(G) \hookrightarrow W_0^{1,1}(G)$.

Another consequence of Korn's inequality is:

Lemma A.2.4. *Let* $\mathbf{u} \in E_0^h(G)$. *Then the field* $\mathbf{w} := \ln(1 + |\mathbf{u}|)\mathbf{u}$ *belongs to the space* $\mathrm{BD}(G)$, *and the total variation* $|\boldsymbol{\varepsilon}(\mathbf{w})|(G)$ *of* $\mathbf{w}$ *is bounded in terms of* $\|\boldsymbol{\varepsilon}(\mathbf{u})\|_{L^h(G)}$, *i.e. we have*

$$|\boldsymbol{\varepsilon}(\mathbf{w})|(G) \le C\left(\|\mathbf{u}\|_{E_0^h(G)}\right). \tag{A.2.10}$$

Proof. Consider first the case $\mathbf{u} \in C_0^\infty(G)$. Then it holds

$$\boldsymbol{\varepsilon}(\mathbf{w}) = \ln(1 + |\mathbf{u}|)\boldsymbol{\varepsilon}(\mathbf{u}) + \frac{1}{2}\left(\mathbf{u}^i\partial_j \ln(1 + |\mathbf{u}|) + \mathbf{u}^j\partial_i \ln(1 + |\mathbf{u}|)\right)_{1\le i,j\le n},$$

hence

$$|\boldsymbol{\varepsilon}(\mathbf{w})| \le \ln(1 + |\mathbf{u}|)|\boldsymbol{\varepsilon}(\mathbf{u})| + c(n)\frac{|\mathbf{u}|}{1 + |\mathbf{u}|}|\nabla\mathbf{u}|.$$

From Young's inequality for N-functions we get for $s,\, t \ge 0$

$$h'(t)s \le \tilde{h}(h'(t)) + h(s),$$

$\tilde{h}$ denoting the conjugate function of h. Moreover we have

$$\tilde{h}(h'(t)) = th'(t) - h(t) \le h(t).$$

These inequalities imply

$$\ln(1 + |\mathbf{u}|)|\boldsymbol{\varepsilon}(\mathbf{u})| \le h'(|\mathbf{u}|)|\boldsymbol{\varepsilon}(\mathbf{u})| \le h(|\mathbf{u}|) + h(|\boldsymbol{\varepsilon}(\mathbf{u})|,$$

hence

$$\int_G |\boldsymbol{\varepsilon}(\mathbf{w})|\,\mathrm{d}x \le \int_\Omega h(|\mathbf{u}|)\,\mathrm{d}x + \int_G h(|\boldsymbol{\varepsilon}(\mathbf{u})|)\,\mathrm{d}x + c(n)\int_G |\nabla\mathbf{u}|\,\mathrm{d}x.$$

The quantity $\int_G h(|\boldsymbol{\varepsilon}(\mathbf{u})|)\,\mathrm{d}x$ can be estimated in terms of $\|\boldsymbol{\varepsilon}(\mathbf{u})\|_{L^h(G)}$ (and vice versa), to $\int_G |\nabla\mathbf{u}|\,\mathrm{d}x$ we apply Lemma A.2.1, and finally observe that $\int_G h(|\mathbf{u}|)\,\mathrm{d}x$ is bounded e.g. by $\int_G |\mathbf{u}|^{d/d-1}\,\mathrm{d}x$ and this integral can be handled via (A.2.5). Altogether we have (A.2.10) for the smooth case.

If $\mathbf{u} \in E_0^h(G)$ is arbitrary, then we choose $\mathbf{u}_\nu \in C_0^\infty(G)$ such that $\|\mathbf{u} - \mathbf{u}_\nu\|_{E_0^h(G)} \to 0$ as $\nu \to \infty$. This in particular gives $\|\mathbf{u}_\nu\|_{E_0^h} \to \|\mathbf{u}\|_{E_0^h(G)}$, and (A.2.10) shows that

$$\sup_\nu \int_G |\boldsymbol{\varepsilon}(\mathbf{w}_\nu)|\,\mathrm{d}x < \infty, \qquad \mathbf{w}_\nu := \ln(1 + |\mathbf{u}_\nu|)\mathbf{u}_\nu. \tag{A.2.11}$$

If we apply (A.2.5) to $\mathbf{u}_\nu - \mathbf{u}$, we get $\mathbf{u}_\nu \to \mathbf{u}$ in $L^{d/(d-1)}(G)$, and for a suitable subsequence it holds $\mathbf{u}_\nu \to \mathbf{u}$ a.e., and therefore $\mathbf{w}_\nu \to \mathbf{w}$ a.e. By (A.2.11) and (A.2.5) we see that $\{\mathbf{w}_\nu\}$ is bounded sequence in $\mathrm{BD}(G)$, thus there is a strongly convergent subsequence in $L^1(G)$ (see Lemma A.2.1) which means that there exists $\widetilde{\mathbf{w}} \in \mathrm{BD}(G)$ such that $\mathbf{w}_\nu \to \widetilde{\mathbf{w}}$ in $L^1(G)$. The finiteness of $|\boldsymbol{\varepsilon}(\widetilde{\mathbf{w}})|(G)$ follows by lower semi-continuity, i.e.

$$|\boldsymbol{\varepsilon}(\widetilde{\mathbf{w}})|(G) \le \liminf_{\nu\to\infty} \int_G |\boldsymbol{\varepsilon}(\mathbf{w}_\nu)|\,\mathrm{d}x\,. \tag{A.2.12}$$

Clearly we have $\widetilde{\mathbf{w}} = \mathbf{w}$, and (A.2.10) for $\mathbf{w}$ follows from (A.2.12) and the version of (A.2.10) for $\mathbf{w}_\nu$. □

Now we can prove the main result of this section:

Theorem A.2.47. *The embedding $E_0^h(G) \hookrightarrow L^{d/(d-1)}(G)$ is compact. More precisely, if $\mathbf{u}_\nu$ denotes a bounded sequence in $E_0^h(G)$, then there exists a subsequence $\mathbf{u}_\nu$ (not relabelled) and a function $\mathbf{u} \in E_0^h(G)$ such that $\mathbf{u}_\nu \to \mathbf{u}$ in $L^{d/(d-1)}(\Omega)$ and $\boldsymbol{\varepsilon}(\mathbf{u}_\nu) \rightharpoonup \boldsymbol{\varepsilon}(\mathbf{u})$ in $L^1(G)$ for $\nu \to \infty$.*

Proof. Suppose that $\sup_{\nu\in\mathbb{N}} \|\mathbf{u}_\nu\|_{E_0^h(G)} < \infty$. From Lemma A.2.4 we deduce the existence of a field $\mathbf{u} \in L^1(G)$ such that

$$\mathbf{u}_\nu \to \mathbf{u} \quad \text{in } L^1(G) \quad \text{and a.e.}, \tag{A.2.13}$$

where here and in what follows we will pass to subsequences whenever this is necessary. According to the De La Vallée Poussin criterion for weak compactness in L^1 or by a theorem of Dunford and Pettis (cf. [9], Theorem 1.38) we get from

$$\sup_{\nu\in\mathbb{N}} \int_G |\boldsymbol{\varepsilon}(\mathbf{u}_\nu)| \ln(1 + |\boldsymbol{\varepsilon}(\mathbf{u}_\nu)|)\,\mathrm{d}x < \infty$$

that $\boldsymbol{\varepsilon}(\mathbf{u}_\nu) \rightharpoonup: \boldsymbol{\sigma}$ in $L^1(G)$, and clearly $\boldsymbol{\sigma} = \boldsymbol{\varepsilon}(\mathbf{u})$. Moreover, by lower semi-continuity it holds

$$\int_\Omega h(|\boldsymbol{\varepsilon}(\mathbf{u})|)\,\mathrm{d}x \le \liminf_{\nu\to\infty} \int_G h(|\boldsymbol{\varepsilon}(\mathbf{u}_\nu)|)\,,$$

so that $\mathbf{u}$ is an element of the space $E^{1,h}(G)$. In order to show $\mathbf{u} \in E_0^h(G)$, we follow the arguments of Frehse and Seregin [81]: since $\boldsymbol{\varepsilon}(\mathbf{u}_\nu) \rightharpoonup \boldsymbol{\varepsilon}(\mathbf{u})$ in $L^1(G)$ we can find a sequence $\{\boldsymbol{\sigma}_\mu\}$, $\boldsymbol{\sigma}_\mu$ being an element of the convex hull of $\{\boldsymbol{\varepsilon}(\mathbf{u}_\nu) : \nu \geq \mu\}$, such that $\boldsymbol{\sigma}_\mu \to \boldsymbol{\varepsilon}(\mathbf{u})$ in $L^1(G)$. This follows from the well-known Banach–Saks lemma. We have

$$\boldsymbol{\sigma}_\mu = \sum_{\nu=\mu}^{N(\mu)} \lambda_\nu^\mu \boldsymbol{\varepsilon}(\mathbf{u}_\nu), \ \sum_{\nu=\mu}^{N(\mu)} \lambda_\nu^\mu = 1, \ 0 \leq \lambda_\nu^\mu \leq 1$$

with suitable coefficients λ_ν^μ and integers $N(\mu) \geq \mu$. Let

$$\overline{\mathbf{u}}_\mu := \sum_{\nu=\mu}^{N(\mu)} \lambda_\nu^\mu \, \mathbf{u}_\nu .$$

These functions belong to $E_0^h(G)$ and satisfy

$$\|\overline{\mathbf{u}}_\mu - \mathbf{u}\|_{L^1(G)} \leq \sum_{\nu=\mu}^{N(\mu)} \lambda_\nu^\mu \|\mathbf{u}_\nu - \mathbf{u}\|_{L^1(G)} \to 0, \quad \mu \to \infty,$$

which is a consequence of (A.2.13). Moreover it holds

$$\int_G |\boldsymbol{\varepsilon}(\overline{\mathbf{u}}_\mu)| \, dx = \int_G |\boldsymbol{\sigma}_\mu| \, dx \to \int_G |\boldsymbol{\varepsilon}(\mathbf{u})| \, dx, \ \mu \to \infty,$$

and according to Lemma A.2.2 b) these two convergences imply the L^1-convergence of the traces of $\mathbf{u}_\mu$ towards the trace of $\mathbf{u}$. In conclusion $\mathbf{u}|_{\partial G} = 0$, hence $\mathbf{u} \in E_0^h(G)$, and it remains to show that

$$\mathbf{u}_\nu \to \mathbf{u} \quad \text{in } L^{d/(d-1)}(G) \tag{A.2.14}$$

holds. From our assumption combined with (A.2.10) we get

$$\sup_{\nu \in \mathbb{N}} \int_\Omega |\boldsymbol{\varepsilon}(\mathbf{w}_\nu)| < \infty , \tag{A.2.15}$$

$\mathbf{w}_\nu := \ln(1 + |\mathbf{u}_\nu|)\mathbf{u}_\nu$, and (A.2.15) together with the first part of Lemma A.2.4 gives

$$\sup_{\nu \in \mathbb{N}} \|\mathbf{w}_\nu\|_{L^{d/(d-1)}(G)} < \infty . \tag{A.2.16}$$

Let $\Gamma(t) := h\left(t^{\frac{d-1}{d}}\right)^{d/(d-1)}$, $t \geq 0$. Then

$$\frac{\Gamma(t)}{t} = \left[\frac{h\left(t^{\frac{d-1}{d}}\right)}{t^{\frac{d-1}{d}}}\right]^{\frac{d}{d-1}} \longrightarrow \infty, \ t \to \infty , \tag{A.2.17}$$

and (compare (A.2.16))

$$\int_{\Omega} \Gamma\left(|\mathbf{u}_\nu|^{\frac{d}{d-1}}\right) \mathrm{d}x = \int_{G} h(|\mathbf{u}_\nu|)^{\frac{d}{d-1}} \mathrm{d}x = \int_{\Omega} |\mathbf{w}_\nu|^{\frac{d}{d-1}} \mathrm{d}x \leq c < \infty\,, \qquad \text{(A.2.18)}$$

therefore $|\mathbf{u}_\nu|^{d/(d-1)} \rightharpoonup: g$ weakly in $L^1(G)$ by quoting the De La Vallée Poussin criterion one more time. By (A.2.13) we must have $g = |\mathbf{u}|^{d/(d-1)}$, since $|\mathbf{u}_\nu|^{d/(d-1)} \to |\mathbf{u}|^{d/(d-1)}$ a.e. on Ω. This in particular implies

$$\|\mathbf{u}_\nu\|_{L^{d/(d-1)}(G)} \to \|\mathbf{u}\|_{L^{d/(d-1)}(G)}, \;\; \nu \to \infty\,,$$

where we combined (A.2.17) and (A.2.18) with Vitali's Theorem. At the same time it follows from

$$\sup_{\nu \in \mathbb{N}} \|\mathbf{u}_\nu\|_{L^{d/(d-1)}(G)} < \infty$$

and (A.2.13), that $\mathbf{u}_\nu \rightharpoonup \mathbf{u}$ in $L^{d/(d-1)}(G)$. Putting both convergences together, the Radon–Riesz lemma (cf. [89], p. 47, Proposition 3) gives our claim (A.2.14), and Theorem A.2.47 is proved. □

In the setting of Prandtl–Eyring fluids we have to work in the space $E^h_{0,\mathrm{div}}(G)$ which according to Lemma 4.1.6 in [77] is the closure of $C^\infty_{0,\mathrm{div}}(G)$ in the class $E^h(G)$ w.r.t. the norm $\|\cdot\|_{E^h(G)}$. From Theorem A.2.47 it follows

Corollary A.2.1. *The statement of Theorem A.2.47 remains valid, if the space $E^h_0(G)$ is replaced by the subclass $E^h_{0,\mathrm{div}}(G)$.*

APPENDIX B

The $\mathcal{A}$-Stokes system

The aim of this section is to present regularity results for the (non-stationary) $\mathcal{A}$-Stokes system depending on the right hand side (in divergence form). Let us fix for this section a bounded domain $G \subset \mathbb{R}^3$ with C^2-boundary and a time interval $(0, T)$. Moreover, let $\mathcal{A} : \mathbb{R}^{3\times 3} \to \mathbb{R}^{3\times 3}$ be an elliptic tensor.

B.1 THE STATIONARY PROBLEM

The $\mathcal{A}$-Stokes problem (in the pressure-free formulation) with right hand side $\mathbf{f} \in L^1(G)$ reads as: find $\mathbf{v} \in \mathrm{LD}_{0,\mathrm{div}}(G)$ such that

$$\int_G \mathcal{A}(\boldsymbol{\varepsilon}(\mathbf{v}), \boldsymbol{\varepsilon}(\boldsymbol{\varphi}))\,\mathrm{d}x = \int_G \mathbf{f}\cdot\boldsymbol{\varphi}\,\mathrm{d}x \quad \text{for all} \quad \boldsymbol{\varphi} \in C^\infty_{0,\mathrm{div}}(G). \tag{B.1.1}$$

The right hand side can also be given in divergence form, i.e.

$$\int_G \mathcal{A}(\boldsymbol{\varepsilon}(\mathbf{v}), \boldsymbol{\varepsilon}(\boldsymbol{\varphi}))\,\mathrm{d}x = \int_G \mathbf{F} : \nabla\boldsymbol{\varphi}\,\mathrm{d}x \quad \text{for all} \quad \boldsymbol{\varphi} \in C^\infty_{0,\mathrm{div}}(G) \tag{B.1.2}$$

for $\mathbf{F} \in L^1(G)$. For certain purposes it is convenient to discuss the problem with a fixed divergence. To be precise for $g \in L^1_\perp(G)$ we are seeking for a function $\mathbf{v} \in \mathrm{LD}_0(G)$ with $\operatorname{div}\mathbf{v} = g$ satisfying (B.1.1) or (B.1.2). We have the following L^q-estimates.

Lemma B.1.1. *Let $G \subset \mathbb{R}^3$ be a bounded C^2-domain and $1 < q < \infty$.*

a) *Let $\mathbf{f} \in L^q(G)$ and $g \in W^{1,q}(G)$ with $\int_G g\,\mathrm{d}x = 0$. Then there is a unique solution $\mathbf{w} \in W^{2,q} \cap W^{1,q}_0(G)$ to (B.1.1) such that $\operatorname{div}\mathbf{v} = g$ and*

$$\fint_B |\nabla^2\mathbf{w}|^q\,\mathrm{d}x \le c\fint_G |\mathbf{f}|^q\,\mathrm{d}x + c\fint_G |\nabla g|^q\,\mathrm{d}x,$$

where c only depends on $\mathcal{A}$ and q.

b) *Let $\mathbf{F} \in L^q(G)$ and $g \in L^q_0(G)$. Then there is a unique solution $\mathbf{w} \in W^{1,q}_0(G)$ to (B.1.2) such that $\operatorname{div}\mathbf{v} = g$ and*

$$\fint_G |\nabla\mathbf{w}|^q\,\mathrm{d}x \le c\fint_G |\mathbf{F}|^q\,\mathrm{d}x + c\fint_G |g|^q\,\mathrm{d}x,$$

where c only depends on $\mathcal{A}$ and q.

Existence Theory for Generalized Newtonian Fluids.
DOI: http://dx.doi.org/10.1016/B978-0-12-811044-7.00023-9

In case $\mathcal{A} = I$ both parts follow from [10], Thm. 4.1. However, the main tool in [10] is the theory from [6,7] where very general linear systems are investigated. Hence it is clear that the results also hold in case of an arbitrary elliptic tensor $\mathcal{A}$.

Corollary B.1.1. *Let the assumptions of Lemma B.1.1 be satisfied. Assume further that* $\mathbf{f} = 0$ *and*

$$g = \operatorname{div} \mathbf{g} + g_0 \tag{B.1.3}$$

with $\mathbf{g} \in W^{1,q}(G)$ *and* $g_0 \in L_0^q(G)$ *with* $\operatorname{spt} g_0 \Subset G$. *Then we have*

$$\int_G |\mathbf{w}|^q \, dx \leq c \int_G |\mathbf{g}|^q \, dx + c \int_{\partial G} |\mathbf{g} \cdot \mathcal{N}|^q \, d\mathcal{H}^2 + c\,\mathcal{L}^3(\operatorname{spt} g_0)^\beta \int_G |g_0|^q \, dx$$

for some $\beta > 0$.

Proof. We follow the lines of [133, Thm. 2.4]. Let $\boldsymbol{\psi} \in L^{q'}(G)$ be arbitrary. In accordance to Lemma B.1.1 a) there is a unique solution $\mathbf{u} \in W^{2,q'} \cap W^{1,q'}_{0,\operatorname{div}}(G)$ to the $\mathcal{A}$-Stokes problem with right-hand-side $\boldsymbol{\psi}$ such that

$$\int_G |\nabla^2 \mathbf{u}|^{q'} \, dx \leq c \int_G |\boldsymbol{\psi}|^{q'} \, dx.$$

By De Rahm's theorem there is a unique $\vartheta \in L_\perp^{q'}(G)$ such that

$$\int_G \mathcal{A}(\boldsymbol{\varepsilon}(\mathbf{u}), \boldsymbol{\varepsilon}(\boldsymbol{\varphi})) \, dx = \int_G \vartheta \operatorname{div} \boldsymbol{\varphi} + \int_G \boldsymbol{\psi} \cdot \boldsymbol{\varphi} \, dx \quad \text{for all } \boldsymbol{\varphi} \in W_0^{1,q}(G).$$

We can conclude that $\vartheta \in W^{1,q'}(G)$ with

$$\fint_G |\nabla \vartheta|^{q'} \, dx \leq c\left(\fint_G |\nabla^2 \mathbf{u}|^{q'} \, dx + \fint_G |\boldsymbol{\psi}|^{q'} \, dx\right) \leq c \fint_G |\boldsymbol{\psi}|^{q'} \, dx. \tag{B.1.4}$$

We proceed by

$$\begin{aligned}
\int_G \mathbf{w} \cdot \boldsymbol{\psi} \, dx &= \int_G \mathcal{A}(\boldsymbol{\varepsilon}(\mathbf{u}), \boldsymbol{\varepsilon}(\mathbf{w})) \, dx - \int_G \vartheta \operatorname{div} \mathbf{w} \, dx = -\int_G \vartheta \operatorname{div} \mathbf{w} \, dx \\
&= -\int_G \vartheta \operatorname{div} \mathbf{g} \, dx - \int_G g_0 \, \vartheta \, dx \\
&= \int_G \nabla \vartheta \cdot \mathbf{g} \, dx - \int_{\partial G} \mathbf{g} \cdot \mathcal{N} \vartheta \, d\mathcal{H}^2 - \int_G g_0 \, \vartheta \, dx.
\end{aligned}$$

We finish the proof by estimating this integrals separately using (B.1.4). For the first one we obtain

$$\begin{aligned}
\left|\int_G \nabla \vartheta \cdot \mathbf{g} \, dx\right| &\leq \left(\int_G |\nabla \vartheta|^{q'} \, dx\right)^{\frac{1}{q'}} \left(\int_G |\mathbf{g}|^q \, dx\right)^{\frac{1}{q}} \\
&\leq \left(\int_G |\boldsymbol{\psi}|^{q'} \, dx\right)^{\frac{1}{q'}} \left(\int_G |\mathbf{g}|^q \, dx\right)^{\frac{1}{q}}.
\end{aligned}$$

Similarly, we estimate using the trace theorem

$$\begin{aligned}\left|\int_{\partial G} \vartheta \mathbf{g}\cdot\mathcal{N}\,\mathrm{d}\mathcal{H}^2\right| &\le \left(\int_{\partial G}|\vartheta|^{q'}\,\mathrm{d}\mathcal{H}^2\right)^{\frac{1}{q'}}\left(\int_{\partial G}|\mathbf{g}\cdot\mathcal{N}|^q\,\mathrm{d}\mathcal{H}^2\right)^{\frac{1}{q}}\\ &\le \left(\int_{G}|\nabla\vartheta|^{q'}\,\mathrm{d}x\right)^{\frac{1}{q'}}\left(\int_{\partial G}|\mathbf{g}\cdot\mathcal{N}|^q\,\mathrm{d}\mathcal{H}^2\right)^{\frac{1}{q}}\\ &\le \left(\int_{G}|\boldsymbol{\psi}|^{q'}\,\mathrm{d}x\right)^{\frac{1}{q'}}\left(\int_{\partial G}|\mathbf{g}\cdot\mathcal{N}|^q\,\mathrm{d}\mathcal{H}^2\right)^{\frac{1}{q}}.\end{aligned}$$

In case $q' < 3$ we obtain for the third integral

$$\begin{aligned}\left|\int_{G} g_0\,\vartheta\,\mathrm{d}x\right| &\le \left(\int_{G}|\vartheta|^{\frac{3q'}{3-q'}}\,\mathrm{d}x\right)^{\frac{3-q'}{dq'}}\left(\int_{G}|g_0|^{\frac{3q'}{4q'-3}}\,\mathrm{d}x\right)^{\frac{4q'-3}{3q'}}\\ &\le \left(\int_{G}|\nabla\vartheta|^{q'}\,\mathrm{d}x\right)^{q'}\mathcal{L}^3(\operatorname{spt} g_0)^{\beta}\left(\int_{G}|g_0|^q\,\mathrm{d}x\right)^{\frac{1}{q}}\\ &\le \left(\int_{G}|\boldsymbol{\psi}|^{q'}\,\mathrm{d}x\right)^{\frac{1}{q'}}\mathcal{L}^3(\operatorname{spt} g_0)^{\beta}\left(\int_{G}|g_0|^q\,\mathrm{d}x\right)^{\frac{1}{q}},\end{aligned}$$

where $\beta = \frac{1}{3}$. If $q' \ge 3$ we can modify the proof by replacing $\frac{3q'}{3-q'}$ with an arbitrary number $r > q'$ and choose $\beta = 3r'$. As $\boldsymbol{\psi}$ is arbitrary, the claim follows. □

B.2 THE NON-STATIONARY PROBLEM

Now we turn to the parabolic problem and the first result is a local L^q-estimate for weak solutions. In case of the $\mathcal{A}$-heat system this follows from the continuity of the corresponding semigroup (see [132]). It is also known for the non-stationary Stokes-system (see [133] and [88]) but not in our setting.

Theorem B.2.48. *Let* $\mathbf{f} \in L^q(Q_0)$ *for some* $q > 2$, *where* $Q_0 := (0, T) \times \mathbb{R}^3$ *and let* $\mathbf{v} \in L^2((0, T); W^{1,2}_{0,\mathrm{div}}(\mathbb{R}^3))$ *be the unique weak solution to*

$$\int_{Q_0} \mathbf{v}\cdot\partial_t\boldsymbol{\varphi}\,\mathrm{d}x\,\mathrm{d}t - \int_{Q_0} \mathcal{A}(\boldsymbol{\varepsilon}(\mathbf{v}), \boldsymbol{\varepsilon}(\boldsymbol{\varphi}))\,\mathrm{d}x = \int_{Q_0} \mathbf{f}\cdot\boldsymbol{\varphi}\,\mathrm{d}x\,\mathrm{d}t \tag{B.2.5}$$

for all $\boldsymbol{\varphi} \in C^{\infty}_{0,\mathrm{div}}([0, T) \times \mathbb{R}^3)$. *Then we have* $\nabla^2\mathbf{v} \in L^q_{loc}(Q_0)$ *and there holds*

$$\int_0^T\int_B |\nabla^2\mathbf{v}|^q\,\mathrm{d}x\,\mathrm{d}t \le c_B\int_{Q_0}|\mathbf{f}|^q\,\mathrm{d}x\,\mathrm{d}t$$

for all balls $\mathcal{B} \subset \mathbb{R}^3$.

Proof. The main ingredient is the proof of the following auxiliary result which has been used in a similar version in [42].

i) We start with interior estimates. Let

$$Q_r := Q_r(x_0, t_0) := (t_0 - r^2, t_0 + r^2) \times \mathcal{B}_r(x_0)$$

be a parabolic cylinder such that $4Q_r \subset Q_0$. We claim the following: There is a constant $N_1 > 0$ such that for every $\varepsilon > 0$ there is $\delta = \delta(\varepsilon) > 0$ such that (since $\mathbf{f} \in L^2(0, T; L^2_{loc}(\mathbb{R}^3))$ the standard interior regularity theory implies $\nabla^2\mathbf{v} \in L^2(0, T; L^2_{loc}(\mathbb{R}^3))$ and $\partial_t\mathbf{v} \in L^2(0, T; L^2_{loc}(\mathbb{R}^3))$)

$$\begin{aligned} &\mathcal{L}^4\big(Q_r \cap \{\mathcal{M}(|\mathbf{f}|)^2 > N_1^2\}\big) \geq \varepsilon \mathcal{L}^4(Q_r) \\ &\Rightarrow Q_r \subset \{\mathcal{M}(|\nabla^2\mathbf{v}|^2) > 1\} \cup \{\mathcal{M}(|\mathbf{f}|^2) > \delta^2\}. \end{aligned} \tag{B.2.6}$$

Let us assume for simplicity that $r = 1$. In fact, we will establish (B.2.6) by showing

$$\begin{aligned} &Q_1 \cap \{\mathcal{M}(|\nabla^2\mathbf{v}|^2) \leq 1\} \cap \{\mathcal{M}(|\mathbf{f}|^2) \leq \delta^2\} \neq \emptyset \\ &\Rightarrow \mathcal{L}^4\big(Q_1 \cap \{\mathcal{M}(|\nabla\mathbf{v}|)^2 > N_1^2\}\big) < \varepsilon \mathcal{L}^4(Q_1) \end{aligned} \tag{B.2.7}$$

and applying a simple scaling argument. In order to show (B.2.7) we compare $\mathbf{v}$ with a solution to a homogeneous problem on

$$Q_4 = (t_0^4 - 4^2, t_0^4 + 4^2) \times \mathcal{B}_4(x_0^4) \subset Q_0$$

which is smooth in the interior. So let us define $\mathbf{h}$ as the unique solution to

$$\begin{cases} \partial_t\mathbf{h} - \operatorname{div}\mathcal{A}(\boldsymbol{\varepsilon}(\mathbf{h})) = \nabla\pi_{\mathbf{h}} & \text{in } Q_4, \\ \operatorname{div}\mathbf{h} = 0 & \text{in } Q_4, \\ \mathbf{h} = \mathbf{v} & \text{on } I_4 \times \partial\mathcal{B}_4, \\ \mathbf{h}(t_0^4, \cdot) = \mathbf{v}(t_0^4, \cdot) & \text{in } \mathcal{B}_4. \end{cases} \tag{B.2.8}$$

We test the difference of both equations with $\mathbf{v} - \mathbf{h}$. This yields by the ellipticity of $\mathcal{A}$

$$\begin{aligned} &\sup_{t \in I_4} \int_{\mathcal{B}_4} |\mathbf{v}(t) - \mathbf{h}(t)|^2 \,\mathrm{d}x + \int_{Q_4} |\boldsymbol{\varepsilon}(\mathbf{v}) - \boldsymbol{\varepsilon}(\mathbf{h})|^2 \,\mathrm{d}x\,\mathrm{d}t \\ &\leq c \int_{Q_4} |\mathbf{f}|^2 \,\mathrm{d}x\,\mathrm{d}t + c \int_{Q_4} |\mathbf{v} - \mathbf{h}|^2 \,\mathrm{d}x\,\mathrm{d}t. \end{aligned}$$

An application of Korn's inequality and Gronwall's lemma implies

$$\sup_{t \in I_4} \int_{\mathcal{B}_4} |\mathbf{v}(t) - \mathbf{h}(t)|^2 \,\mathrm{d}x + \int_{Q_4} |\nabla\mathbf{v} - \nabla\mathbf{h}|^2 \,\mathrm{d}x\,\mathrm{d}t \leq c \int_{Q_4} |\mathbf{f}|^2 \,\mathrm{d}x\,\mathrm{d}t. \tag{B.2.9}$$

First we insert $\partial_t(\mathbf{v}-\mathbf{h})$ which yields similarly

$$\int_{Q_4} |\partial_t(\mathbf{v}-\mathbf{h})|^2\,\mathrm{d}x + \sup_{t\in I_4}\int_{\mathcal{B}_4} |\nabla\mathbf{v}-\nabla\mathbf{h}|^2\,\mathrm{d}x \le c\int_{Q_4} |\mathbf{f}|^2\,\mathrm{d}x\,\mathrm{d}t. \tag{B.2.10}$$

We can introduce the pressure terms $\pi_{\mathbf{v}}, \pi_{\mathbf{h}} \in L^2(I_4, L^2_0(\mathcal{B}_4))$ in the equations for $\mathbf{v}$ and $\mathbf{h}$ and show

$$\int_{Q_4} |\pi_{\mathbf{v}}-\pi_{\mathbf{h}}|^2\,\mathrm{d}x \le c\int_{Q_4} |\mathbf{f}|^2\,\mathrm{d}x\,\mathrm{d}t. \tag{B.2.11}$$

Estimate (B.2.11) can be shown by using the Bogovskiĭ operator, see Section 2.1. Setting $\mathrm{Bog}=\mathrm{Bog}_{\mathcal{B}_4}$ we gain due to (B.2.10) for any $\varphi\in C_0^\infty(Q_4)$ that

$$\begin{aligned}
\int_{Q_4} (\pi_{\mathbf{v}}-\pi_{\mathbf{h}})\varphi\,\mathrm{d}x\,\mathrm{d}t &= \int_{Q_4} (\pi_{\mathbf{v}}-\pi_{\mathbf{h}})\,\mathrm{div}\,\mathrm{Bog}(\varphi-\varphi_{\mathcal{B}_4})\,\mathrm{d}x\,\mathrm{d}t\\
&= \int_{Q_4} \mathcal{A}\big(\boldsymbol{\varepsilon}(\mathbf{v}-\mathbf{h}),\boldsymbol{\varepsilon}\big(\mathrm{Bog}(\varphi-\varphi_{\mathcal{B}_4})\big)\big)\,\mathrm{d}x\,\mathrm{d}t\\
&+ \int_{Q_4} \mathbf{f}\cdot\mathrm{Bog}(\varphi-\varphi_{\mathcal{B}_4})\big)\,\mathrm{d}x\,\mathrm{d}t\\
&- \int_{Q_4} \partial_t(\mathbf{v}-\mathbf{h})\cdot\mathrm{Bog}(\varphi-\varphi_{\mathcal{B}_4})\,\mathrm{d}x\,\mathrm{d}t\\
&\le c\Big(\|\nabla(\mathbf{v}-\mathbf{h})\|_2+\|\mathbf{f}\|_2+\|\partial_t(\mathbf{v}-\mathbf{h})\|_2\Big)\big\|\nabla\mathrm{Bog}(\varphi-(\varphi)_{\mathcal{B}_4})\big\|_2\\
&\le c\Big(\int_{Q_4}|\mathbf{f}|^2\,\mathrm{d}x\,\mathrm{d}t\Big)^{\frac12}\Big(\int_{Q_4}|\varphi|^2\,\mathrm{d}x\,\mathrm{d}t\Big)^{\frac12}.
\end{aligned}$$

Now we choose a cut off function $\eta\in C_0^\infty(\mathcal{B}_4)$ with $0\le\eta\le 1$ and $\eta\equiv 1$ on $\mathcal{B}_3$. We insert $\partial_\gamma(\eta^2\partial_\gamma(\mathbf{v}-\mathbf{h}))$ in the equation for $\mathbf{v}-\mathbf{h}$ and sum over $\gamma\in\{1,2,3\}$. We gain

$$\begin{aligned}
&\sup_{t\in I_4}\int_{\mathcal{B}_4} \eta^2|\nabla(\mathbf{v}-\mathbf{h})|^2\,\mathrm{d}x + \int_{Q_4} \eta^2|\nabla\boldsymbol{\varepsilon}(\mathbf{v})-\nabla\boldsymbol{\varepsilon}(\mathbf{h})|^2\,\mathrm{d}x\,\mathrm{d}t\\
&\le c\int_{Q_4} \mathbf{f}\cdot\partial_\gamma(\eta^2\partial_\gamma(\mathbf{h}-\mathbf{h}))\,\mathrm{d}x\,\mathrm{d}t + c(\nabla\eta)\int_{Q_4} |\nabla\mathbf{v}-\nabla\mathbf{h}|^2\,\mathrm{d}x\,\mathrm{d}t\\
&+ c\int_{Q_4} (\pi-\pi_{\mathbf{h}})\cdot\partial_\gamma(\nabla\eta^2\cdot\partial_\gamma(\mathbf{h}-\mathbf{h}))\,\mathrm{d}x\,\mathrm{d}t.
\end{aligned}$$

We estimate the term involving $\mathbf{f}$ by

$$\int_{Q_4} \mathbf{f}\cdot\partial_\gamma(\eta^2\partial_\gamma(\mathbf{v}-\mathbf{h}))\,\mathrm{d}x\,\mathrm{d}t$$

$$\leq c(\kappa)\int_{Q_4}|\mathbf{f}|^2\,\mathrm{d}x\,\mathrm{d}t+\kappa\int_{Q_4}\eta^2|\nabla^2\mathbf{v}-\nabla^2\mathbf{h}|^2\,\mathrm{d}x\,\mathrm{d}t$$
$$+c(\nabla\eta)\int_{Q_4}|\nabla\mathbf{v}-\nabla\mathbf{h}|^2\,\mathrm{d}x\,\mathrm{d}t,$$

where $\kappa > 0$ is arbitrary. The term involving $\pi - \pi_{\mathbf{h}}$ can be estimated in the same fashion. Choosing $\kappa > 0$ small enough and using the inequality $|\nabla^2\mathbf{u}| \leq c\,|\nabla\boldsymbol{\varepsilon}(\mathbf{u})|$ as well as (B.2.9)–(B.2.11) shows

$$\sup_{t\in I_4}\int_{\mathcal{B}_3}|\nabla(\mathbf{v}-\mathbf{h})(t)|^2\,\mathrm{d}x+\int_{I_4}\int_{\mathcal{B}_3}|\nabla^2\mathbf{v}-\nabla^2\mathbf{h}|^2\,\mathrm{d}x\,\mathrm{d}t\leq c\int_{Q_4}|\mathbf{f}|^2\,\mathrm{d}x\,\mathrm{d}t. \tag{B.2.12}$$

Now, let us assume that $(\mathrm{B.2.7})_1$ holds. Then there is a point $(t_0, x_0) \in Q_1$ such that

$$⨍_{Q_\sigma(t_0,x_0)}|\nabla^2\mathbf{v}|^2\,\mathrm{d}x\,\mathrm{d}t\leq 1,\quad ⨍_{Q_\sigma(t_0,x_0)}|\mathbf{f}|^2\,\mathrm{d}x\,\mathrm{d}t\leq\delta^2 \tag{B.2.13}$$

for all $\sigma > 0$. Since $Q_4 \subset Q_6(t_0, x_0)$ we have

$$\int_{Q_4}|\nabla^2\mathbf{v}|^2\,\mathrm{d}x\,\mathrm{d}t\leq c,\quad \int_{Q_4}|\mathbf{f}|^2\,\mathrm{d}x\,\mathrm{d}t\leq c\delta^2. \tag{B.2.14}$$

As $\mathbf{h}$ is smooth we know that

$$N_0^2 := \sup_{Q_3}|\nabla^2\mathbf{h}|^2 < \infty. \tag{B.2.15}$$

From this we aim to conclude that

$$Q_1\cap\{\mathcal{M}(|\nabla^2\mathbf{v}|^2)>N_1^2\}\subset Q_1\cap\{\mathcal{M}(\chi_{Q_3}|\nabla^2\mathbf{v}-\nabla^2\mathbf{h}|^2)>N_0^2\} \tag{B.2.16}$$

for $N_1^2 := \max\{4N_0^2, 2^5\}$. To establish (B.2.16) suppose that

$$(t,x)\in Q_1\cap\{\mathcal{M}(\chi_{Q_3}|\nabla^2\mathbf{v}-\nabla^2\mathbf{h}|^2)\leq N_0^2\}. \tag{B.2.17}$$

If $\sigma \leq 2$ we have $Q_\sigma(t,x) \subset Q_3$ and gain by (B.2.15)

$$⨍_{Q_\sigma(t,x)}|\nabla^2\mathbf{v}|^2\,\mathrm{d}x\,\mathrm{d}t$$
$$\leq 2⨍_{Q_\sigma(t,x)}\chi_{Q_3}|\nabla^2\mathbf{v}-\nabla^2\mathbf{h}|^2\,\mathrm{d}x\,\mathrm{d}t+2⨍_{Q_\sigma(t,x)}\chi_{Q_3}|\nabla^2\mathbf{h}|^2\,\mathrm{d}x\,\mathrm{d}t$$
$$\leq 4N_0^2.$$

If $\sigma \geq 2$ we have by (B.2.13)

$$⨍_{Q_\sigma(t,x)}|\nabla^2\mathbf{v}|^2\,\mathrm{d}x\,\mathrm{d}t\leq 2^5⨍_{Q_{2\sigma}(t_0,x_0)}|\nabla^2\mathbf{v}|^2\,\mathrm{d}x\,\mathrm{d}t\leq 2^5.$$

Combining the both cases yields (B.2.16). This implies together with the continuity of the maximal function on L^2, (B.2.12) and (B.2.14)

$$\begin{aligned}
&\mathcal{L}^4(Q_1 \cap \{\mathcal{M}(|\nabla^2\mathbf{v}|^2) > N_1^2\}) \\
&\quad \le \mathcal{L}^{d+1}(Q_1 \cap \{\mathcal{M}(\chi_{Q_3}|\nabla^2\mathbf{v} - \nabla^2\mathbf{h}|^2) > N_0^2\}) \\
&\quad \le \frac{c}{N_0^2}\int_{Q_3} |\nabla^2\mathbf{v} - \nabla^2\mathbf{h}|^2 \,\mathrm{d}x\,\mathrm{d}t \\
&\quad \le \frac{c}{N_0^2}\int_{Q_3} |\mathbf{f}|^2 \,\mathrm{d}x\,\mathrm{d}t \le \frac{c}{N_0^2}\delta^2 \\
&\quad = \varepsilon\mathcal{L}^4(Q_1),
\end{aligned}$$

choosing $\delta := c^{-1/2}N_0\sqrt{\varepsilon}$. So we have shown (B.2.7) which yields (B.2.6) by a scaling argument.
If $(\text{B.2.6})_1$ holds then we have

$$\begin{aligned}
&\mathcal{L}^4(Q_r \cap \{\mathcal{M}(|\nabla^2\mathbf{v}|)^2 > N_1^2\}) \\
&\quad \le \varepsilon\mathcal{L}^4\Big(Q_r \cap \{\mathcal{M}(|\nabla^2\mathbf{v}|^2) > 1\} \cup \{\mathcal{M}(|\mathbf{f}|^2) > \delta^2\}\Big) \\
&\quad \le \varepsilon\Big(\mathcal{L}^4(Q_r \cap \{\mathcal{M}(|\nabla^2\mathbf{v}|^2) > 1\}) + \mathcal{L}^4(Q_r \cap \{\mathcal{M}(|\mathbf{f}|^2) > \delta^2\})\Big).
\end{aligned}$$

Multiplying the equation for $\mathbf{v}$ by some small number $\varrho = \varrho(\|\mathbf{f}\|_q, \|\nabla^2\mathbf{v}\|_2)$ we can assume that

$$\mathcal{L}^4(Q_r \cap \{\mathcal{M}(|\nabla^2\mathbf{v}|^2) > N_1^2\}) < \varepsilon. \tag{B.2.18}$$

By induction we can establish that

$$\begin{aligned}
&\mathcal{L}^4(Q_r \cap \{\mathcal{M}(|\nabla^2\mathbf{v}|^2) > N_1^{2k}\}) \\
&\quad \le \varepsilon^k\mathcal{L}^4(Q_r \cap \{\mathcal{M}(|\nabla^2\mathbf{v}|^2) > 1\} \\
&\quad + c\sum_{i=1}^{k}\varepsilon^i\mathcal{L}^4\Big(Q_r \cap \big\{\mathcal{M}(|\mathbf{f}|^2) > \delta^2 N_1^{2(k-i)}\big\}\Big).
\end{aligned}$$

In the induction step one has to introduce $\mathbf{v}_1 := \frac{\mathbf{v}}{N_1}$ which is a solution to the $\mathcal{A}$-Stokes problem with right hand side $\mathbf{f}_1 := \frac{\mathbf{f}}{N_1}$. Now we will show $\nabla^2\mathbf{v} \in L^q(Q_r)$. For this we use the equivalence for $1 \le q_0 < \infty$

$$\sum_{k=1}^{\infty} L^{q_0 k}\mathcal{L}^4\Big(Q_r \cap \big\{(t,x) : |u(t,x)| > \theta L^k\big\}\Big) < \infty \quad \Leftrightarrow \quad u \in L^{q_0}(Q_r),$$

which holds for any measurable function u, see [44]. Here $L > 1$ and $\theta > 0$ are arbitrary. So we aim to prove that

$$\sum_{k=1}^{\infty}(N_1^2)^{\frac{q}{2}k}\mathcal{L}^4(Q_r \cap \{\mathcal{M}(|\nabla^2\mathbf{v}|^2) > N_1^{2k}\}) < \infty \tag{B.2.19}$$

to conclude $\mathcal{M}(|\nabla^2\mathbf{v}|^2) \in L^{\frac{q}{2}}(Q_r)$ and hence $\nabla^2\mathbf{v} \in L^q(Q_r)$. Since $\mathbf{f} \in L^q(Q_r)$ we have $\mathcal{M}(|\mathbf{f}|^2) \in L^{q/2}(Q_r)$ (recall (5.2.4)) and we have

$$\sum_{k=1}^{\infty}(N_1^2)^{\frac{q}{2}k}\mathcal{L}^4\big(Q_r \cap \{\mathcal{M}(|\mathbf{f}|^2) > \delta^2 N_1^{2k}\}\big) < \infty.$$

We obtain

$$\begin{aligned}
&\sum_{k=1}^{\infty}(N_1)^{qk}\mathcal{L}^4\big(Q_r \cap \{\mathcal{M}(|\nabla^2\mathbf{v}|^2) > N_1^{2k}\}\big)\\
&\quad\leq c\sum_{k=1}^{\infty}(N_1)^{qk}\varepsilon^k\mathcal{L}^4\big(Q_r \cap \{\mathcal{M}(|\nabla^2\mathbf{v}|^2) > 1\}\big)\\
&\quad+ c\sum_{k=1}^{\infty}(N_1)^{qk}\sum_{i=1}^{k}\varepsilon^i\mathcal{L}^4\big(Q_r \cap \{\mathcal{M}(|\mathbf{f}|^2) > \delta^2 N_1^{2(k-i)}\}\big)\\
&\quad\leq c_r\sum_{k=1}^{\infty}(\varepsilon N_1)^{qk}\\
&\quad+ c\sum_{i=1}^{\infty}\varepsilon^i(N_1)^{qi}\sum_{k=i}^{\infty}(N_1)^{q(k-i)}\mathcal{L}^4\big(Q_r \cap \{\mathcal{M}(|\mathbf{f}|^2) > \delta^2 N_1^{2(k-i)}\}\big)\\
&\quad\leq c_r\sum_{k=1}^{\infty}(\varepsilon N_1^q)^k.
\end{aligned}$$

If we choose $\varepsilon N_1^q < 1$ the sum in (B.2.19) is converging and we have $\nabla^2\mathbf{v} \in L^q(Q_r)$. Since the mapping $\mathbf{f} \mapsto \nabla^2\mathbf{v}$ is linear we gain the desired estimate

$$\int_{Q_r}|\nabla^2\mathbf{v}|^q\,\mathrm{d}x\,\mathrm{d}t \leq c_r\int_{Q_0}|\mathbf{f}|^q\,\mathrm{d}x\,\mathrm{d}t. \tag{B.2.20}$$

ii) Now let Q_1 be a cylinder such that $4Q_1 \cap (-\infty, 0] \times \mathbb{R}^3 \neq \emptyset$. Moreover, assume that $Q_1 \cap Q_0 \neq \emptyset$. We consider the solution $\tilde{\mathbf{h}}$ to

$$\begin{cases}
\partial_t\tilde{\mathbf{h}} - \operatorname{div}\mathcal{A}(\boldsymbol{\varepsilon}(\tilde{\mathbf{h}})) = \nabla\pi_{\tilde{\mathbf{h}}} & \text{in } \tilde{Q}_4,\\
\operatorname{div}\tilde{\mathbf{h}} = 0 & \text{in } \tilde{Q}_4,\\
\tilde{\mathbf{h}} = \mathbf{v} & \text{on } \tilde{I}_4 \times \partial\mathcal{B}_4,\\
\tilde{\mathbf{h}}(t_0^4, \cdot) = 0 & \text{in } \mathcal{B}_4,
\end{cases} \tag{B.2.21}$$

where $\tilde{I}_m := I_m \cap (0, T)$ and $\tilde{Q}_m := \tilde{I}_m \times \mathcal{B}_m$. We can establish a variant of (B.2.12) on $\tilde{Q}_4$. Now we have $\sup_{\tilde{Q}_3}|\nabla^2\tilde{\mathbf{h}}|^2 < \infty$ due to the smooth initial datum of $\tilde{\mathbf{h}}$ (recall that $\mathbf{v}(0, \cdot) = 0$ a.e.). So we can finish the proof as before and gain $\nabla^2\mathbf{v} \in L^q(Q_1)$. This implies again (B.2.20).

iii) The situation $4Q_1 \cap [T, \infty) \times \mathbb{R}^3 \neq \emptyset$ is uncritical again and we can assume that ii) and iii) do not occur for the same cylinder (by choosing sufficiently small cubes).

Covering the set $(0, T) \times \mathcal{B}$ by smaller cylinders and combing i)–iii) yield the desired estimate. □

Corollary B.2.1. *Under the assumptions of Theorem B.2.48 we have for all balls $\mathcal{B} \subset \mathbb{R}^3$ the following estimates for some constant $c_{\mathcal{B}}$ which does not depend on T.*

a) *The following holds*

$$\int_0^T \int_{\mathcal{B}} \left(\left| \frac{\mathbf{v}}{T} \right|^q + \left| \frac{\nabla \mathbf{v}}{\sqrt{T}} \right|^q + |\nabla^2 \mathbf{v}|^q \right) \mathrm{d}x\,\mathrm{d}t \leq c_{\mathcal{B}} \int_{Q_0} |\mathbf{f}|^q \,\mathrm{d}x\,\mathrm{d}t.$$

b) *We have $\partial_t \mathbf{v} \in L^q(0, T; L^q_{loc}(\mathbb{R}^3))$ together with*

$$\int_0^T \int_{\mathcal{B}} |\partial_t \mathbf{v}|^q \,\mathrm{d}x\,\mathrm{d}t \leq c_{\mathcal{B}} \int_{Q_0} |\mathbf{f}|^q \,\mathrm{d}x\,\mathrm{d}t.$$

c) *There is $\pi \in L^q((0, T), W^{1,q}_{loc}(\mathbb{R}^3))$ such that*

$$\int_{Q_0} \mathbf{v} \cdot \partial_t \boldsymbol{\varphi} \,\mathrm{d}x\,\mathrm{d}t - \int_{Q_0} \mathcal{A}(\boldsymbol{\varepsilon}(\mathbf{v}), \boldsymbol{\varepsilon}(\boldsymbol{\varphi}))\,\mathrm{d}x$$
$$= \int_{Q_0} \pi \operatorname{div} \boldsymbol{\varphi} \,\mathrm{d}x\,\mathrm{d}t + \int_{Q_0} \mathbf{f} \cdot \boldsymbol{\varphi} \,\mathrm{d}x\,\mathrm{d}t$$

for all $\boldsymbol{\varphi} \in C_0^\infty([0, T) \times \mathbb{R}^3)$.

d) *The following holds*

$$\int_0^T \int_{\mathcal{B}} \left(\left| \frac{\pi}{\sqrt{T}} \right|^q + |\nabla \pi|^q \right) \mathrm{d}x\,\mathrm{d}t \leq c_{\mathcal{B}} \int_{Q_0} |\mathbf{f}|^q \,\mathrm{d}x\,\mathrm{d}t.$$

Proof. The estimate in a) is a simple scaling argument. Having a solution $\mathbf{v}$ defined on $(0, T) \times \mathbb{R}^3$ we gain a solution $\widehat{\mathbf{v}}$ on $(0, 1) \times \mathbb{R}^3$ by setting

$$\widehat{\mathbf{v}}(s, x) := \frac{1}{T} \mathbf{v}(Ts, \sqrt{T}x).$$

Now we apply Theorem B.2.48 to $\widehat{\mathbf{v}}$. The constant which appears is independent of T. Transforming back to $\mathbf{v}$ yields the claimed inequality.

b) For $\boldsymbol{\varphi} \in C_0^\infty(Q_0)$ with $\boldsymbol{\varphi}(t, x) = \tau(t)\boldsymbol{\psi}(x)$ where $\boldsymbol{\varphi} \in C_0^\infty(G)$ ($G \in \mathbb{R}^3$ a bounded Lipschitz domain) we have

$$\int_{Q_0} \mathbf{v} \cdot \partial_t \boldsymbol{\varphi} \,\mathrm{d}x\,\mathrm{d}t = \int_0^T \partial_t \tau \int_{\mathbb{R}^3} \mathbf{v} \cdot \left(\boldsymbol{\psi}_{\mathrm{div}} + \nabla \boldsymbol{\Psi} \right) \mathrm{d}x\,\mathrm{d}t$$

$$= \int_0^T \partial_t \tau \int_{\mathbb{R}^3} \mathbf{v} \cdot \boldsymbol{\psi}_{\mathrm{div}} \, \mathrm{d}x \, \mathrm{d}t$$
$$= \int_{Q_0} \mathbf{v} \cdot \partial_t \boldsymbol{\varphi}_{\mathrm{div}} \, \mathrm{d}x \, \mathrm{d}t$$

where $\boldsymbol{\psi}_{\mathrm{div}} := \boldsymbol{\psi} - \nabla \Delta_G^{-1} \operatorname{div} \boldsymbol{\psi}$ and $\boldsymbol{\Psi} := \Delta_G^{-1} \operatorname{div} \boldsymbol{\psi}$. (Here Δ_G^{-1} is the solution operator to the Laplace equation with zero boundary datum on ∂G.) Here we took into account $\Psi|_{\partial G} = 0$ as well as $\operatorname{div} \mathbf{v} = 0$. Using $\nabla^2 \mathbf{v} \in L^2(0, T; L^2_{loc}(\mathbb{R}^3))$ we proceed by

$$\begin{aligned}
\int_{Q_0} \mathbf{v} \cdot \partial_t \boldsymbol{\varphi} \, \mathrm{d}x \, \mathrm{d}t &= \int_0^T \int_G \big(\mathbf{f} - \operatorname{div} \mathcal{A}\boldsymbol{\varepsilon}(\mathbf{v})\big) \cdot \boldsymbol{\varphi}_{\mathrm{div}} \, \mathrm{d}x \, \mathrm{d}t \\
&\leq c \bigg(\int_0^T \int_G \big(|\nabla^2 \mathbf{v}|^q + |\mathbf{f}|^q\big) \, \mathrm{d}x \, \mathrm{d}t \bigg)^{\frac{1}{q}} \bigg(\int_0^T \int_G |\boldsymbol{\varphi}_{\mathrm{div}}|^{q'} \, \mathrm{d}x \, \mathrm{d}t \bigg)^{\frac{1}{q'}} \\
&\leq c \bigg(\int_{Q_0} |\mathbf{f}|^q \, \mathrm{d}x \, \mathrm{d}t \bigg)^{\frac{1}{q}} \bigg(\int_0^T \int_G |\boldsymbol{\varphi}|^{q'} \, \mathrm{d}x \, \mathrm{d}t \bigg)^{\frac{1}{q'}}.
\end{aligned}$$

In the last step we used the estimate from Theorem B.2.48 and continuity of $\nabla \Delta_G^{-1} \operatorname{div}$ on $L^{q'}(G)$. Duality implies $\partial_t \mathbf{v} \in L^q(0, T; L^q_{loc}(\mathbb{R}^3))$ and we can introduce the pressure function $\pi \in L^q(0, T; L^q_{loc}(\mathbb{R}^3))$ as claimed in b) by De Rahm's Theorem. Using the equation for $\mathbf{v}$ and the estimates in a) and b) we gain

$$\int_Q |\nabla \pi|^q \, \mathrm{d}x \, \mathrm{d}t \leq c \int_{Q_0} |\mathbf{f}|^q \, \mathrm{d}x \, \mathrm{d}t.$$

The estimate for π in d) follows again by scaling. □

Corollary B.2.2. *Let* $\mathbf{f} \in L^q(Q_0^+)$ *for some* $q > 2$ *where* $Q_0^+ := (0, T) \times \mathbb{R}^3_+$, $\mathbb{R}^3_+ = \mathbb{R}^3 \cap [x_3 > 0]$, *and let* $\mathbf{v} \in L^2((0, T); W^{1,2}_{0,\mathrm{div}}(\mathbb{R}^3_+))$ *with* $\mathbf{v}|_{x_3=0} = 0$ *be the unique weak solution to*

$$\int_{Q_0^+} \mathbf{v} \cdot \partial_t \boldsymbol{\varphi} \, \mathrm{d}x \, \mathrm{d}t - \int_{Q_0^+} \mathcal{A}(\boldsymbol{\varepsilon}(\mathbf{v}), \boldsymbol{\varepsilon}(\boldsymbol{\varphi})) \, \mathrm{d}x = \int_{Q_0^+} \mathbf{f} \cdot \boldsymbol{\varphi} \, \mathrm{d}x \, \mathrm{d}t \qquad \text{(B.2.22)}$$

for all $\boldsymbol{\varphi} \in C^\infty_{0,\mathrm{div}}([0, T) \times \mathbb{R}^3_+)$. *Then the results from Theorem B.2.48 and Corollary B.2.1 hold for* $\mathbf{v}$ *for all half balls* $\mathcal{B}^+(z) = \mathcal{B}(z) \cap [x_3 > 0] \subset \mathbb{R}^3$ *with* $z_3 = 0$.

Proof. We will show a variant of the L^q-estimate from Theorem B.2.48 on half balls $\mathcal{B}^+$, i.e.

$$\int_0^T \int_{\mathcal{B}^+} |\nabla^2 \mathbf{v}|^q \, \mathrm{d}x \, \mathrm{d}t \leq c_{\mathcal{B}} \int_0^T \int_{Q^+} |\mathbf{f}|^q \, \mathrm{d}x \, \mathrm{d}t. \qquad \text{(B.2.23)}$$

From this we can follow estimates in the fashion of Corollary B.2.1 as done there. In order to establish (B.2.23) we will proceed as in the proof of Theorem B.2.48 replacing all balls with half ball. So let $Q_1 \subset \mathbb{R}^4$ such that $4Q_1 \subset Q_0$ (the other situation can be shown along the modifications indicated at the end of the proof of Theorem B.2.48). Moreover, assume that $Q_1 = I_1 \times \mathcal{B}_1(z)$ where $z_3 = 0$. We compare $\mathbf{v}$ with the unique solution $\mathbf{h}^+$ to

$$\begin{cases} \partial_t \mathbf{h}^+ - \operatorname{div} \mathcal{A}(\boldsymbol{\varepsilon}(\mathbf{h}^+)) = \nabla \pi_{\mathbf{h}^+} & \text{in } Q_4^+, \\ \operatorname{div} \mathbf{h}^+ = 0 & \text{in } Q_4^+, \\ \mathbf{h}^+ = \mathbf{v} & \text{on } I_4 \times \partial \mathcal{B}_4^+, \\ \mathbf{h}^+(0, \cdot) = \mathbf{v}(0, \cdot) & \text{in } \mathcal{B}_4^+. \end{cases} \tag{B.2.24}$$

We gain a version of the estimate (B.2.9) and (B.2.10) on half-balls. In fact, there holds

$$\sup_{t \in I_4} \int_{\mathcal{B}_4^+} |\mathbf{v} - \mathbf{h}^+|^2 \, dx + \int_{Q_4^+} |\nabla \mathbf{v} - \nabla \mathbf{h}^+|^2 \, dx \, dt \le c \int_{Q_4^+} |\mathbf{f}|^2 \, dx \, dt, \tag{B.2.25}$$

$$\int_{Q_4^+} |\partial_t (\mathbf{v} - \mathbf{h}^+)|^2 \, dx + \sup_{t \in I_4} \int_{\mathcal{B}_4^+} |\nabla \mathbf{v} - \nabla \mathbf{h}^+|^2 \, dx \le c \int_{Q_4^+} |\mathbf{f}|^2 \, dx \, dt. \tag{B.2.26}$$

We can introduce the pressure terms $\pi_{\mathbf{v}}, \pi_{\mathbf{h}^+} \in L^2(I_4, L^2_0(\mathcal{B}_4^+))$ in the equations for $\mathbf{v}$ and $\mathbf{h}^+$ and show

$$\int_{Q_4^+} |\pi_{\mathbf{v}} - \pi_{\mathbf{h}^+}|^2 \, dx \le c \int_{Q_4^+} |\mathbf{f}|^2 \, dx \, dt. \tag{B.2.27}$$

This can be done as in the proof of (B.2.11) using the Bogovskiĭ operator on $\mathcal{B}_4^+$. Estimate (B.2.27) can be shown by using the Bogovskiĭ operator introduced in [24]. Now we insert $\partial_\gamma(\eta^2 \partial_\gamma(\mathbf{v} - \mathbf{h}))$ for $\gamma \in \{1, 2\}$ in the equation for $\mathbf{v}$ $\mathbf{h}$. Here we choose $\eta \in C_0^\infty(\mathcal{B}_4)$ with $0 \le \eta \le 1$ and $\eta \equiv 1$ on $\mathcal{B}_3$. This yields together with (B.2.25)–(B.2.27)

$$\int_{Q_3^+} |\tilde{\nabla} \nabla (\mathbf{v} - \mathbf{h})|^2 \, dx \le c \int_{Q_4^+} |\mathbf{f}|^2 \, dx \, dt, \tag{B.2.28}$$

where $\tilde{\nabla} := (\partial_1, \partial_2)$. Finally, the only term which is missing is $\partial_3^2(\mathbf{v} - \mathbf{h})$. On account of $\operatorname{div}(\mathbf{v} - \mathbf{h}) = 0$ we have (cf. [21])

$$|\partial_3^2 (\mathbf{v} - \mathbf{h})| \le c \big(|\tilde{\nabla}(\pi_{\mathbf{v}} - \pi_{\mathbf{h}^+})| + |\tilde{\nabla} \nabla (\mathbf{v} - \mathbf{h})| + |\mathbf{f}| \big). \tag{B.2.29}$$

So we have to estimate derivatives of the pressure. In fact we have

$$\int_{Q_3^+} |\tilde{\nabla}(\pi_{\mathbf{v}} - \pi_{\mathbf{h}^+})|^2 \, dx \le c \int_{Q_4^+} |\mathbf{f}|^2 \, dx \, dt. \tag{B.2.30}$$

We can show this similarly to the proof of (B.2.27) replacing φ by $\partial_\gamma \varphi$ and using (B.2.28). Combining (B.2.28)–(B.2.30) implies

$$\int_{Q_3^+} |\nabla^2 \mathbf{v} - \nabla^2 \mathbf{h}|^2 \, dx \le c \int_{Q_4^+} |\mathbf{f}|^2 \, dx \, dt.$$

Moreover, we know $\sup_{Q_3^+} |\nabla^2 \mathbf{h}^+|^2 < \infty$. Note that $\mathbf{h}^+ = 0$ on $Q_3 \cap [x_3 = 0]$. This allows to show $\nabla^2 \mathbf{v} \in L^q(Q_1^+)$ as in the proof of Theorem B.2.48. □

Corollary B.2.3. *Let $\mathbf{f} \in L^q(Q_0^{\nu,\xi})$ for some $q > 2$ where $Q_0^{\nu,\xi} := (0, T) \times \mathbb{R}^3_{\nu,\xi}$, $\mathbb{R}^3_{\nu,\xi} = \mathbb{R}^3 \cap [(x - \xi) \cdot \nu > 0]$ for some $\nu, \xi \in \mathbb{R}^3$. Let $\mathbf{v} \in L^2(0, T; W^{1,2}_{0,\mathrm{div}}(\mathbb{R}^3_{\nu,\xi}))$ with $\mathbf{v}|_{(x-\xi)\cdot\nu=0} = 0$ be the unique weak solution to*

$$\int_{Q_0^{\nu,\xi}} \mathbf{v} \cdot \partial_t \boldsymbol{\varphi} \, \mathrm{d}x \, \mathrm{d}t - \int_{Q_0^{\nu,\xi}} \mathcal{A}(\boldsymbol{\varepsilon}(\mathbf{v}), \boldsymbol{\varepsilon}(\boldsymbol{\varphi})) \, \mathrm{d}x = \int_{Q_0^{\nu,\xi}} \mathbf{f} \cdot \boldsymbol{\varphi} \, \mathrm{d}x \, \mathrm{d}t \qquad \text{(B.2.31)}$$

for all $\boldsymbol{\varphi} \in C^\infty_{0,\mathrm{div}}([0, T) \times \mathbb{R}^3_{\nu,\xi})$. Then the results from Theorem B.2.48 and Corollary B.2.1 hold for $\mathbf{v}$ for all half balls $\mathcal{B}^{\nu,\xi}(z) = \mathcal{B}(z) \cap [(x - \xi) \cdot \nu > 0] \subset \mathbb{R}^3$ with $(z - \xi) \cdot \nu = 0$.

Proof. The proof follows easily from Corollary B.2.2 by rotation of the coordinate system. There is an orthogonal matrix $V \in \mathbb{R}^{3\times3}$ such that the mapping $z = V(x - \xi)$ transforms $\mathbb{R}^3_{\nu,\xi}$ to $\mathbb{R}^3_+$. We define

$$\tilde{\mathbf{v}}(t, x) = V^{-1}\mathbf{v}(t, V(x - \xi)), \quad \tilde{\mathbf{f}}(t, x) = V^{-1}\mathbf{f}(t, V(x - \xi)),$$

as well as the bilinear form $\tilde{\mathcal{A}}$ by

$$\tilde{\mathcal{A}}(\boldsymbol{\zeta}, \boldsymbol{\xi}) := \mathcal{A}\big(V\boldsymbol{\zeta}V^{-1}, V\boldsymbol{\xi}V^{-1}\big), \quad \boldsymbol{\zeta}, \boldsymbol{\xi} \in \mathbb{R}^{3\times3}.$$

Note that the ellipticity constants of $\mathcal{A}$ and $\tilde{\mathcal{A}}$ coincide as V is an orthogonal matrix. Now it is easy to see that $\tilde{\mathbf{v}} \in L^2(0, T; W^{1,2}_{0,\mathrm{div}}(\mathbb{R}^3_+))$ satisfies $\tilde{\mathbf{v}}|_{x_3=0} = 0$ and is a solution to (B.2.22) with right hand side $\tilde{\mathbf{f}}$ and bilinear form $\tilde{\mathcal{A}}$. Hence Corollary B.2.3. □

Theorem B.2.49. *Let $Q := (0, T) \times G$ with a bounded domain $G \subset \mathbb{R}^3$ having a C^2-boundary. Let $\mathbf{f} \in L^q(Q)$ for some $q > 2$. Then there is a unique weak solution $\mathbf{v} \in L^\infty(0, T; L^2(G)) \cap L^q(0, T; W^{1,q}_{0,\mathrm{div}}(G))$ to*

$$\int_Q \mathbf{v} \cdot \partial_t \boldsymbol{\varphi} \, \mathrm{d}x \, \mathrm{d}t - \int_Q \mathcal{A}(\boldsymbol{\varepsilon}(\mathbf{v}), \boldsymbol{\varepsilon}(\boldsymbol{\varphi})) \, \mathrm{d}x = \int_Q \mathbf{f} \cdot \boldsymbol{\varphi} \, \mathrm{d}x \, \mathrm{d}t \qquad \text{(B.2.32)}$$

for all $\boldsymbol{\varphi} \in C^\infty_{0,\mathrm{div}}([0, T) \times G)$ such that $\nabla^2 \mathbf{v} \in L^q(Q)$. Moreover, we have

$$\int_Q |\nabla^2 \mathbf{v}|^q \, \mathrm{d}x \, \mathrm{d}t \le c \int_Q |\mathbf{f}|^q \, \mathrm{d}x \, \mathrm{d}t.$$

Proof. Due to the local L^q-theory for the whole space problem and the half-space problem which follow from Corollary B.2.1 and Corollary B.2.2 (with the right scaling in T) the proof is similar to [133], Thm. 4.1, in the case $\mathcal{A} = I$. Note that L^q-estimates for the stationary problem on bounded domains with given divergence are stated in Lemma B.1.1. We want to invert the operator

$$\begin{aligned}\mathscr{L} : \mathscr{Y} &\to L^q(I; L^q_{\mathrm{div}}(G)),\\ \mathbf{v} &\mapsto \mathcal{P}_q\big(\partial_t \mathbf{v} - \operatorname{div} \mathcal{A}\boldsymbol{\varepsilon}(\mathbf{v})\big).\end{aligned}$$

The space $\mathscr{Y}$ is given by

$$\mathscr{Y} := L^q(I; W^{1,q}_{0,\mathrm{div}} \cap W^{2,q}(G)) \cap W^{1,q}(I; L^q(G)) \cap \{\mathbf{v}(0,\cdot) = 0,\ \mathbf{v}_{\partial G} = 0\}$$

and $\mathcal{P}_q$ is the Helmholtz projection from $L^q(G)$ into $L^q_{\mathrm{div}}(G)$. The latter one is defined by

$$L^q_{\mathrm{div}}(G) := \overline{C^\infty_{0,\mathrm{div}}(G)}^{\|\cdot\|_q}.$$

The Helmholtz-projection $\mathcal{P}_q\mathbf{u}$ of a function $\mathbf{u} \in L^q(G)$ can be defined as $\mathcal{P}_q\mathbf{u} := \mathbf{u} - \nabla h$, where h is the solution to the Neumann-problem

$$\begin{cases} \Delta h = \operatorname{div} \mathbf{u} & \text{on} \quad G,\\ \mathcal{N}_B \cdot (\nabla h - \mathbf{u}) = 0 & \text{on} \quad \partial G.\end{cases}$$

We will try to find an operator $\mathcal{R} : L^q(I; L^q_{\mathrm{div}}(G)) \to \mathscr{Y}$ such that

$$\mathscr{L} \circ \mathscr{R} = I + \tau \tag{B.2.33}$$

with $\|\tau\| < 1$. The range of $\mathscr{L} \circ \mathscr{R}$ (which then is $L^q(0, T; L^q_{\mathrm{div}}(G))$) is contained in the range of $\mathscr{L}$. So $\mathscr{L}$ is onto.

Let G_k, $0 = 1, \ldots, N$ be a covering of G such that $G_0 \Subset G$ and G_k covers a (small) boundary strip of G. There are local coordinates

$$z = \mathscr{Z}_k(y^k) = (y_1^k, y_2^k, y_3^k - F_k(y_1^k, y_2^k)), \quad y^k = (y_1^k, y_2^k) \in \mathcal{B}_\lambda(\xi_k)$$

where F_k is a C^2-function and $0 < \lambda \le \lambda_0 < 1$. Here ξ_k denotes the center point of $S_k = \partial G \cap G_k$ and ν_k the outer unit normal of S_k at the point ξ_k. In this coordinate system we have a flat boundary which is contained in the plane $\{(x - \xi_k) \cdot \nu_k = 0\}$. We consider a decomposition of unity $(\zeta_k)_{k=0}^N \subset C_0^\infty(\mathbb{R}^3)$ with respect to G_k such that $\operatorname{spt}\zeta_k \subset G_k$. We can assume that $|\nabla^l \zeta_k| \le c\lambda^{-l}$ for $l = 1, 2$ and that the multiplicity of the covering of G by the domains G_k does not depend on λ.

Furthermore, $\mathscr{M}_0 f$ and $\mathscr{M} f$ is the extension of a function f (by zero) to the whole space or the half space respectively. Note that if $\mathbf{f} \in L^q_{\mathrm{div}}(G)$, then $\mathscr{M}_0\mathbf{f} \in L^q_{\mathrm{div}}(\mathbb{R}^3)$. Finally, we denote by $\mathscr{U}^k\mathbf{f}$ (for $\mathbf{f} \in L^q(\mathbb{R}^3_+)$) and $\mathscr{U}_0\mathbf{f}$ (for $\mathbf{f} \in L^q(\mathbb{R}^3)$) the solution on the half space (corresponding to the plane $\{(x - \xi_k) \cdot \nu_k = 0\}$) and the whole space respectively (see Corollary B.2.1

and Corollary B.2.3). By $\mathscr{Q}^k\mathbf{f}$ we denote the pressure corresponding to $\mathscr{U}^k\mathbf{f}$. Now we define the operators

$$\mathscr{R}_0\mathbf{f} := \zeta_0\mathscr{U}_0\mathscr{M}_0\mathbf{f} + \sum_{k=1}^{N}\zeta_k\mathscr{Z}_k^{-1}\mathscr{U}^k\mathscr{M}\mathscr{Z}_k\mathbf{f},$$

$$\mathscr{P}_0\mathbf{f} := \sum_{k=1}^{N}\mathscr{Z}_k^{-1}\sigma\mathscr{Q}^k\mathscr{M}\mathscr{Z}_k\mathbf{f}.$$

The idea is in the interior to extend the force to the whole space, compute the whole space solution and localize again. At the boundary it is more tricky since we have to flatten the problem before considering the half space problem (and of course we have to transform back after solving it). Due to the involved cut-off functions $\mathscr{R}_0\mathbf{f}$ is in general not divergence-free. This will be corrected in the following way: we set

$$\mathscr{R}\mathbf{f} = \mathscr{R}_0\mathbf{f} + \mathscr{R}_1\mathbf{f}, \quad \mathscr{P}\mathbf{f} = \mathscr{P}_0\mathbf{f} + \mathscr{P}_1\mathbf{f}.$$

Here $\mathscr{R}_1\mathbf{f} = \mathbf{w}$ and $\mathscr{P}_1\mathbf{f} = s$, where $(\mathbf{w}, s)$ is the unique solution to the stationary problem

$$-\operatorname{div}\mathcal{A}\boldsymbol{\varepsilon}(\mathbf{w}) + \nabla s = 0, \quad \operatorname{div}\mathbf{w} = \operatorname{div}\mathscr{R}_0\mathbf{f}, \quad \mathbf{w}|_{\partial G} = 0,$$

cf. Lemma B.1.1. Now we clearly have $\mathscr{R}\mathbf{f} \in \mathscr{Y}$. We need to establish (B.2.33). We abbreviate

$$\mathbf{u}_k := \mathscr{Z}_k^{-1}\mathscr{U}^k\mathscr{M}\mathscr{Z}_k\mathbf{f},$$
$$\pi_k := \mathscr{Z}_k^{-1}\mathscr{Q}^k\mathscr{M}\mathscr{Z}_k\mathbf{f}.$$

We define $\tilde{\nabla}_k = \nabla_x + \mathscr{Z}_k\nabla F_k\partial_{z_3}$ and accordingly $\widetilde{\operatorname{div}}_k$ as well as $\tilde{\boldsymbol{\varepsilon}}_k$. We gain on Q_k^T

$$\partial_t\mathbf{u}_k - \widetilde{\operatorname{div}}_k\mathcal{A}\big(\tilde{\boldsymbol{\varepsilon}}_k(\mathbf{u}_k)\big) + \tilde{\nabla}_k\pi_k = \mathscr{Z}_k^{-1}\mathscr{Z}_k\mathbf{f} = \mathbf{f}.$$

There holds

$$\partial_t\mathscr{R}\mathbf{f} - \operatorname{div}\mathcal{A}\boldsymbol{\varepsilon}(\mathscr{R}\mathbf{f}) + \nabla\mathscr{P}\mathbf{f} = \mathbf{f} + \mathscr{S}\mathbf{f} + \partial_t\mathscr{R}_1\mathbf{f}, \tag{B.2.34}$$

$$\begin{aligned}\mathscr{S}\mathbf{f} = &-\mathcal{A}\boldsymbol{\varepsilon}(\mathscr{U}_0\mathscr{M}_0\mathbf{f})\nabla\zeta_0 - \operatorname{div}\mathcal{A}\big(\zeta_0\odot\mathscr{U}_0\mathscr{M}_0\mathbf{f}\big) - \sum_{k=1}^{N}\mathcal{A}\boldsymbol{\varepsilon}(\mathbf{u}_k)\nabla\zeta_k\\ &-\sum_{k=1}^{N}\operatorname{div}\mathcal{A}\big(\nabla\zeta_k\odot\mathbf{u}_k\big) + \sum_{k=1}^{N}\nabla\zeta_k\pi_k - \sum_{k=1}^{N}\zeta_k(\tilde{\nabla}_k - \nabla)\pi_k \qquad \text{(B.2.35)}\\ &+\sum_{k=1}^{N}\zeta_k\Big(\widetilde{\operatorname{div}}_k\mathcal{A}\big(\tilde{\boldsymbol{\varepsilon}}_k(\cdot)\big) - \operatorname{div}\mathcal{A}\big(\boldsymbol{\varepsilon}(\cdot)\big)\Big)\mathbf{u}_k.\end{aligned}$$

From (B.2.34) it follows

$$\mathscr{L}\mathscr{R}\mathbf{f}=\mathbf{f}+\mathcal{P}_q\mathscr{S}\mathbf{f}+\mathcal{P}_q\partial_t\mathscr{R}_1\mathbf{f}$$

i.e. (B.2.33) with $\tau=\mathcal{P}_q\mathscr{S}+\mathcal{P}_q\partial_t\mathscr{R}_1$. We need to estimate the norms of the operators $\mathscr{S}$ and $\partial_t\mathscr{R}_1$. If we choose T small enough the first two terms in (B.2.35) are small in accordance to Corollary B.2.1. The same is true for the first three sums as a consequence of Corollary B.2.3. All together we have

$$\sum_{i=1}^{5}\|T_i\|_q\leq\delta(\lambda,T)\|\mathbf{f}\|_q$$

with $\delta(\lambda,T)\to 0$ for $T\to 0$ (and any fixed λ). Note that $\delta(\lambda,T)$ does not depend on N on account of the localization. We will argue similarly for the next three sums assuming that the gradients of the F_k are small (meaning that λ is small). Here, we gain

$$\|T_6\|_q+\|T_7\|_q\leq\kappa(\lambda)\|\mathbf{f}\|_q$$

with $\kappa(\lambda)\to 0$ for $\lambda\to 0$. Note that $\kappa(\lambda)$ does neither depend on T nor on N. By choosing first λ small enough such that $\kappa(\lambda)\leq\frac{1}{8}$ and then T small enough such that $\delta(\lambda,T)\leq\frac{1}{8}$ we can follow

$$\|\mathscr{S}\mathbf{f}\|_q\leq\tfrac{1}{4}\|\mathbf{f}\|_q. \tag{B.2.36}$$

Now we are going to show the same for $\partial_t\mathscr{R}_1$. In order to achieve this we consider the function $\mathbf{w}'=\partial_t\mathscr{R}_1\mathbf{f}$ which is the solution to

$$-\operatorname{div}\mathcal{A}\boldsymbol{\varepsilon}(\partial_t\mathbf{w}')+\nabla s'=0,\quad \operatorname{div}\mathbf{w}'=\operatorname{div}\partial_t\mathscr{R}_0\mathbf{f},\quad \mathbf{w}'|_{\partial G}=0.$$

We have the identity

$$\begin{aligned}\operatorname{div}\partial_t\mathscr{R}_0\mathbf{f}&=\sum_{k=0}^{N}\nabla\zeta_k\cdot\partial_t\mathbf{u}_k=\sum_{k=0}^{N}\tilde{\nabla}_k\zeta_k\cdot\partial_t\mathbf{u}_k+\sum_{k=0}^{N}(\nabla-\tilde{\nabla}_k)\zeta_k\cdot\partial_t\mathbf{u}_k\\&=\sum_{k=1}^{N}\tilde{\nabla}_k\zeta_k\cdot\Big(\widetilde{\operatorname{div}}_k\mathcal{A}\big(\tilde{\boldsymbol{\varepsilon}}_k(\mathbf{u}_k)\big)-\tilde{\nabla}_k\pi_k\Big)+\sum_{k=0}^{N}(\tilde{\nabla}_k-\nabla)\zeta_k\cdot\mathbf{f}\\&\quad+\sum_{k=1}^{N}(\nabla-\tilde{\nabla}_k)\zeta_k\cdot\partial_t\mathbf{u}_k=:\mathscr{T}_1+\mathscr{T}_2+\mathscr{T}_3,\end{aligned}$$

where $\mathbf{u}_0=\mathscr{U}_0\mathscr{M}_0\mathbf{f}$ and $\tilde{\nabla}_0=\nabla$. We use the formula

$$\mathscr{T}_1=\sum_{k=1}^{N}\widetilde{\operatorname{div}}_k\Big(\tilde{\nabla}_k\zeta_k\mathcal{A}\big(\tilde{\boldsymbol{\varepsilon}}_k(\mathbf{u}_k)\big)-\tilde{\nabla}_k\zeta_k\pi_k\Big)$$

$$-\sum_{k=1}^{N}\Big(\tilde{\nabla}_k^2\zeta_k\mathcal{A}\big(\tilde{\boldsymbol{\varepsilon}}_k(\mathbf{u}_k)\big)-\tilde{\Delta}_k\zeta_k\pi_k\Big)=:\mathscr{T}_1^1+\mathscr{T}_1^2,$$

as well as

$$\begin{aligned}(\tilde{\nabla}_k-\nabla)g^k&=\mathscr{Z}_k\nabla F_k\partial_{z_3}g^k=\mathscr{Z}_k\partial_{z_3}\big(\nabla F_kg^k\big)\\&=\partial_{\nu_k}\big(\mathscr{Z}_k\nabla F_kg^k\big)=\nu_k\cdot\nabla\big(\mathscr{Z}_k\nabla F_kg^k\big)\\&=\operatorname{div}\Big(\nu_k\mathscr{Z}_k^{-1}\big(\nabla F_kg^k\big)\Big).\end{aligned}$$

Now we can apply Corollary B.1.1 with $\operatorname{div}\mathbf{g}=\mathscr{T}_1^1$ and $g_0=\mathscr{T}_1^2+\mathscr{T}_2+\mathscr{T}_3$. By Corollary B.2.2 there are constants $\delta'(\lambda,T),\delta''(\lambda,T)$ with $\delta'(\lambda,T)\to 0$ and $\delta''(\lambda,T)\to 0$ for $T\to 0$ (and any fixed λ) such that

$$\|\mathbf{g}\|_q\leq\delta'(\lambda,T)\|\mathbf{f}\|_q,\quad \|g_0\|_q\leq c(1+\delta''(\lambda,T))\,\|\mathbf{f}\|_q.$$

As a consequence of Corollary B.1.1 this yields

$$\|\partial_t\mathscr{R}_1\mathbf{f}\|_q\leq c\big(\lambda^\beta+\delta'(\lambda,T)+\delta'(\lambda,T)\big)\|\mathbf{f}\|_q.$$

Choosing first λ and then T small enough we gain

$$\|\partial_t\mathscr{R}_1\mathbf{f}\|_q\leq\tfrac{1}{4}\|\mathbf{f}\|_q. \tag{B.2.37}$$

Combining (B.2.36) and (B.2.37) implies $\|\tau\|\leq\frac{1}{2}$. Hence $\mathscr{L}$ is onto (recall (B.2.33)). This means we have shown the claim for T sufficiently small, say $T=T_0\ll 1$. It is easy to extend it to the whole interval. Let $(\mathbf{v},\pi)$ be the solution on $[0,T_0]$. We extend it in an even manner to the interval $[0,2T_0]$. On the interval $[T_0,2T_0]$ we define $(\mathbf{v}',\pi')$ as solution to the $\mathcal{A}$-Stokes system with right-hand-side

$$\mathbf{f}'(t,x)=\mathbf{f}(t,x)-\mathbf{f}(x,2T_0-t)+2\partial_t\mathbf{v}(x,2T_0-t).$$

If we set $\mathbf{v}'=0,\pi'=0$ on $[0,T_0]$ then $(\mathbf{v}+\mathbf{v}',\pi+\pi')$ is the solution on $[0,2T_0]$. This can be repeated to construct the solution on $[0,T]$. □

B.3 THE NON-STATIONARY PROBLEM IN DIVERGENCE FORM

In order to treat problems with right hand side in divergence form we consider the $\mathcal{A}$-Stokes operator

$$\mathscr{A}_q:=-\mathcal{P}_q\operatorname{div}\mathcal{A}\big(\boldsymbol{\varepsilon}(\cdot)\big).$$

The $\mathcal{A}$-Stokes operator $\mathscr{A}_q$ enjoys the same properties than the Stokes operator $\mathbb{A}_q$ (see for instance [87]).

For the $\mathcal{A}$-Stokes operator it holds $D(\mathscr{A}_q) = W^{1,q}_{0,\mathrm{div}} \cap W^{2,q}(G)$, where D denotes the domain, and

$$\|\mathbf{u}\|_{2,q} \le c_1 \|\mathscr{A}_q \mathbf{u}\|_q \le c_2 \|\mathbf{u}\|_{2,q}, \quad \mathbf{u} \in D(\mathscr{A}_q), \tag{B.3.38}$$

$$\int_G \mathscr{A}_q \mathbf{u} \cdot \mathbf{w}\,\mathrm{d}x = \int_G \mathbf{u} \cdot \mathscr{A}_{q'} \mathbf{w}\,\mathrm{d}x \quad \mathbf{u} \in D(\mathscr{A}_q),\ \mathbf{w} \in D(\mathscr{A}_{q'}). \tag{B.3.39}$$

Inequality (B.3.38) is a consequence of Lemma B.1.1 a) and the continuity of $\mathcal{P}_q$.

Since $\mathscr{A}_q$ is positive its root $\mathscr{A}_q^{\frac{1}{2}}$ is well-defined with $D(\mathscr{A}_q^{\frac{1}{2}}) = W^{1,q}_{0,\mathrm{div}}(G)$ and

$$\|\mathbf{u}\|_{1,q} \le c_1 \|\mathscr{A}_q^{\frac{1}{2}} \mathbf{u}\|_q \le c_2 \|\mathbf{u}\|_{1,q}, \quad \mathbf{u} \in D(\mathscr{A}_q^{\frac{1}{2}}), \tag{B.3.40}$$

$$\int_G \mathscr{A}_q^{\frac{1}{2}} \mathbf{u} \cdot \mathbf{w}\,\mathrm{d}x = \int_G \mathbf{u} \cdot \mathscr{A}_{q'}^{\frac{1}{2}} \mathbf{w}\,\mathrm{d}x \quad \mathbf{u} \in D(\mathscr{A}_q^{\frac{1}{2}}),\ \mathbf{w} \in D(\mathscr{A}_{q'}^{\frac{1}{2}}). \tag{B.3.41}$$

Finally, the inverse operator $\mathscr{A}_q^{-\frac{1}{2}} : L^q_{\mathrm{div}}(G) \to W^{1,q}_{0,\mathrm{div}}(G)$ is defined and it holds

$$\|\nabla \mathscr{A}_q^{-\frac{1}{2}} \mathbf{u}\|_q \le c\|\mathbf{u}\|_q, \quad \mathbf{u} \in \mathcal{D}(\mathscr{A}_q^{-\frac{1}{2}}), \tag{B.3.42}$$

$$\int_G \mathscr{A}_q^{-\frac{1}{2}} \mathbf{u} \cdot \mathbf{w}\,\mathrm{d}x = \int_G \mathbf{u} \cdot \mathscr{A}_{q'}^{-\frac{1}{2}} \mathbf{w}\,\mathrm{d}x \quad \mathbf{u} \in D(\mathscr{A}_q^{-\frac{1}{2}}),\ \mathbf{w} \in D(\mathscr{A}_{q'}^{-\frac{1}{2}}). \tag{B.3.43}$$

From the definition of the square root of a positive self-adjoint operator follows also that

$$\mathscr{A}_{q'}^{\frac{1}{2}} : W^{2,q'} \cap W^{1,q'}_{0,\mathrm{div}}(G) \to W^{1,q'}_{0,\mathrm{div}}(G),$$
$$\mathscr{A}_{q'}^{-\frac{1}{2}} : W^{1,q'}_{0,\mathrm{div}}(G) \to W^{2,q'} \cap W^{1,q'}_{0,\mathrm{div}}(G),$$

together with

$$\|\nabla \mathscr{A}_q^{\frac{1}{2}} \mathbf{u}\|_q \le c\|\mathbf{u}\|_{2,q}, \quad \mathbf{u} \in W^{2,q'} \cap W^{1,q'}_{0,\mathrm{div}}(G), \tag{B.3.44}$$

$$\|\nabla \mathscr{A}_q^{-\frac{1}{2}} \mathbf{u}\|_q \le c\|\mathbf{u}\|_q, \quad \mathbf{u} \in W^{1,q'}_{0,\mathrm{div}}(G). \tag{B.3.45}$$

Finally we state the main result of this section.

Theorem B.3.50. *Let $Q := (0, T) \times G$ with a bounded domain $G \subset \mathbb{R}^3$ having a C^2-boundary. Let $\mathbf{F} \in L^q(Q)$, where $q \in (1, \infty)$. There is a unique solution $\mathbf{w} \in L^q(0, T; W^{1,q}_{0,\mathrm{div}}(G))$ to*

$$\int_Q \mathbf{w} \cdot \partial_t \boldsymbol{\varphi}\,\mathrm{d}x\,\mathrm{d}t - \int_Q \mathcal{A}(\boldsymbol{\varepsilon}(\mathbf{w}), \boldsymbol{\varepsilon}(\boldsymbol{\varphi}))\,\mathrm{d}x\,\mathrm{d}t = \int_Q \mathbf{F} : \nabla \boldsymbol{\varphi} \tag{B.3.46}$$

for all $\boldsymbol{\varphi} \in C^{\infty}_{0,\mathrm{div}}([0,T) \times G)$. *Moreover we have*

$$\fint_Q |\nabla \mathbf{w}|^q \,\mathrm{d}x\,\mathrm{d}t \le c \fint_Q |\mathbf{F}|^q \,\mathrm{d}x\,\mathrm{d}t,$$

where c *only depends on* $\mathcal{A}$ *and* q.

Proof. Let us first assume that $q > 2$. Then Theorem B.2.49 applies. We set $\mathbf{f} := \mathscr{A}_q^{-\frac{1}{2}} \operatorname{div} \mathbf{F}$ which is defined via the duality

$$\int_G \mathscr{A}_q^{-\frac{1}{2}} \operatorname{div} \mathbf{F} \cdot \boldsymbol{\varphi} \,\mathrm{d}x = \int_G \mathbf{F} : \nabla \mathscr{A}_{q'}^{-\frac{1}{2}} \boldsymbol{\varphi} \,\mathrm{d}x, \quad \boldsymbol{\varphi} \in C^{\infty}_{0,\mathrm{div}}(G),$$

using (B.3.43). So we gain $\mathbf{f} \in L^q_{\mathrm{div}}(G)$ with

$$\|\mathbf{f}\|_q \le c\, \|\mathbf{F}\|_q. \tag{B.3.47}$$

We define $\tilde{\mathbf{w}} \in L^q(0,T; W^{1,q}_{0,\mathrm{div}}(G))$ as the unique solution to

$$\int_Q \tilde{\mathbf{w}} \cdot \partial_t \boldsymbol{\varphi} \,\mathrm{d}x\,\mathrm{d}t - \int_Q \mathcal{A}(\boldsymbol{\varepsilon}(\tilde{\mathbf{w}}), \boldsymbol{\varepsilon}(\boldsymbol{\varphi})) \,\mathrm{d}x\,\mathrm{d}t = \int_Q \mathbf{f} \cdot \boldsymbol{\varphi} \,\mathrm{d}x\,\mathrm{d}t \tag{B.3.48}$$

for all $\boldsymbol{\varphi} \in C^{\infty}_{0,\mathrm{div}}([0,T) \times G)$. Theorem B.2.49 yields $\tilde{\mathbf{w}} \in L^q(0,T; W^{2,q}(G))$ and

$$\|\tilde{\mathbf{w}}\|_{2,q} \le c\, \|\mathbf{f}\|_q. \tag{B.3.49}$$

We want to return to the original problem and set $\mathbf{w} := \mathscr{A}_q^{\frac{1}{2}} \tilde{\mathbf{w}}$ thus we have $\mathbf{w} \in L^q(0,T; W^{1,q}_{0,\mathrm{div}}(G))$. Since $\mathscr{A}_{q'}^{\frac{1}{2}} : W^{1,q'}_{0,\mathrm{div}} \cap W^{2,q'}(G) \to W^{1,q'}_{0,\mathrm{div}}(G)$ we can replace $\boldsymbol{\varphi}$ by $\mathscr{A}_{q'}^{\frac{1}{2}} \boldsymbol{\varphi}$ in (B.3.48). This implies using (B.3.41) and the definition of $\mathbf{f}$

$$\int_Q \mathbf{w} \cdot \partial_t \boldsymbol{\varphi} \,\mathrm{d}x\,\mathrm{d}t + \int_Q \operatorname{div} \mathcal{A}(\boldsymbol{\varepsilon}(\tilde{\mathbf{w}})) : \mathscr{A}_{q'}^{\frac{1}{2}} \boldsymbol{\varphi} \,\mathrm{d}x\,\mathrm{d}t = \int_Q \mathbf{F} : \nabla \boldsymbol{\varphi}$$

for all $\boldsymbol{\varphi} \in C^{\infty}_{0,\mathrm{div}}(Q)$. On account of $\mathscr{A}_{q'}^{\frac{1}{2}} \boldsymbol{\varphi} \in W^{1,q'}_{0,\mathrm{div}}(G)$ and $\mathscr{A}_q^{\frac{1}{2}} \tilde{\mathbf{w}} \in W^{1,q}_{0,\mathrm{div}}(G)$ we gain due to (B.3.41)

$$\begin{aligned}
&\int_Q \operatorname{div} \mathcal{A}(\boldsymbol{\varepsilon}(\tilde{\mathbf{w}})) : \mathscr{A}_{q'}^{\frac{1}{2}} \boldsymbol{\varphi} \,\mathrm{d}x\,\mathrm{d}t \\
&= \int_Q \mathscr{A}_q \tilde{\mathbf{w}} : \mathscr{A}_{q'}^{\frac{1}{2}} \boldsymbol{\varphi} \,\mathrm{d}x\,\mathrm{d}t = \int_Q \mathscr{A}_q^{\frac{1}{2}} \tilde{\mathbf{w}} : \mathscr{A}_{q'} \boldsymbol{\varphi} \,\mathrm{d}x\,\mathrm{d}t \\
&= \int_Q \mathbf{w} \cdot \operatorname{div} \mathcal{A}(\boldsymbol{\varepsilon}(\boldsymbol{\varphi})) \,\mathrm{d}x\,\mathrm{d}t = - \int_Q \mathcal{A}(\boldsymbol{\varepsilon}(\mathbf{w}), \boldsymbol{\varepsilon}(\boldsymbol{\varphi})) \,\mathrm{d}x\,\mathrm{d}t
\end{aligned}$$

using (B.3.41) and $\mathbf{w} \in W^{1,q}_{0,\mathrm{div}}(G)$. This shows that $\mathbf{w}$ is the unique solution to (6.2.27). Moreover, we obtain the desired regularity estimate via

$$\begin{aligned}\int_Q |\nabla \mathbf{w}|^q \,\mathrm{d}x\,\mathrm{d}t &\leq c \int_Q |\mathscr{A}_q^{\frac{1}{2}} \mathbf{w}|^q \,\mathrm{d}x\,\mathrm{d}t = c \int_Q |\mathscr{A}_q \tilde{\mathbf{w}}|^q \,\mathrm{d}x\,\mathrm{d}t \\ &\leq c \int_Q |\nabla^2 \tilde{\mathbf{w}}|^q \leq c \int_Q |\mathbf{f}|^q \,\mathrm{d}x\,\mathrm{d}t \\ &\leq c \int_Q |\mathbf{F}|^q \,\mathrm{d}x\,\mathrm{d}t\end{aligned}$$

as a consequence of (B.3.40), the definition of $\mathbf{w}$, (B.3.38), (B.3.49), and (B.3.47). A simple scaling argument shows that the inequality is independent of the diameter of I and B. So we have shown the claim for $q > 2$.

The case $q = 2$ follows easily from a priori estimates and Korn's inequality. So let us assume that $q < 2$. Duality arguments show that

$$\frac{1}{q} \fint_Q |\nabla \mathbf{w}|^q \,\mathrm{d}x\,\mathrm{d}t = \sup_{\mathbf{G} \in L^{q'}(Q)} \left[\fint_Q \nabla \mathbf{w} : \mathbf{G} \,\mathrm{d}x\,\mathrm{d}t - \frac{1}{q'} \fint_Q |\mathbf{G}|^{q'} \,\mathrm{d}x\,\mathrm{d}t \right].$$

For a given $\mathbf{G} \in L^{q'}(Q)$ let $\mathbf{z}_\mathbf{G}$ be the unique $L^{q'}(0, T; W^{1,q'}_{0,\mathrm{div}}(G))$-solution to

$$\fint_Q \mathbf{z} \cdot \partial_t \boldsymbol{\xi} \,\mathrm{d}x\,\mathrm{d}t + \int_Q \mathcal{A}(\boldsymbol{\varepsilon}(\mathbf{z}), \boldsymbol{\varepsilon}(\boldsymbol{\xi})) \,\mathrm{d}x\,\mathrm{d}t = \int_Q \mathbf{G} : \nabla \boldsymbol{\xi} \,\mathrm{d}x\,\mathrm{d}t \qquad \text{(B.3.50)}$$

for all $\boldsymbol{\xi} \in C^\infty_{0,\mathrm{div}}((0, T] \times G)$. This is a backward parabolic equation with end datum zero. We have that $\partial_t \mathbf{z}_\mathbf{G} \in L^{q'}(0, T; W^{-1,q'}_{\mathrm{div}}(G))$ such that test-functions can be chosen from the space $L^q(0, T; W^{1,q}_{0,\mathrm{div}}(G))$. Due to $q' > 2$ the first part of the proof (applied to $\tilde{\mathbf{z}}_{\tilde{\mathbf{G}}}(t, \cdot) = \mathbf{z}_\mathbf{G}(T - t, \cdot)$, where $\tilde{\mathbf{G}}(t, \cdot) = \mathbf{G}(T - t, \cdot)$) yields

$$\fint_Q |\nabla \mathbf{z}_\mathbf{G}|^{q'} \,\mathrm{d}x\,\mathrm{d}t \leq c \fint_Q |\mathbf{G}|^{q'} \,\mathrm{d}x\,\mathrm{d}t.$$

This and $\mathbf{w}(0, \cdot) = 0$ implies (using $\mathbf{w}$ as a test-function in (B.3.50))

$$\begin{aligned}&\fint_Q |\nabla \mathbf{w}|^q \,\mathrm{d}x\,\mathrm{d}t \\ &\leq c \sup_{\mathbf{G} \in L^{q'}(Q)} \left[\fint_Q \mathcal{A}(\boldsymbol{\varepsilon}(\mathbf{w}), \boldsymbol{\varepsilon}(\mathbf{z}_\mathbf{G})) \,\mathrm{d}x\,\mathrm{d}t - \fint_Q \partial_t \mathbf{z}_\mathbf{G} \cdot \mathbf{w} \,\mathrm{d}x\,\mathrm{d}t - \fint_Q |\nabla \mathbf{z}_\mathbf{G}|^{q'} \,\mathrm{d}x\,\mathrm{d}t \right] \\ &\leq c \sup_{\boldsymbol{\xi} \in C^\infty_{0,\mathrm{div}}(Q)} \left[\fint_Q \mathcal{A}(\boldsymbol{\varepsilon}(\mathbf{w}), \boldsymbol{\varepsilon}(\boldsymbol{\xi})) \,\mathrm{d}x\,\mathrm{d}t - \fint_Q \mathbf{w} \cdot \partial_t \boldsymbol{\xi} \,\mathrm{d}x\,\mathrm{d}t - \fint_Q |\nabla \boldsymbol{\xi}|^{q'} \,\mathrm{d}x\,\mathrm{d}t \right].\end{aligned}$$

The equation for **w** and Young's inequality finally give

$$\begin{aligned}\fint_Q |\nabla\mathbf{w}|^q \,\mathrm{d}x\,\mathrm{d}t &\leq c \sup_{\boldsymbol{\xi}\in C^\infty_{0,\mathrm{div}}(Q)} \left[\fint_Q \mathbf{F} : \nabla\boldsymbol{\xi} \,\mathrm{d}x\,\mathrm{d}t - \fint_Q |\nabla\boldsymbol{\xi}|^{q'} \,\mathrm{d}x\,\mathrm{d}t\right] \\ &\leq c \fint_Q |\mathbf{F}|^q \,\mathrm{d}x\,\mathrm{d}t\end{aligned}$$

and hence the claim. □

APPENDIX C

Itô's formula in infinite dimensions

In this appendix we establish a version of Itô's formula which holds for weak solutions of SPDEs on a probability space $(\Omega, \mathcal{F}, \mathbb{P})$. Let $\mathbf{u}$ be a solution to the system

$$\begin{aligned}\int_G \mathbf{u}(t) \cdot \boldsymbol{\varphi}\,\mathrm{d}x &= \int_G \mathbf{u}_0 \cdot \boldsymbol{\varphi}\,\mathrm{d}x + \int_0^t \int_G \mathbf{H} : \nabla\boldsymbol{\varphi}\,\mathrm{d}x\,\mathrm{d}\sigma \\ &\quad + \int_0^t \int_G \mathbf{h} \cdot \boldsymbol{\varphi}\,\mathrm{d}x\,\mathrm{d}t + \int_G \int_0^t \boldsymbol{\varphi} \cdot \Phi\,\mathrm{d}\mathbf{W}_\sigma\,\mathrm{d}x\end{aligned} \tag{C.0.1}$$

for all $\varphi \in C_0^\infty(G)$, where $\mathbf{W}$ is given by (9.1.1). We assume

(I1) $\mathbf{u}_0 \in L^2(\Omega, \mathcal{F}_0, \mathbb{P}; L^2(G))$;
(I2) $\mathbf{H} \in L^{p'}(\Omega, \mathcal{F}, \mathbb{P}; L^{p'}(Q))$ adapted to $(\mathcal{F}_t)_{t\geq 0}$;
(I3) $\mathbf{h} \in L^{q'}(\Omega, \mathcal{F}, \mathbb{P}; L^{q'}(Q))$ adapted to $(\mathcal{F}_t)_{t\geq 0}$;
(I4) $\Phi \in L^2(\Omega, \mathcal{F}, \mathbb{P}; L^2(0, T; L_2(U, L^2(G))))$ progressively measurable.

The function has to be taken from the function space

$$\begin{aligned}\mathcal{W}_{p,q} &:= L^q(\Omega \times Q; \mathbb{P} \otimes \mathcal{L}^{d+1}) \cap L^p(\Omega, \mathcal{F}, \mathbb{P}; L^p(0, T; W_0^{1,p}(G))) \\ &\quad \cap \left\{\mathbf{w} \in L^1(\Omega \times Q; \mathbb{P} \otimes \mathcal{L}^{d+1}) : \mathbb{E}\left[\sup_{t\in(0,T)} \int_G |\mathbf{w}|^2\,\mathrm{d}x\right] < \infty\right\}.\end{aligned}$$

Lemma C.0.1 (Itô's Lemma). *Let*

$$f : L^2(G) \to \mathbb{R}, \quad \mathbf{v} \mapsto \int_G F(x, \mathbf{v})\,\mathrm{d}x,$$

where $F \in C^2(G \times \mathbb{R}^d)$ has the following properties: for all $(x, \boldsymbol{\xi}) \in G \times \mathbb{R}^d$ we have

- $|D_{\boldsymbol{\xi}} F(x, \boldsymbol{\xi})| \leq c(1 + |\boldsymbol{\xi}|)$ *and* $|D^2_{\boldsymbol{\xi}}(x, \boldsymbol{\xi})| \leq c$;
- $|D_x D_{\boldsymbol{\xi}} F(x, \boldsymbol{\xi})| \leq c(1 + |\boldsymbol{\xi}|)$ *and* $|D_x D^2_{\boldsymbol{\xi}}(x, \boldsymbol{\xi})| \leq c$.

Let $\mathbf{u} \in \mathcal{W}_{p,q}$ with $p, q \in (1, \infty)$ be progressively measurable. Assume that (C.0.1) is satisfied and that (I1)–(I4) *hold. Then we have*

$$\begin{aligned}f(\mathbf{u}(t)) &= f(\mathbf{u}(0)) + \int_0^t f'(\mathbf{u})\,\mathrm{d}\mathbf{u} + \frac{1}{2}\int_0^t f''(\mathbf{u})\,\mathrm{d}\langle\langle \mathbf{u}\rangle\rangle_\sigma \\ &= \int_G F(\cdot, \mathbf{u}(0))\,\mathrm{d}x + \int_0^t \int_G \mathbf{H} : \nabla\{D_{\boldsymbol{\xi}} F(\cdot, \mathbf{u})\}\,\mathrm{d}x\,\mathrm{d}\sigma\end{aligned}$$

Existence Theory for Generalized Newtonian Fluids.
DOI: http://dx.doi.org/10.1016/B978-0-12-811044-7.00024-0

$$+\int_0^t\int_G \mathbf{h}\cdot D_\xi F(\cdot,\mathbf{u})\,\mathrm{d}x\,\mathrm{d}\sigma+\int_G\int_0^t D_\xi F(\cdot,\mathbf{u})\cdot\Phi\,\mathrm{d}\mathbf{W}_\sigma\,\mathrm{d}x$$
$$+\frac{1}{2}\sum_{k=1}^{\infty}\int_0^t\int_G D_\xi^2 F(\cdot,\mathbf{u})(\Phi\mathbf{e}_k,\Phi\mathbf{e}_k)\,\mathrm{d}x\,\mathrm{d}\sigma.$$

Proof. We follow the ideas of [57], Prop. 1. We replace $\boldsymbol{\varphi}$ with the mollification $\boldsymbol{\varphi}_\varrho$, where $\varrho<\operatorname{dist}(\operatorname{spt}(\boldsymbol{\varphi}),\partial G)$. This yields

$$\mathbf{u}_\varrho(t)=(\mathbf{u}_0)_\varrho-\int_0^t \operatorname{div}\mathbf{H}_\varrho\,\mathrm{d}\sigma+\int_0^t \mathbf{h}_\varrho\,\mathrm{d}\sigma+\sum_k\int_0^t(\Phi_k)_\rho\,\mathrm{d}\beta_k,$$

where $\Phi_k:=\Phi\mathbf{e}_k$ a.e. on G. So we can apply the common finite-dimensional Itô formula (see Theorem 8.3.34) to the real-valued process $t\mapsto f(\mathbf{u}_\varrho(t))$ and gain

$$f(\mathbf{u}_\varrho(t))=f(\mathbf{u}_\varrho(0))+\int_0^t f'(\mathbf{u}_\varrho)\,\mathrm{d}\mathbf{u}_\varrho+\frac{1}{2}\int_0^t f''(\mathbf{u}_\varrho)\,\mathrm{d}\langle\langle\mathbf{u}_\varrho\rangle\rangle_\sigma. \qquad \text{(C.0.2)}$$

Here we have for $\mathbf{w}\in L^2(G)$

$$f'(\mathbf{w})\boldsymbol{\varphi}=\int_G D_\xi F(x,\mathbf{w})\cdot\boldsymbol{\varphi}\,\mathrm{d}x,\quad \boldsymbol{\varphi}\in L^2(G),$$
$$f''(\mathbf{w})(\boldsymbol{\varphi},\boldsymbol{\psi})=\int_G D_\xi^2 F(x,\mathbf{w})(\boldsymbol{\varphi},\boldsymbol{\psi})\,\mathrm{d}x,\quad \boldsymbol{\varphi},\boldsymbol{\psi}\in L^2(G).$$

Now we have to pass to the limit in ϱ. It is easy to see that $f(\mathbf{u}_\varrho(t))$ converges to $f(\mathbf{u}(t))$ $\mathbb{P}$-a.s. for a.e. t. This is a consequence of the properties of the convolution. Similarly we have $f((\mathbf{u}_0)_\varrho)\to f(\mathbf{u}_0)$, so

$$f(\mathbf{u}_\varrho(t))\longrightarrow(\mathbf{u}(t)),\quad f((\mathbf{u}_0)_\varrho)\longrightarrow(\mathbf{u}_0), \qquad \text{(C.0.3)}$$

for $\varrho\to0$. Let us now consider the first integral which can be written as

$$\int_0^t f'(\mathbf{u}_\varrho)\,\mathrm{d}\mathbf{u}_\varrho$$
$$=\int_0^t\int_G \mathbf{H}_\varrho:D_\xi^2F(\cdot,\mathbf{u}_\varrho)\nabla\mathbf{u}_\varrho\,\mathrm{d}x\,\mathrm{d}\sigma+\int_0^t\int_G \mathbf{H}_\varrho:D_\xi D_x F(\cdot,\mathbf{u}_\varrho)\,\mathrm{d}x\,\mathrm{d}\sigma$$
$$+\int_0^t\int_G \mathbf{h}_\varrho\cdot D_\xi F(\cdot,\mathbf{u}_\varrho)\,\mathrm{d}x\,\mathrm{d}\sigma+\sum_k\int_0^t\int_G(\Phi_k)_\rho\cdot D_\xi F(\cdot,\mathbf{u}_\varrho)\,\mathrm{d}x\,\mathrm{d}\beta_k$$
$$=(I)_\varrho+(II)_\varrho+(III)_\varrho+(IV)_\varrho.$$

We have to show that all the integrals converge to their counterparts denoted by (I)–(IV). By the properties of the convolution and the assumptions on F we have $\mathbb{P}$-a.s.

$$\begin{aligned}\mathbf{H}_\varrho&\longrightarrow\mathbf{H}\ \text{ in }\ L^{p'}(Q), & \mathbf{h}_\varrho&\longrightarrow\mathbf{h}\ \text{ in }\ L^{q'}(Q),\\ \nabla\mathbf{u}_\varrho&\longrightarrow\nabla\mathbf{u}\ \text{ in }\ L^p(Q), & \mathbf{u}_\varrho&\longrightarrow\mathbf{u}\ \text{ in }\ L^q(Q).\end{aligned} \qquad \text{(C.0.4)}$$

So we gain due $|D_\xi^2 F| \le c$ and (C.0.4) $\mathbb{P}$-a.s.

$$\begin{aligned}
|(I)_\varrho - (I)| \le & \left| \int_0^t \int_G \big(\mathbf{H}_\varrho - \mathbf{H}\big) : D_\xi^2 F(\cdot, \mathbf{u}_\varrho) \nabla \mathbf{u}_\varrho \,\mathrm{d}x\,\mathrm{d}\sigma \right| \\
& + \left| \int_0^t \int_G \mathbf{H} : D_\xi^2 F(\cdot, \mathbf{u}_\varrho) \big(\nabla \mathbf{u}_\varrho - \nabla \mathbf{u}\big) \,\mathrm{d}x\,\mathrm{d}\sigma \right| \\
& + \left| \int_0^t \int_G \mathbf{H} : \Big(D_\xi^2 F(\cdot, \mathbf{u}_\varrho) - D_\xi^2 F(\cdot, \mathbf{u})\Big) \nabla \mathbf{u} \,\mathrm{d}x\,\mathrm{d}\sigma \right| \\
& \longrightarrow 0, \quad \varrho \to 0,
\end{aligned}$$

for a.e. t. Similarly, the following holds

$$\begin{aligned}
|(II)_\varrho - (II)| \le & \left| \int_0^t \int_G \big(\mathbf{H}_\varrho - \mathbf{H}\big) : D_x D_\xi F(\cdot, \mathbf{u}_\varrho) \,\mathrm{d}x\,\mathrm{d}\sigma \right| \\
& + \left| \int_0^t \int_G \mathbf{H} : \big(D_x D_\xi F(\cdot, \mathbf{u}_\varrho) - D_x D_\xi F(\cdot, \mathbf{u})\big) \,\mathrm{d}x\,\mathrm{d}\sigma \right| \\
& \longrightarrow 0, \quad \varrho \to 0.
\end{aligned}$$

Here we took into account that $|D_x D_\xi^2 F| \le c$ implies

$$\big|D_x D_\xi F(\cdot, \mathbf{u}_\varrho) - D_x D_\xi F(\cdot, \mathbf{u})\big| \le c\,|\mathbf{u}_\varrho - \mathbf{u}|.$$

Analogously, we can prove that $\lim_\varrho (III)_\varrho = (III)$.

For the stochastic integral we have by the Burkholder–Davis–Gundi inequality

$$\begin{aligned}
& \mathbb{E}\Big[\sup_{(0,T)} |(IV)_\varrho - (IV)|\Big] \\
& \le c\,\mathbb{E}\bigg[\int_0^T \sum_k \Big(\int_G \big\{(\Phi_k)_\rho \cdot D_\xi F(\cdot, \mathbf{u}_\varrho) - (\Phi_k) \cdot D_\xi F(\cdot, \mathbf{u})\big\} \,\mathrm{d}x\Big)^2\bigg]^{\frac{1}{2}} \\
& \le c\,\mathbb{E}\bigg[\int_0^T \|D_\xi F(\cdot, \mathbf{u}_\delta) - D_\xi F(\cdot, \mathbf{u})\|_2^2 \|\Phi_\varrho\|_{L_2(U, L^2(G))}^2\bigg]^{\frac{1}{2}} \\
& + c\,\mathbb{E}\bigg[\int_0^T \|D_\xi F(\cdot, \mathbf{u})\|_2^2 \|\Phi - \Phi_\varrho\|_{L_2(U, L^2(G))}^2\bigg]^{\frac{1}{2}} \longrightarrow 0, \quad \varrho \to 0.
\end{aligned}$$

Here we used that $\|\Phi_\varrho\|_{L_2(U,L^2(G))} \le \|\Phi\|_{L_2(U,L^2(G))}$ and $\|\Phi - \Phi_\varrho\|_{L_2(U,L^2(G))} \to 0$ for a.e. (ω, t) and $|D_\xi F(\cdot, \mathbf{u}_\delta) - D_\xi F(\cdot, \mathbf{u})| \le |\mathbf{u}_\delta - \mathbf{u}|$. Plugging all together we have shown

$$\int_0^t f'(\mathbf{u}_\varrho)\,\mathrm{d}\mathbf{u}_\varrho \longrightarrow \int_0^t f'(\mathbf{u})\,\mathrm{d}\mathbf{u} \tag{C.0.5}$$

for $\varrho \to 0$.

What remains is the convergence of the correction term. Here we have

$$(V)_\varrho := \frac{1}{2}\int_0^t f''(\mathbf{u}_\varrho)\,\mathrm{d}\langle\langle \mathbf{u}_\varrho\rangle\rangle_\sigma = \sum_k \int_0^t \int_G D_\xi^2(\cdot,\mathbf{u}_\varrho)(\Phi\mathbf{e}_k)_\varrho^2\,\mathrm{d}x\,\mathrm{d}\sigma$$

and so

$$\begin{aligned}
|(V)_\varrho - (V)| &\le \sum_k \int_0^t \int_G |D_\xi^2(\cdot,\mathbf{u}_\varrho)(\Phi\mathbf{e}_k)_\varrho^2 - D_\xi^2(\cdot,\mathbf{u}_\varrho)(\Phi\mathbf{e}_k)^2|\,\mathrm{d}x\,\mathrm{d}\sigma\\
&\le \sum_k \int_0^t \int_G |D_\xi^2(\cdot,\mathbf{u}_\varrho)\big((\Phi\mathbf{e}_k)_\varrho^2 - (\Phi\mathbf{e}_k)^2\big)|\,\mathrm{d}x\,\mathrm{d}\sigma\\
&\quad + \sum_k \int_0^t \int_G |\big(D_\xi^2(\cdot,\mathbf{u}_\varrho) - D_\xi^2(\cdot,\mathbf{u}_\varrho)\big)(\Phi\mathbf{e}_k)^2|\,\mathrm{d}x\,\mathrm{d}\sigma.
\end{aligned}$$

This means we have

$$\frac{1}{2}\int_0^t f''(\mathbf{u}_\varrho)\,\mathrm{d}\langle\langle \mathbf{u}_\varrho\rangle\rangle_\sigma \longrightarrow \frac{1}{2}\int_0^t f''(\mathbf{u})\,\mathrm{d}\langle\langle \mathbf{u}\rangle\rangle_\sigma. \tag{C.0.6}$$

Combining (C.0.3)–(C.0.6) implies the claim. □

REFERENCES

1. E. Acerbi, N. Fusco, Semicontinuity problems in the calculus of variations, Arch. Ration. Mech. Anal. 86 (2) (1984) 125–145.
2. E. Acerbi, N. Fusco, A regularity theorem for minimizers of quasiconvex integrals, Arch. Ration. Mech. Anal. 99 (3) (1987) 261–281.
3. E. Acerbi, N. Fusco, An approximation lemma for $W^{1,p}$ functions, in: Material Instabilities in Continuum Mechanics, Edinburgh, 1985–1986, in: Oxford Sci. Publ., Oxford Univ. Press, New York, 1988, pp. 1–5.
4. E. Acerbi, G. Mingione, Regularity results for stationary electro-rheological fluids, Arch. Ration. Mech. Anal. 164 (2002) 213–259.
5. R.A. Adams, Sobolev Spaces, Pure Appl. Math., vol. 65, Academic Press, Inc., New York, 1975.
6. S. Agmon, A. Douglis, L. Nirenberg, Estimates near the boundary for solutions of elliptic partial differential equations satisfying general boundary conditions. I, Commun. Pure Appl. Math. 12 (1959) 623–727.
7. S. Agmon, A. Douglis, L. Nirenberg, Estimates near the boundary for solutions of elliptic partial differential equations satisfying general boundary conditions. II, Commun. Pure Appl. Math. 17 (1964) 35–92.
8. H. Amann, Compact embeddings of vector-valued Sobolev and Besov spaces, Glass. Mat., III. Ser. 35 (55) (2000) 161–177.
9. L. Ambrosio, N. Fusco, D. Pallara, Functions of Bounded Variation and Free Discontinuity Problems, Oxf. Math. Monogr., The Clarendon Press, Oxford University Press, New York, 2000.
10. C. Amrouche, V. Girault, Decomposition of vector spaces and application to the Stokes problem in arbitrary dimension, Czechoslov. Math. J. 44 (1994) 109–140.
11. G. Anzellotti, M. Giaquinta, Existence of the displacements field for an elasto plastic body subject to Hencky's law and von Mises yield condition, Manuscr. Math. 32 (1980) 101–136.
12. L. Arnold, Stochastic Differential Equations: Theory and Applications, J. Wiley & Sons, New York, 1973.
13. G. Astarita, G. Marrucci, Principles of Non-Newtonian Fluid Mechanics, McGraw–Hill, London–New York, 1974.
14. J.-P. Aubin, Un théorème de compacité, C. R. Acad. Sci. Paris 256 (1963) 5042–5044.
15. R.J. Bagby, D.S. Kurtz, A rearranged good λ inequality, Trans. Am. Math. Soc. 293 (1986) 71–81.
16. J.M. Ball, Convexity conditions and existence theorems in nonlinear elasticity, Arch. Ration. Mech. Anal. 63 (1977) 337–403.
17. M.T. Barlow, One dimensional stochastic differential equation with no strong solution, J. Lond. Math. Soc. (2) 26 (1982) 335–347.
18. C. Bennett, K. Rudnick, On Lorentz–Zygmund spaces, Diss. Math. 185 (1980).
19. C. Bennett, R. Sharpley, Interpolation of Operators, Academic Press, Boston, 1988.
20. A. Bensoussan, R. Temam, Équations stochastiques du type Navier–Stokes, J. Funct. Anal. 13 (1973) 195–222 (in French).
21. H. Beirão da Veiga, P. Kaplický, M. Růžička, Boundary Regularity of Shear Thickening Flows, J. Math. Fluid Mech. 13 (2011) 387–404.

22. M. Bildhauer, M. Fuchs, Variants of the Stokes problem: the case of anisotropic potentials, J. Math. Fluid Mech. 5 (2003) 364–402.
23. R. Bird, R. Armstrong, O. Hassager, Dynamics of Polymeric Liquids, vol. 1: Fluid Mechanics, second edition, John Wiley, 1987.
24. M.E. Bogovskiĭ, Solutions of some problems of vector analysis, associated with the operators div and grad, in: Theory of Cubature Formulas and the Application of Functional Analysis to Problems of Mathematical Physics, Akad. Nauk SSSR Sibirsk. Otdel. Inst. Mat., Novosibirsk, 1980, pp. 5–40, 149 (in Russian).
25. D. Breit, Analysis of generalized Navier–Stokes equations for stationary shear thickening flows, Nonlinear Anal. 75 (2012) 5549–5560.
26. D. Breit, Existence Theory for Generalized Newtonian Fluids, Postdoctoral thesis, LMU Munich, Department of Mathematics, 2013.
27. D. Breit, Existence theory for stochastic power law fluids, J. Math. Fluid Mech. 17 (2015) 295–326.
28. D. Breit, The $\mathcal{A}$-Stokes approximation for non-stationary problems, Quart. J. Math. 67 (2016) 201–231.
29. D. Breit, Existence theory for generalized Newtonian fluids, in: Recent Advances in Partial Differential Equations and Applications, in: Contemp. Math., vol. 666, Amer. Math. Soc., Providence, RI, 2016, pp. 99–110.
30. D. Breit, A. Cianchi, Negative Orlicz–Sobolev norms and strongly nonlinear systems in fluid mechanics, J. Differ. Equ. 259 (2015) 48–83.
31. D. Breit, A. Cianchi, L. Diening, Trace-free Korn inequalities in Orlicz spaces, preprint at arXiv:1605.01006v1.
32. D. Breit, L. Diening, Sharp conditions for Korn inequalities in Orlicz spaces, J. Math. Fluid Mech. 14 (2012) 565–573.
33. D. Breit, L. Diening, M. Fuchs, Solenoidal Lipschitz truncation and applications in fluid mechanics, J. Differ. Equ. 253 (2012) 1910–1942.
34. D. Breit, L. Diening, S. Schwarzacher, Solenoidal Lipschitz truncation for parabolic PDEs, Math. Models Methods Appl. Sci. 23 (2013) 2671–2700.
35. D. Breit, M. Fuchs, The nonlinear Stokes problem with general potentials having superquadratic growth, J. Math. Fluid Mech. 13 (2011) 371–385.
36. D. Breit, M. Hofmanová, Stochastic Navier–Stokes equations for compressible fluids, Indiana Univ. Math. J. 165 (4) (2016).
37. S.C. Brenner, L.R. Scott, The Mathematical Theory of Finite Element Methods, Texts Appl. Math., vol. 15, Springer-Verlag, New York, 1994.
38. Z. Brzeźniak, M. Ondreját, Strong solutions to stochastic wave equations with values in Riemannian manifolds, J. Funct. Anal. 253 (2007) 449–481.
39. M. Bulíček, J. Málek, K.R. Rajagopal, Navier's slip and evolutionary Navier–Stokes-like systems with pressure and shear-rate dependent viscosity, Indiana Univ. Math. J. 56 (2007) 61–86.
40. M. Bulíček, S. Schwarzacher, Existence of very weak solutions to elliptic systems of p-Laplacian type, Calc. Var. Partial Differ. Equ. 55 (2016) 52.
41. W.G. Burgers, J.M. Burgers, First report on viscosity and plasticity, Verh. Akad. Amsterdam (I) 15 (1935), Chapter V.
42. S.-S. Byun, L. Wang, L^p estimates for parabolic equations in Reifenberg domains, J. Funct. Anal. 223 (2005) 44–85.
43. A.P. Calderón, A. Zygmund, Singular integrals and periodic functions, Stud. Math. 14 (1954) 249–271.

44. L.A. Caffarelli, X. Cabré, Fully Nonlinear Elliptic Equations, Colloq. Publ. – Am. Math. Soc., vol. 43, American Mathematical Society, Providence, RI, 1995, vi+104 pp.
45. J. Chen, Z.-M. Chen, Stochastic non-Newtonian fluid motion equations of a nonlinear bipolar viscous fluid, J. Math. Anal. Appl. 369 (2) (2010) 486–509.
46. A. Cianchi, A sharp embedding theorem for Orlicz–Sobolev spaces, Indiana Univ. Math. J. 45 (1996) 39–65.
47. A. Cianchi, Strong and weak type inequalities for some classical operators in Orlicz spaces, J. Lond. Math. Soc. 60 (1999) 187–202.
48. A. Cianchi, Korn type inequalities in Orlicz spaces, J. Funct. Anal. 267 (2014) 2313–2352.
49. Ph. Clément, G. Sweers, Uniform anti-maximum principles for polyharmonic equations, Proc. Am. Math. Soc. 129 (2000) 467–474.
50. Ph. Clément, G. Sweers, Uniform anti-maximum principles, J. Differ. Equ. 164 (2000) 118–154.
51. R.R. Coifman, C. Fefferman, Weighted norm inequalities for maximal functions and singular integrals, Stud. Math. 51 (1974) 241–250.
52. S. Conti, D. Faraco, F. Maggi, A new approach to counterexamples to L1 estimates: Korn's inequality, geometric rigidity, and regularity for gradients of separately convex functions, Arch. Ration. Mech. Anal. 175 (2005) 287–300.
53. R. Courant, D. Hilbert, Methoden der mathematischen Physik, vol. 2, Springer, Berlin, 1937.
54. G. Dal Maso, F. Murat, Almost everywhere convergence of gradients of solutions to nonlinear elliptic systems, Nonlinear Anal. 31 (1998) 405–412.
55. G. Da Prato, J. Zabczyk, Stochastic Equations in Infinite Dimensions, Encycl. Math. Appl., vol. 44, Cambridge University Press, Cambridge, 1992.
56. A. Debussche, N. Glatt-Holtz, R. Temam, Local martingale and pathwise solutions for an abstract fluids model, Phys. D, Nonlinear Phenom. 240 (14–15) (2011) 1123–1141.
57. A. Debussche, M. Hofmanova, J. Vovelle, Degenerate parabolic stochastic partial differential equations: quasilinear case, Ann. Probab. 44 (3) (2016) 1916–1955.
58. E. De Giorgi, Frontiere orientate di misura minima, in: Seminario di Matematica della Scuola Normale Superiore di Pisa, Editrice Tecnico Scientica, Pisa, 1961.
59. L. Diening, P. Kaplicky, L^q theory for a generalized Stokes System, Manuscr. Math. 141 (2013) 333–361.
60. L. Diening, Ch. Kreuzer, E. Süli, Finite element approximation of steady flows of incompressible fluids with implicit power-law-like rheology, SIAM J. Numer. Anal. 51 (2) (2013) 984–1015.
61. L. Diening, B. Stroffolini, A. Verde, Partial regularity for minimizers of quasiconvex functionals with general growth, SIAM J. Math. Anal. 44 (2012) 3594–3616.
62. L. Diening, J. Málek, M. Steinhauer, On Lipschitz truncations of Sobolev functions (with variable exponent) and their selected applications, ESAIM Control Optim. Calc. Var. 14 (2008) 211–232.
63. L. Diening, M. Růžička, Interpolation operators in Orlicz Sobolev spaces, Numer. Math. 107 (1) (2007) 107–129.
64. L. Diening, M. Růžička, K. Schumacher, A decomposition technique for John domains, Ann. Acad. Sci. Fenn., Math. 35 (2010) 87–114.

65. L. Diening, M. Růžička, J. Wolf, Existence of weak solutions for unsteady motions of generalized Newtonian fluids, Ann. Sc. Norm. Super. Pisa, Cl. Sci. (5) IX (2010) 1–46.
66. G. Duvaut, J.L. Lions, Inequalities in Mechanics and Physics, Springer Grundlehren, vol. 219, Springer, Berlin, 1976.
67. F. Duzaar, G. Mingione, Harmonic type approximation lemmas, J. Math. Anal. Appl. 352 (2009) 301–335.
68. F. Duzaar, G. Mingione, Second order parabolic systems, optimal regularity, and singular sets of solutions, Ann. Inst. Henri Poincaré, Anal. Non Linéaire 22 (2005) 705–751.
69. H.J. Eyring, Viscosity, plasticity, and diffusion as example of absolute reaction rates, J. Chem. Phys. 4 (1936) 283–291.
70. D.M. Èidus, On a mixed problem of the theory of elasticity, Dokl. Akad. Nauk SSSR 76 (1951) 181–184.
71. H. Federer, Geometric Measure Theory, Grundlehren Math. Wiss., vol. 153, Springer Verlag, Berlin, 1969.
72. F. Flandoli, An introduction to 3D stochastic fluid dynamics, in: SPDE in Hydrodynamic: Recent Progress and Prospects, in: Lect. Notes Math., vol. 1942, Springer, Berlin, 2008, pp. 51–150.
73. F. Flandoli, D. Gątarek, Martingale and stationary solutions for stochastic Navier–Stokes equations, Probab. Theory Relat. Fields 102 (1995) 367–391.
74. M. Fuchs, On stationary incompressible Norton fluids and some extensions of Korn's inequality, Z. Anal. Anwend. 13 (2) (1994) 191–197.
75. M. Fuchs, Korn inequalities in Orlicz spaces, Ir. Math. Soc. Bull. 65 (2010) 5–9.
76. M. Fuchs, G. Seregin, Variational methods for fluids of Prandtl–Eyring type and plastic materials with logarithmic hardening, Math. Methods Appl. Sci. 22 (1999) 317–351.
77. M. Fuchs, G. Seregin, Variational Methods for Problems from Plasticity Theory and for Generalized Newtonian Fluids, Lect. Notes Math., vol. 1749, Springer Verlag, Berlin–Heidelberg–New York, 2000.
78. J. Frehse, J. Málek, M. Steinhauer, An existence result for fluids with shear dependent viscosity—steady flows, Nonlinear Anal. 30 (1997) 3041–3049.
79. J. Frehse, J. Málek, M. Steinhauer, On analysis of steady flows of fluids with shear-dependent viscosity based on the Lipschitz truncation method, SIAM J. Math. Anal. 34 (5) (2003) 1064–1083.
80. J. Frehse, J. Málek, M. Steinhauer, On existence results for fluids with shear dependent viscosity—unsteady flows, in: Partial Differential Equations, Praha, 1998, in: Chapman & Hall/CRC Res. Notes Math., vol. 406, Chapman & Hall/CRC, Boca Raton, 2000, pp. 121–129.
81. J. Frehse, G. Seregin, Regularity of solutions to variational problems of the deformation theory of plasticity with logarithmic hardening, Proc. St. Petersburg Math. Soc. 5 (1998/1999) 184–222; English translation: Amer. Math. Soc. Transl. II 193 (1999) 127–152.
82. A. Friedman, Stochastic Differential Equations and Applications I, Academic Press, New York, 1975.
83. A. Friedman, Stochastic Differential Equations and Applications II, Academic Press, New York, 1976.
84. K. Friedrichs, On the boundary value problems of the theory of elasticity and Korn's inequality, Ann. Math. 48 (2) (1947) 441–471.

85. G.P. Galdi, An introduction to the mathematical theory of the Navier–Stokes equations, vol. I, Springer Tracts Nat. Philos., vol. 38, Springer, Berlin–New York, 1994.
86. G.P. Galdi, An Introduction to the Mathematical Theory of the Navier–Stokes Equations, vol. II, Springer Tracts Nat. Philos., vol. 39, Springer, Berlin–New York, 1994.
87. G.P. Galdi, C.G. Simader, H. Sohr, A class of solutions to stationary Stokes and Navier–Stokes equations with boundary data in $W^{-1/q,q}$, Math. Ann. 331 (2005) 41–74.
88. M. Giaquinta, Multiple Integrals in the Calculus of Variations and Nonlinear Elliptic Systems, Ann. Math. Stud., vol. 105, Princeton University Press, Princeton, New Jersey, 1983.
89. M. Giaquinta, G. Modica, J. Souček, Cartesian Currents in the Calculus of Variations. I: Cartesian Currents, Ergeb. Math. Grenzgeb. (3), vol. 37, Springer-Verlag, Berlin, 1998.
90. D. Gilbarg, N.S. Trudinger, Elliptic Partial Differential Equations of Second Order, 2nd edition, revised 3rd printing, Grundlehren Math. Wiss., vol. 224, Springer, Berlin–Heidelberg–New York, 1998.
91. J. Gobert, Une inéquation fondamentale de la théorie de l'élasticité, Bull. Soc. R. Sci. Liège 3–4 (1962) 182–191.
92. J. Gobert, Sur une inégalité de coercivité, J. Math. Anal. Appl. 36 (1971) 518–528.
93. L. Grafakos, Classical and Modern Fourier Analysis, Pearson Education, Inc., Upper Saddle River, NJ, 2004.
94. M. Hofmanová, Degenerate parabolic stochastic partial differential equations, Stoch. Process. Appl. 123 (12) (2013) 4294–4336.
95. M. Hofmanová, J. Seidler, On weak solutions of stochastic differential equations, Stoch. Anal. Appl. 30 (1) (2012) 100–121.
96. N. Ikeda, S. Watanabe, Stochastic Differential Equations and Diffusion Processes, 2nd ed., N.-Holl. Math. Libr., vol. 24, North-Holland, Amsterdam, 1989.
97. T. Iwaniec, Projections onto gradient fields and L^p-estimates for degenerated elliptic operators, Stud. Math. 75 (3) (1983) 293–312.
98. T. Iwaniec, p-harmonic tensors and quasiregular mappings, Ann. Math. (2) 136 (3) (1992) 589–624.
99. A. Jakubowski, The almost sure Skorokhod representation for subsequences in nonmetric spaces, Teor. Veroâtn. Primen. 42 (1) (1997/1998) 209–216; translation in Theory Probab. Appl. 42 (1) (1998) 167–174.
100. I. Karatzas, S.E. Shreve, Brownian Motion and Stochastic Calculus, Springer, 1998.
101. J. Kinnunen, J.L. Lewis, Very weak solutions of parabolic systems of p-Laplacian type, Ark. Mat. 40 (1) (2002) 105–132.
102. B. Kirchheim, S. Müller, V. Svérak, Studying nonlinear pde by geometry in matrix space, in: S. Hildebrandt, H. Karcher (Eds.), Geometric Analysis and Nonlinear Partial Differential Equations, Springer, 2003, pp. 347–395.
103. V. Kokilashvili, M. Krbec, Weighted Inequalities in Lorentz and Orlicz Spaces, World Scientific Publishing, Singapore, 1991.
104. A. Korn, Über einige Ungleichungen, welche in der Theorie der elastischen und elektrischen Schwingungen eine Rolle spielen, in: Classe des Sciences Mathématiques et Naturels, 9, Novembre, Bull. Internat. Acad. Sci. Cracovie (1909) 705–724.
105. N.V. Krylov, B.L. Rozovskii, Stochastic evolution equations, Itogi Nauki i Tekh. Ser. Sovrem. Probl. Mat., vol. 14, VINITI, Moscow, 1979, pp. 71–146; English transl.: J. Sov. Math. 16 (4) (1981) 1233–1277.

106. O.A. Ladyzhenskaya, The Mathematical Theory of Viscous Incompressible Flow, Gorden and Breach, New York, 1969.
107. O.A. Ladyzhenskaya, On some new equations describing dynamics of incompressible fluids and on global solvability of boundary value problems to these equations, Trudy Steklov's Math. Inst. 102 (1967) 85–104.
108. O.A. Ladyzhenskaya, On some modifications of the Navier–Stokes equations for large gradients of velocity, Zap. Nauč. Semin. Leningrad. Otdel. Mat. Inst. Steklov (LOMI) 7 (1968) 126–154.
109. J.L. Lions, Quelques méthodes de résolution des problèmes aux limites non linéaires, Dunod, Gauthier-Villars, Paris, 1969.
110. H. Matthies, G. Strang, E. Christiansen, The saddle point of a differential program, in: R. Glowinski, E. Rodin, O.C. Zienkiewicz (Eds.), Energy Methods in Finite Element Analysis, Volume Dedicated to Professor Veubeke, Whiley, New York, 1978.
111. J. Málek, J. Necǎs, M. Rokyta, M. Růžička, Weak and Measure Valued Solutions to Evolutionary PDEs, Chapman & Hall, London–Weinheim–New York, 1996.
112. J. Malý, W.P. Ziemer, Fine Regularity of Solutions of Elliptic Partial Differential Equations, American Mathematical Society, 1997.
113. G. Mingione, Towards a nonlinear Calderon–Zygmund theory, Quad. Mat. 23 (2009) 371–458.
114. R. von Mises, Mechanik der festen Körper in plastisch-deformablem Zustand, Götti. Nachr., Math.-Phys. Klasse (1913) 582–592, Berlin.
115. C.B. Morrey, Quasi-convexity and the semicontinuity of multiple integrals, Pac. J. Math. 2 (1952) 25–53.
116. P.P. Mosolov, V.P. Mjasnikov, On the correctness of boundary value problems in the mechanics of continuous media, Math. USSR Sb. 17 (2) (1972) 257–267.
117. R. Müller, Das schwache Dirichletproblem in L^q für den Bipotentialoperator in beschränkten Gebieten und in Außengebieten, PhD thesis, Universität Bayreuth, Bayreuth / Bayreuth. Math. Schr. 49, 1994/1995, pp. 115–211.
118. S. Müller, Variational models for microstructure and phase transitions, in: Calculus of Variations and Geometric Evolution Problems, in: F. Bethuel (Ed.), Springer Lect. Notes Math., vol. 1713, Springer, Berlin, 1999, pp. 85–210.
119. J. Nečas, Sur les normes équivalentes dans $W_p^{(k)}(\Omega)$ et sur la coecivité des formes formellement positives, in: Séminaire des mathématiques supérieures, in: Equ. Dériv. Partielles, vol. 19, Les Presses de l'Université de Montréal, Montréal, 1965, pp. 102–128.
120. D. Ornstein, A non-inequality for differential operators in the L_1 norm, Arch. Ration. Mech. Anal. 11 (1964) 40–49.
121. W. Prager, Finite plastic deformation, in: Rheology. Theory and Applications, vol. 1, F.R. Reich (Ed.), Academic Press, New York, 1956, pp. 63–96.
122. W. Prager, Einführung in die Kontinumsmechanik, Birkhäuser Verlag, Basel–Stuttgart, 1961/1973; Engl. transl.: Introduction to the Mechanics of Continua, Dover Publ., New York, 1973.
123. W. Prager, On slow visco-plastic flow, in: G. Birkhoff, et al. (Eds.), Studies in Math. Mech. Presented to R. v. Mises., Academic Press, New York, 1954, pp. 208–216.
124. C. Prévôt, M. Röckner, A concise course on stochastic partial differential equations, Lect. Notes Math., vol. 1905, Springer, Berlin, 2007.
125. M.M. Rao, Z.D. Ren, Theory of Orlicz Spaces, Pure Appl. Math., vol. 146, Marcel Dekker, Inc., New York–Basel–Hong Kong, 1991.

126. Y.G. Reshetnyak, Estimates for certain differential operators with finite-dimensional kernel, Sib. Math. J. 11 (2) (1970) 315–326.
127. B.L. Rozovskii, Stochastic evolution systems. Linear theory and applications to non-linear filtering, Math. Appl., Sov. Ser., vol. 35, Kluwer Academic Publishers, Dordrecht etc., 1990, xviii.
128. T. Roubíček, A generalization of the Lions–Temam compact imbedding theorem, Časopic Pro Pěstování Mathematiky 115 (1990) 338–342.
129. M. Růžička, Electrorheological Fluids: Modeling and Mathematical Theory, Lect. Notes Math., vol. 1748, Springer, Berlin, 2000.
130. S.S. Samko, A.A. Kilbas, O.I. Marichev, Fractional Integrals and Derivatives, Gordon and Breach, Yverdon, 1993.
131. J. Simon, Démonstration constructive d'un théorème de G. de Rahm, C. R. Acad. Sci. Paris 316 (1993) 1167–1172.
132. Z. Shen, Resolvent Estimates in L^p for eliptic systems in Lipschitz domains, J. Funct. Anal. 133 (1995) 224–251.
133. V.A. Solonnikov, Estimates for solutions of nonstationary Navier–Stokes equations, J. Sov. Math. 8 (1977) 467–528.
134. E.M. Stein, Harmonic analysis: real-variable methods, orthogonality, and oscillatory integrals, in: Monographs in Harmonic Analysis, III, Princeton University Press, Princeton, NJ, 1993, with the assistance of Timothy S. Murphy.
135. G. Strang, R. Temam, Functions of bounded deformation, Arch. Ration. Mech. Anal. 75 (1981) 7–21.
136. P. Suquet, Existence et régularité des solutions des equations de la plasticité parfaite, Thèse de 3e Cycle, Université de Paris-IV; Also C. R. Acad. Sci. Paris, Ser. D 286 (1978) 1201–1204.
137. R. Temam, Navier–Stokes Equations, North-Holland, Amsterdam, 1977.
138. R. Teman, Problèmes mathémathiques en plasticité, Gaulliers-Villars, Paris, 1985.
139. Y. Terasawa, N. Yoshida, Stochastic power-law fluids: existence and uniqueness of weak solutions, Ann. Appl. Probab. 21 (5) (2011) 1827–1859.
140. J. Wolf, Existence of weak solutions to the equations of nonstationary motion of non-Newtonian fluids with shear-dependent viscosity, J. Math. Fluid Mech. 9 (2007) 104–138.
141. B. Yan, On a reverse estimate for Hodge decompositions of p-Laplacian type operators, J. Differ. Equ. 173 (2001) 160–177.
142. N. Yoshida, Stochastic shear thickenning fluids: strong convergence of the Galerkin approximation and the energy inequality, Ann. Appl. Probab. 22 (3) (2012) 1215–1242.
143. E. Zeidler, Nonlinear Functional Analysis II/B–Nonlinear Monotone Operators, vol. 120, Springer Verlag, Berlin–Heidelberg–New York, 1990.

INDEX

Y

Printed in the United States
By Bookmasters